$\mu_{\bar{x}}$	population mean of distribution of $\bar{X}$; $E(\bar{X})$
n	number of elements in the value set of a random variable
N	number of observations upon a random variable; number of trials
ν	(lowercase Greek nu) number of degrees of freedom
p	probability; probability of success in a binomial experiment
q	probability of failure in a binomial experiment; $1 - p$
r	sample correlation coefficient
ρ	(lowercase Greek rho) population correlation coefficient
s^2	sample variance
s	sample standard deviation; $\sqrt{s^2}$
Σ	(capital Greek sigma) summation (see Appendix I)
σ^2	(lowercase Greek sigma) population variance; $V(X)$
$\hat{\sigma}^2$	unbiased estimator of σ^2; $s^2(N/N - 1)$
$\sigma_{\bar{x}}^2$	variance of the sample mean; $V(\bar{X}) = \sigma^2/N$
$\hat{\sigma}_{\bar{x}}^2$	unbiased estimator of $\sigma_{\bar{x}}^2$; $\hat{\sigma}^2/N = s^2/(N - 1)$
σ	population standard deviation; $\sqrt{\sigma^2}$
$t(\nu)$	distribution of Student's t for ν degrees of freedom
t	any random variable distributed as Student's t
ϕ	(lowercase Greek phi) probability density
$V(X)$	variance of a random variable X; the population variance of X (see Appendix II)
X	random variable defined by the set $\{x_1, x_2, \ldots x_n\}$
x (or x_i)	particular value of the random variable X (midpoint of interval Δx_i)
x_a	lower real limit of interval Δx_i
x_b	upper real limit of interval Δx_i
$\tilde{x}$	the sample median; the median of a particular sample
$\bar{x}$	the sample mean; the mean of a particular sample
$\bar{x}_c$	critical value of the sample mean
$\bar{X}$	the sample mean; random variable defined by the set $\{\bar{x}_1, \bar{x}_2, \ldots \bar{x}_n\}$
$\chi^2(\nu)$	(lowercase Greek chi) distribution of chi square for ν degrees of freedom
χ^2	any random variable distributed as chi square
$\tilde{\chi}^2$	Pearson's chi square
Z	any random variable in standard form
z	particular value of Z

William A. Maeser
28 January 1975

Introduction to
Probability and Statistics

Introduction to Probability and Statistics: Concepts and Principles

Harry Frank
University of Michigan
Flint

John Wiley & Sons, Inc.
New York, London, Sydney, Toronto

Library of Congress Cataloging in Publication Data:

Frank, Harry, 1942–
   Introduction to probability and statistics.

   1. Statistics. 2. Probabilities. I Title.

HA29.F64   519.2   73-16147
ISBN 0-471-27500-X

Printed in the United States of America

10 9 8 7 6 5 4 3 2 1

To four teachers
Howard Rhine who taught writing
Sharmon Nash who taught thinking
the late Jacobus tenBroek who taught discipline
O. J. Harvey who put it all together

and Alfa Romeo, Inc.
who inspired many of the examples

Preface

To the student

This textbook is intended for an introductory statistics course in the social or biological sciences. Therefore, you are probably an undergraduate working toward a degree in biology, psychology, political science, or some related field. Irrespective of your major, these thoughts are addressed to you.

In the last 25 years, research in the behavioral and natural sciences has become increasingly complex. To some degree, of course, this complexity simply reflects the maturation of these disciplines and their emerging sense of scientific parity with the more established physical sciences. Second, much of this complexity results from rapid advances in other areas, such as applied mathematics, computer technology, and philosophy of science. For the most part, however, it appears to be a social phenomenon. Society today is faced with terribly complex problems and has called on the scientific community for solutions. Whatever its causes, complexity in research demands sophisticated methods of collecting, organizing, and interpreting scientific data; introductory statistics has therefore become an almost universal part of academic curricula in anthropology, biology, economics, environmental science, genetics, geography, political science, psychology, and sociology.

Unfortunately, this growing emphasis on quantitative methods has not met with unequivocal student approval. Although some students approach statistics with a surprising degree of interest and anticipation, these are significantly outnumbered by those who enroll with hostile reluctance. Statistics, they argue, is simply another bothersome and irrelevant requirement, an ordeal to be completed as quickly and as painlessly as possible. Despite current rhetoric to the contrary, I suggest to these students that not all things need be justified on the grounds of relevance to be interesting or even enjoyable. Chocolate ice cream, scotch whiskey, and Baroque music are cases in point. If you meet statistics (or anything else) with an open mind, you might find it intrinsically rewarding. As the late golf pro, Walter Hagen, once said, "We pass this way only once, so don't hurry, don't worry, and be sure to smell the flowers on the way."

Another typical student reaction, particularly among those who have experienced difficulty with mathematics in high school or college, is fear, which may range from mild apprehension to downright terror. For those to whom this applies, I can offer some substantive assurance. This is *not* a mathematical statistics book. A number of my students have received top marks with no more preparation than one year of high school algebra. These students, however, had more mathematical *maturity* than their formal mathematical *training* might indicate. That is, they had a facility for symbolic representation and manipulation of abstract concepts and relationships. This skill, rather than mathematical background per se, is the real requirement for the present text. (Indeed, several students have suggested that a course in logic would provide better preparation than any course in mathematics.) Except for the unusually sophisticated, therefore, students are not advised to confront this book with less than two years of high school mathematics. And, the average student is advised to complete the equivalent of two full years of high school algebra. If you now suspect that your background may be deficient, but extrinsic considerations preclude your taking a preparatory course, there are a number of fine algebra review books in paperback designed specifically for introductory statistics students in your position. One that I generally recommend to my students is *Mathematics of Introductory Statistics: A Programmed Review* by Andrew R. Baggaley.

Goals of the Book

Book titles are not generally very informative, and we therefore tend to ignore them. In this case, however, there has been an effort to convey some definite information. The emphasis of this book is on the *concepts* and *principles* of probability and statistics. Specific applications and techniques are of secondary concern. There are several reasons for this. First, it is my unabashed prejudice that routine application without a thorough understanding of concepts and principles is simply ritual and catechism. More important, one semester is not sufficient time to make each of you a competent statistician. But it is fervently hoped that a proper foundation in the fundamental principles of statistics will enable you to learn the statistical techniques that your own work may require from any good statistical reference book. Finally, more and more routine statistical application is being performed by computer. Therefore, there is diminishing usefulness in teaching the student to do what a computer can do faster and more accurately. It becomes far more important that the student have an understanding of underlying principles so that he may competently select the computer program that is most appropriate to his particular task.

Presentation and Structure

In discussing any presentation of mathematical content, it is customary to distinguish between *rigorous* and intuitive approaches. I feel that it is sometimes more appropriate to distinguish between *formal* and intuitive approaches. This distinction implies that nonmathematical presentations can be as conceptually rigorous as formally deductive proofs and derivations. The key to rigor is not conformity to the canons of Aristotelian convention, but the degree to which a student comprehends the justification for any conclusion presented by a teacher or author. It is not necessary that this understanding be articulated in the language of pure mathematics.

In this textbook, therefore, most of the major concepts are developed in much the same way that children learn basic quantitative principles before school teachers convince them that mathematics is an arcane art, incomprehensible to all but a select few. Each new topic is generally introduced with one or two problems that the student can solve either intuitively or with only minimal recourse to mathematical conventions. We then examine our solutions and abstract the basic strategy or operations that the solutions have in common. Finally, we summarize the operations in formal mathematical notation and thereby derive complex mathematical expressions, which are somehow less formidable when developed in this manner. In some sense, therefore, our basic approach is inductive and thus departs from the conventional deductive paradigm for mathematical derivation.

Finally, there has been a concerted effort to organize the material coherently; each new principle and concept is presented as a development of earlier concepts and principles. Thus, no topic is introduced until it is required for an understanding of a topic soon to follow. It also means that when new topics are introduced they are developed without deferring necessary explanations to later sections. It is hoped that this sort of orderly presentation and continual building on earlier material will provide the reader with a logical framework that will make his task a bit easier.

Harry Frank

Acknowledgments

At the end of his preface, an author is granted a few paragraphs under the nebulous heading "acknowledgments" to humbly express gratitude to the people "without whom this book would not have been possible," and so forth. Thus, a brief chronology: several years ago, I took an undergraduate statistics course at the University of California from a softspoken and somewhat vague old gentleman named Rheem Jarrett. Some years later at the University of Colorado I completed my sojourn into statistics under the sharp tongue and baleful eye of another of Jarrett's former students, Daniel Bailey. The general orientation of this book undoubtedly derives in large measure from this somewhat incestuous academic geneology.

Orientation, however important, must nonetheless be distinguished from substantive assistance, and in this regard, many persons contributed generously of their time. First among these are R. Shantaram and Joanne Snider, who spent many patient hours explaining and correcting the more blatant of my mathematical indiscretions. In addition, many confusing passages were eliminated by the careful scrutiny of students who used the manuscript in my introductory statistics classes. Also in order is a serious note of thanks to Dan Bailey. His text, *Probability and Statistics: Models for Research*, was often the final arbiter in matters of rigor and precision. Similarly, I would thank a former Michigan colleague, whom by great misfortune I have never met: Dean William Hays. His book, *Statistics*, convinced me that a statistics book could be literate. I should like to thank Colleen Jaskiewicz; the speed and accuracy of her statistical typing clearly deserve at least a line in the *Guiness Book of Records*.

Unlike other authors, I shall not acknowledge my wife for her patience, assistance, or perceptive criticism. Patience is a virtue with which she was born and can, therefore, assume no credit. Moreover, she was of almost no assistance and, although an excellent literary stylist of considerable mathematical ability, patently refused to read the manuscript, much less type or edit it. As for criticism, who can honestly give thanks for that, most *especially* if it is perceptive? She was, however, most helpful at keeping the wolf from the door while I toiled. I speak here of our pet wolf, Samantha, and the door to my study.

Finally, I am indebted to the Literary Executor of the late Sir Ronald A. Fisher, F.R.S., to Dr. Frank Yates, F.R.S., and to Longman Group Ltd., London, for permission to reprint Table **D** and part of Table **E** from their book *Statistical Tables for Biological, Agricultural and Medical Research.*

H.F.

Contents

Part II   Properties of Distributions   71

introduction

What is Statistics?

If statistics could really be defined in a few short sentences, much of this book would be unnecessary. However, even an overly simplified definition can be helpful if sufficiently vague. In this spirit we define statistics as a branch of applied mathematics that scientists use to extract information from data. Practically, statistics is simply one tool that an investigator uses to make sense of what he observes. His observations may be collected in a precisely controlled laboratory where all conditions are held constant except for the variables under consideration. Or, they may be made under far less precise conditions where the data are subject to the simultaneous influence of many factors. In either case the scientist must do two things. First, he must organize his data into some manageable form and summarize them; second, he must interpret his data. These tasks correspond to the major emphases of statistics.

Descriptive statistics is a body of highly conventionalized numerical indices that are used to summarize data. Averages, frequencies, and percentages are statistics that are familiar to all of us as summary statements that describe some property of a collection of data. In any pool of observations there are a number of properties that may be quantitatively described in this manner. Suppose, for example, that a college instructor administers an examination with 100 possible points. If he is conscientious he will probably be concerned that his test is pitched at an appropriate level of difficulty for this particular class. It

1

would therefore be important for him to know whether the majority of his students scored toward the low end of the total possible range, somewhere in the middle, or toward the high end. This information would be summarized in a statistic that describes the *central tendency*, or *location*, of a set of scores.

Location, however, is not the only descriptive property that might be of interest to a teacher. The purpose of any examination is to distinguish among students with greater or lesser command of the subject matter and thus provide a basis for the assignment of grades. If, in this regard, our instructor found that his scores ran from the teens to the nineties, he might at least feel confident that his assignment of extreme grades reflected wide differences in test performance. If, however, he discovered that the highest and lowest examination scores were separated by only a few points, he should seriously question the meaningfulness of *any* grades he might assign. In evaluating an examination, therefore, an instructor might very conceivably wish to determine the extent to which his test scores are spread across their total possible range of values. To assess this property, he would compute a statistic that describes *dispersion*, or *variability*.

Finally, the instructor might be concerned with whether or not these tests results agreed with student performance on the term paper. That is, did students who did well on the term paper also do well on the examination? In this regard he might believe that the two tasks involved different skills or even incompatible skills. This information would be summarized in a statistic describing the property of *covariation*.

The importance of description notwithstanding, the scientific philosphy of our times places much greater emphasis on explanation, prediction, and generalization than on pure description. Throughout most of his adult life, the Dutch astronomer, Tycho Brahe, carefully recorded the positions of the known planets. These observations, however, were only empirical curiosities until Johann Kepler incorporated them into an explanatory model of the solar system that permitted prediction of subsequent planetary motion and generalization to planets not yet discovered. In this spirit, most contemporary statistics texts devote more space to statistical *inference* than to statistical *description*. **Inferential statistics** is a body of quantitative techniques that enable the scientist to make appropriate generalizations from limited observations. Not all scientific generalization requires the application of statistics, however, so before we can define this branch of statistics more precisely, we must examine the conditions that define an inferential statistics problem.

Every problem in statistical inferences involves four common characteristics. First, there is a real or conceptualized *population* about which we wish to draw a conclusion. Some typical populations are college students (past, present, and future), women, white mice of a particular genetic strain, working class male adults, and institutionalized schizophrenics.

Second, the individuals comprising our population, be they students, automobile crankshafts, blood marrow cells, or mineral specimens, are different from one another. That is, they exhibit *variability*. Some of this variability may reflect *systematic* differences among our subjects. In a population of dogs, for example, individual differences in body weight are at least partially determined by breed. Chihuahuas typically weigh less than Siberian Huskies. Moreover, males of any breed are generally larger than females, and one particular set of parents may produce unusually large progeny of either sex. Even among animals of the same breed, sex, and pedigree, however, we might still find some differences in body weight. This variation would probably represent *random* differences in diet, metabolism, environmental stress, exercise, and so forth. In this context, the word "random" does not necessarily imply that such differences are either unknown or uncontrollable. It simply means that they are not subject to systematic manipulation at the time of observation.

Third, we cannot observe the entire population, so we examine a small (but hopefully representative) portion of the population we call a *sample*. If our sample is selected properly, we can use our sample observations to make inferences about the population from which they are drawn.

If random variation contributes to differences among the *individuals* in our population we would expect that different *samples*, even if drawn from precisely the same population, will also differ from one another. In any inference problem, therefore, repeated sampling from the parent population generally produces different outcomes. Take, for example, the population of male students at the University of Michigan. If we selected a sample of 25 students and found their average height to be 5 feet 9 inches, we should not be surprised if the average height for a second sample were, say, 5 feet 8 inches. Furthermore, if samples differ among themselves, it is apparent that we can never be completely certain of the fidelity with which any particular sample reflects the characteristics of the parent population. Thus, no matter how carefully we select our sample, there is always some *residual uncertainty* as to how far we may generalize these observations. If, however, the experimenter can specify, or perhaps control, the extent of this uncertainty, his generalizations may nonetheless be very useful.

Having briefly described the inference problem, we may now define inferential statistics as a body of mathematical conventions and techniques that permit us to make generalizations about populations from sample data and to determine the residual uncertainty intrinsic to such generalizations. In the purely descriptive problem, the investigator is only concerned about the properties of the sample that he observes. But in the inferential problem, the investigator tries to go beyond his immediate observations and draw meaningful conclusions about the world at large. Furthermore, these conclusions must frequently be derived from data that are inherently variable and therefore imprecise. The concepts and methodology employed in this venture and the over-riding scientific philosophy that governs their application are the province of statistical inference.

part one
probability theory

Introductory statistics students are frequently puzzled when they discover that a preliminary exposure to probability is considered essential to the study of statistics. It seems, at the outset, to be an exercise in pure mathematics that is far removed from concerns of scientific research. Nothing could be further from the truth. The techniques and operations that we call statistics were developed explicitly to bring the mathematical rigor of probability theory into the area of scientific inquiry. Thus, probability theory is the foundation on which statistics is built.

If different samples from the same population are expected to differ from one another, we may reasonably infer that even a highly representative sample will never precisely reflect the characteristics of the parent population. Therefore, we can never be 100 percent sure of the conclusions, or inferences, that we make about the population from which the sample is drawn. We therefore reply on probability to specify the *degree* of confidence we place in our conclusions. In the previous example we found that the average height in our first sample was 5 feet 9 inches. Probability theory would permit us to define the limits within which we could generalize this finding to the parent population. That is, we can make statements of the following sort. We are 95 percent confident that the average height of the parent population is between 5 feet $7\frac{1}{2}$ inches and 5 feet $10\frac{1}{2}$ inches.

Other sorts of statistics problems involve making inferences about differences between two populations. Let us suppose that we drew one

5

sample of 25 male students and one sample of 25 female students. If the college women averaged 5 feet 6 inches in height, our two samples would differ by 3 inches. Probability theory permits us to numerically specify the likelihood that this difference reflected random variation rather than a characteristic difference between our male and female student populations. The form of the statement would be as follows. The probability is less than 5 percent that a difference this large could have occurred by random chance.

It is not overstating the case, then, to say that statistics is largely a collection of techniques for transforming data in order that they may be interpreted in probability terms. Without a thorough familiarity with probability theory, therefore, statistics becomes a rather meaningless compendium of algebraic manipulations.

chapter one

the concept of probability

Probability theory was formulated in the seventeenth century as a model for games of chance, and any of you who have played poker, roulette, or shot craps have been exposed to formal probability considerations in terms of "odds." Since the seventeenth century, however, probability has seen much more general application. And, even the nongamblers among you are familiar with statements concerning the probability of rain, the National Safety Council's warnings about the increased probability of serious injury when driving without seat belts, and the American Cancer Society's publicity indicating that the probability of contracting lung cancer is considerably greater for smokers than for nonsmokers.

In these contexts it is intuitively understood that probability is simply a number that indicates the likelihood of some event. If we say that an event has a probability of .50, this means that it is just as likely to happen as not to happen. That is, there is a 50 percent chance that the event will occur. If we say that the probability is 0, the event has a 0 percent chance of occurring; it is certain *not* to happen. And, if the probability is given as 1.0, the event has a 100 percent chance of occurrence. That is, it is *certain* to happen. We see, then, that the probability of an event can take on any value from 0 to 1.0.

9

A. The Classical Definition of Probability

If you were asked the probability of obtaining a head on a single toss of an ordinary coin, you would probably answer .50. (Actually, you would probably answer 50 percent or one-half, but let us assume the formal nomenclature.) Similarly, if you were asked the probability of drawing a club on one random draw from a well-shuffled deck of cards you are likely to say .25. And finally, if you were asked the probability of throwing a 6 on one toss of a fair die you would probably answer with some impatience by this time, $\frac{1}{6}$. (You *should* answer .16667, but most of us prefer the more rational fraction.) While these answers may seem intuitively obvious, it was precisely this sort of "intuitively obvious" situation that gave probability theory its start. Note that all of the above examples have four characteristics in common.

In every instance the number of possible results is *known* at the outset. A toss of a single coin has only *two* possible results: a head or a tail. In selecting a card only four possible suits can be drawn: clubs, diamonds, hearts, and spades. And in throwing a die, only six results can transpire. One can obtain a 1, 2, 3, 4, 5, or 6.

Second, in each example the occurrence of one particular result precludes the occurrence of any other. One cannot, for example, obtain *both* a head and a tail simultaneously. One cannot draw a spade and a club on the *same* draw, and one cannot roll a 6 and a 2 on the *same* throw. Thus, the results are said to be *mutually exclusive*.

Third, all of the results have precisely the same probability of occurring. With an ordinary coin one is as likely to toss a head as a tail. A complete deck of cards contains 13 cards in each of the four suits, so one is as likely to draw a club as he is to draw a diamond or a heart or a spade. And with a fair die, each face is equally likely to land face up.

Finally, we know precisely how many of our possible results would success-fully meet the definition of any outcome that we might specify. Were we concerned with the outcome "draw a red card," we know that two of our possible results, hearts and diamonds, would satisfy this criterion. If on the roll of a single die we defined the outcome "roll an even number," we know that three of our possible results, that is, 2, 4, and 6, would be considered successful.

These considerations are all drawn together in the classical definition of probability:

> If an experiment can produce m different and mutually exclusive results, all of which are equally likely, and if f of these are considered successful, the probability of a success is f/m.

B. The Relative Frequency Definition of Probability

The classical definition of probability was an elegant development and is intellectually satisfying for a number of reasons. It does, however, present some difficulties. First, the expression "equally likely" produces a bit of formal circularity. What does "equally likely" mean in terms of formal mathematics? It is generally taken to mean *equally probable*. Thus, we have probability defined partially in terms of probability, and any logician would tell us very quickly that this is an unequivocal no-no. Even if we are inclined to ignore formal logic as undue pendantry, the constraint that all results be equally likely seriously limits the usefulness of our definition. The logic and symmetry inherent in most games of chance generally permit assumption of the four characteristics listed above. However, we are frequently confronted with situations where we should like to apply probability considerations, but where not all of these conditions are met.

Let us suppose, for example, that we have a very large urn filled with balls of various colors. (*Note.* An urn is a vaselike vessel much beloved by Romantic poets, punsters, and statisticians.) Let us further suppose that we know (because someone peeked) that at least some of the balls are red. We do not know, however, how many *other* colors are represented. Finally, let us imagine that for some obscure reason we are interested in the probability of *not* obtaining a red ball on a single random draw from the urn.

Because we do not know how many different colors are in our collection, we do not know the number of possible results. We cannot, therefore, know how many results should be considered successful (i.e., nonred). Finally, we do not know that all of the colors (however many there be) are represented in equal numbers. This means that we have no justification for assuming that all results are equally likely.

This is the type of situation that is most frequently encountered by scientists, and it is apparent that the classical definition of probability cannot be used. In an effort to determine the probability, therefore, we must take an alternative tack. One likely strategy would be to draw a sample of N balls from the urn, count the number of nonred balls in the sample (call this number f) and use the ratio

$$\frac{f}{N}$$

to approximate the probability of obtaining a nonred ball.

This ratio is called the relative frequency. The number of observed successes (e.g., the number of balls in our sample that were not red) is symbolized as f, and the total number of observations (e.g., the total number of balls that were drawn) is symbolized by the letter N.

To estimate the probability of drawing a nonred ball, let us now suppose that we select 10 balls from our urn and compute the appropriate relative frequency. To assure that our sampling does not affect the proportion of red and nonred balls in the urn, let us further stipulate that the balls are drawn one at a time and that each is replaced before drawing the next. We then repeat this procedure, drawing samples of 100, 1000, 10,000, and 100,000 balls. Data from our five samples might look like that shown in Table 1.1.

Table 1.1 Number and Relative Frequency of Nonred Balls Obtained in N Draws

Sample Size (N)	Number of Nonred Balls (f)	Relative Frequency (f/N)
10	3	.3
100	39	.39
1,000	361	.361
10,000	3,733	.3733
100,000	36,900	.36900

Although our five relative frequencies are all different, we can see that these differences decrease as our samples become larger. Thus, our first two values, .3 and .39, differ by .09; the second and third values differ by just under .03, and the last two values differ by only about .004. And on a purely intuitive basis we can see that these differences should disappear altogether as our samples become *infinitely* large. If so, it must necessarily be true that relative frequencies computed for ever larger samples converge on some *fixed value* as the sample size N approaches infinity. In mathematics such a fixed value is called a *limit*, and as the reader may have anticipated the limit in the present example is the probability of observing a nonred ball. This illustrates a general relationship between the probability of any event E and the relative frequency with which it is observed in a sample of N observations: As any sample becomes infinitely large, the relative frequency of event E approaches as a limit the probability of that event. This relationship is frequently introduced as "the law of large numbers" and may be reworded slightly to provide an alternative definition of probability:

If event E is observed f times in a sample of N observations, then the probability of event E is equal to the limit of

$$\frac{f}{N}$$

As N approaches infinity.

This same definition may be expressed more succinctly as follows:

The probability of event E is equal to

$$\lim_{N \to \infty} \frac{f}{N}$$

where N is the total number of observations and f is the number of times event E is observed.

The notation

$$\lim_{N \to \infty}$$

is borrowed from the calculus and is simply a shorthand representation of the phase "the limit as N approaches infinity of . . . ."

Although the relative frequency definition of probability overcomes a number of the undesirable limitations inherent in the classical definition, it is not entirely free of difficulty. By this definition it should be apparent that the probability is only precisely determined when the sample size is either infinitely large or exhausts the source from which we are sampling. Drawing an infinitely large sample and computing the relative frequency associated with such a sample would be extremely tedious, and we must therefore generally be satisfied with the more or less precise approximation that may be derived from a finite sample.

C. The Mathematical Definition of Probability

The principal objection to both the classical and relative frequency definitions of probability is that they carry restrictive assumptions that are totally unrelated to the formal, mathematical properties of probability. Under the classical definition, for example, it is necessary to assume that all results are equally likely. To use the relative frequency definition, it must be possible, in principle at least, to physically draw a sample of observations. This is because both definitions were formulated after the fact from existing methods for computing probability, which developed to meet specific practical needs.

In point of fact, there are only two mathematical constraints on probability. First, the probability assigned to each possible result must be no greater than 1.0 and no less than 0. Second, the sum of the probabilities of all possible results must be exactly 1.0. The mathematical definition of probability considers only these requirements, and it therefore has fewer and less restrictive limitations than either the classical or the relative frequency definition. Before we can discuss this fairly recent development, however, it is necessary to introduce the concept of a sample space and related notions.

1. Sample Spaces, Outcomes, and Results

Those of you who have taken courses in sciences know that a visual presentation of information, such as a graph or diagram, is often more comprehensible than a verbal explanation. The structure of ordinary language (as opposed to computer language) makes it an inefficient vehicle for the communication of information. Visual representations, however, are more condensed than verbal presentations and therefore permit us to handle simultaneously many more "bits" of information than a verbal communication of the same complexity. In statistics one of the most basic visual representations is the *sample space*. Quite simply, a sample space is a conventionalized representation of all possible results of an experiment. For example, let us suppose that we conduct the following simple experiment. Toss a coin. An exhaustive list of possible results results would be:

Result 1. The coin lands "heads."
Result 2. The coin lands "tails."

In conventional sample space notation our list of results is set off in some geometrically closed figure, such as a circle, oval, or rectangle. Thus, our sample space S for the present experiment might be represented as shown in Figure 1.1.

Heads

Tails

Figure 1.1   Sample space for one toss of a coin.

or, more simply (Figure 1.2)

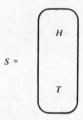

$S =$

H

T

Figure 1.2   Sample space for one toss of a coin.

In this example our experiment has only two possible results, heads and tails. When represented in a sample space, such experimental results are various called sample points, points, elementary events, simple events, or elements. To express the idea that a particular sample point belongs to, or is part of, a particular sample space S, we use the lowercase Greek letter *epsilon*, ϵ. Thus, we can express as follows the statement that obtaining a head is a member of the sample space S defined by our experiment:

$$H \in S$$

which is read "H is an element of S."

In addition to sample points, which are the most elementary units of any sample space, we may define subgroups of points that have some *specific* property in common. These collections are called events or outcomes. To illustrate the relationships among results, outcomes, and the sample space, let us suppose that instead of tossing a single coin we toss two coins. Our experiment would then have four possible results.

Result 1. The first coin lands "heads," and the second coin lands "heads."
Result 2. The first coin lands "heads," and the second coin lands "tails."
Result 3. The first coin lands "tails," and the second coin lands "heads."
Result 4. The first coin lands "tails," and the second coin lands "tails."

The sample space for this experiment (Figure 1.3) would be

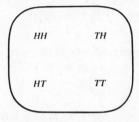

Figure 1.3   Sample space for toss of two coins.

Now let us define an event, E_1, as obtaining one and only one tail. In the present experiment this condition is satisfied by only two results *HT* and *TH*. In the sample space shown in Figure 1.4, therefore, E_1 is represented by the quadrilateral that bounds the subgroup of points *HT* and *TH*.

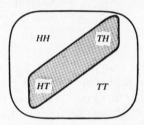

Figure 1.4   Sample space for toss of two coins and event defined by observation of one and only one tail.

The relationship "is an element of," indicated by ϵ, is also applicable to outcomes. Thus, the following notation expresses the statement that observing a tail for the first coin and a head for the second is an element of our event E_1:

$$TH \; \epsilon \; E_1$$

Not all problems in probability involve coin tosses, however, so we need some notation that is more general than that provided by *H*'s and *T*'s. Therefore, when speaking of sample points in general, be they faces of coins, colors of balls drawn from an urn, hands of cards dealt in a poker game, or faces of a die, we use the lowercase letter *e*. Thus, if we wish to make some statement about those sample points that are elements of the set *S*, we simply refer to

$$e \; \epsilon \; S$$

This avoids the cumbersome job of selecting some representative point, say *HT*, and explaining that we are using *HT* only as an example and that our statement is applicable to all points in the space.

Those of you who may have had some exposure to set theory will recognize sample spaces, events, and results as simply a specialized application of basic set notions. Thus, our sample space is the *set* of all possible results of an experiment. Any given event is a *subset* of results that share some defined property, and each result is an *element* of the set.

Set Theory Terminology	*Sample Space Terminology*
1. Set	Sample space
2. Subset	Event, outcome
3. Element	Sample point, elementary event, simple event, result, point

In set notation the possible results of the first experiment would be represented as

$$S = \{H, T\}$$

where S stands for set (instead of sample space), and the complete list of elements (results) is enclosed in braces. The set theory representation of our second experiment would be

$$S = \{HH, HT, TH, TT\}$$

For typographical reasons we shall frequently find it more convenient to use the set notation than the corresponding sample space notation.

At the beginning of this section we stated that formal mathematics places only two restrictions upon probability. First, the probability of each possible result must be no greater than 1.0 and no less than 0. Second, the sum of the probabilities of all possible results must equal 1.0. If both of these conditions are met, then the probability of any outcome is simply equal to the sum of the probabilities associated with the results that satisfy the definition of the outcome. This is the essence of the mathematical definition of probability, but the definition is usually cast in terms of sample spaces, events, and elements.

> The probability of an event E is equal to the sum of the probabilities of the sample points e that comprise the event, where
> (1) the probability of every element in the sample space is greater than or equal to 0 and less than or equal to 1.0, and
> (2) the sum of the probabilities of all sample points in the space is equal to 1.0.

For those with a subversive fondness for elegant mathematical notation, this definition may be expressed somewhat more compactly as

$$p(E) = \Sigma\, p(e \in E) \text{ such that } 0 \leq p(e \in S) \leq 1.0$$
$$\text{and } \Sigma\, p(e \in S) = 1.0$$

Once again we employ some mathematical shorthand and represent the phrase "the sum of . . ." with the capital Greek letter *sigma*, Σ.

To illustrate how this definition works, let us return to our earlier experiment "throw two coins." Recall that our sample space was comprised of four elements:

$$S = \{HH, HT, TH, TT\}$$

If we can assume that both coins are fair, our four possible results must be equally likely. That is, every possible result has a probability of .25:

$$p(e \,\epsilon\, S) = .25$$

Note, however, that the decision to set the probability of every element in the sample space equal to .25 was based on rational considerations, not mathematical considerations. Just so long as each value is between 0 and 1 and just so long as their sum is equal to 1.0, we have satisfied all of the formal mathematical requirements. And, indeed, we note that

$$0 \leq .25 \leq 1.0$$

and

$$.25 + .25 + .25 + .25 = 1.0$$

If, as before, we define the event E_1 as observing one and only one tail, we may therefore set the probability of E_1 equal to the sum of the probabilities of those results that meet this definition. We note again that these results are HT and TH. That is,

$$HT \,\epsilon\, E_1$$

$$TH \,\epsilon\, E_1$$

Therefore

$$p(E_1) = p(HT) + p(TH)$$
$$= .25 + .25$$
$$= .50$$

D. The Probability of Two Events: $p\,(E_1 \; or \; E_2)$

Let us again consider the experiment, "toss two coins." As before we can express the set of all possible results as shown in Figure 1.5.

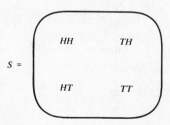

$S =$

HH   TH

HT   TT

Figure 1.5   Sample space for toss of two coins.

Moreover, we know from the mathematical definition of probability that the probability associated with the entire sample space must be 1.0. That is, one of the set of all possible results *must* occur. And, if we may assume that we are tossing fair coins, we also know that all of these results are equally likely. Therefore, the probability of any given result must be equal to .25:

$$p(e \in S) = .25$$

or, more simply,

$$p(e) = .25$$

Now let us further define Event 1, or E_1, as obtaining a head on the first toss (Figure 1.6).

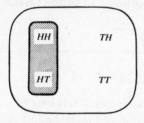

Figure 1.6 *Sample space for toss of two coins and subset of results in which head is obtained on first toss.*

E_1 is expressed as that subset of points bounded by the rectangle, and we can compute the probability of E_1 by adding the probabilities of those points. By the mathematical definition of probability,

$$p(E_1) = p(HH) + p(HT)$$
$$= .25 + .25$$
$$= .50$$

Finally, we shall complicate matters a bit more by defining still a second event, E_2, as obtaining one (and only one) tail (Figure 1.7).

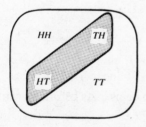

Figure 1.7 *Sample space for toss of two coins and subset of results in which one and only one tail is obtained.*

Once again our event is represented as those elements enclosed by the quadrilateral, and we may employ the mathematical definition of probability to compute the probability of E_2.

$$p(E_2) = p(HT) + p(TH)$$
$$= .25 + .25$$
$$= .50$$

Given these conditions, what is the probability of observing *either* Event 1 or Event 2? That is, what is the probability of observing a result which is an element of *either* the subset E_1 or the subset E_2? We are thus asking for the probability of observing either (1) a head on the first toss or (2) one (and only one) tail. The results that satisfy at least one of these conditions are *HH*, *HT*, and *TH* (Figure 1.8).

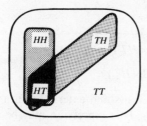

Figure 1.8   Sample space for toss of two coins and two subsets.

Therefore

$$p(E_1 \text{ or } E_2) = p(HH) + p(HT) + p(TH)$$

$$= .25 + .25 + .25$$

$$= .75$$

Similarly, we might ask what is the probability of *both* E_1 and E_2 occurring. In this situation we must obtain some result that is common to *both* events. Thus, the probability of observing both E_1 and E_2 is equal to the sum of the probabilities of all points that are shared by both subsets E_1 and E_2. In the present example there is only one such point, *HT*,

$$p(E_1 \text{ and } E_2) = p(HT)$$

$$= .25$$

1. The Addition Theorem

In a more realistic situation we are likely to have many more than four sample points, and it can become extremely cumbersome to compute $p(E_1 \text{ or } E_2)$ by summing the probabilities of the appropriate sample points. Therefore, we ask, what is $p(E_1 \text{ or } E_2)$ *in terms of* $p(E_1)$, $p(E_2)$, and $p(E_1 \text{ and } E_2)$? Recall that

$$p(E_1 \text{ or } E_2) \quad = .75$$

$$p(E_1) \qquad\quad = .50$$

$$p(E_2) \qquad\quad = .50$$

$$p(E_1 \text{ and } E_2) = .25$$

Therefore

$$p(E_1 \text{ or } E_2) = p(E_1) + p(E_2) - p(E_1 \text{ and } E_2)$$

$$.75 = .50 + .50 - .25$$

This is not as mysterious as it may seem. If we take the simple sum of the probabilities of all sample points in both events, it can be seen that $p(HT)$ would be added in twice:

$$E_1 = \{HH, HT\}$$

$$E_2 = \{HT, TH\}$$

Therefore

$$p(E_1) + p(E_2) = p(HH) + p(HT) + p(HT) + p(TH)$$

$$= p(HH) + p(TH) + 2p(HT)$$

This is because HT is a result common to both events and enters the sum once as an element of E_1 and again as an element of E_2. This expression may be compared with the solution given above for the probability of observing either Event 1 or Event 2:

$$p(E_1 \text{ or } E_2) = p(HH) + p(HT) + p(TH)$$

We see, therefore, that

$$p(E_1) + p(E_2) = p(E_1 \text{ or } E_2) + p(HT)$$

or alternatively that

$$p(E_1 \text{ or } E_2) = p(E_1) + p(E_2) - p(HT)$$

The crux of the matter if that HT is common to both outcomes. Were we considering a problem where more than one result were shared by two events, we would have to subtract out the sum of the probabilities of *all* common points. This sum, as we demonstrated above, is precisely equal to the probability of observing *both* events. Therefore

$$p(E_1 \text{ or } E_2) = p(E_1) + p(E_2) - p(E_1 \text{ and } E_2)$$

This expression involves only the operations of addition and subtraction and is therefore known as the Addition Theorem. This additive property is also reflected in the mathematical notation for the concept "or," which is generally expressed as a plus ($+$) sign. Thus

$$p(E_1 \text{ or } E_2) \qquad \text{and} \qquad p(E_1 + E_2)$$

are interchangeable expressions.

a. A Special Case: Mutually Exclusive Events. When two outcomes have *no* points in common, they are said to be mutually exclusive. In such cases E_1 and E_2 cannot occur simultaneously. Or, alternatively, the occurrence of one event precludes the occurrence of the other. Therefore

$$p(E_1 \text{ and } E_2) = 0$$

and

$$p(E_1 \text{ or } E_2) = p(E_1) + p(E_2) - 0$$
$$= p(E_1) + p(E_2)$$

E. The Probability of Two Events: $p\,(E_1\ and\ E_2)$

In the last section we developed the Addition Theorem for determining the probability of observing either of two events, E_1 or E_2. In the course of this presentation we had occasion to compute and use the probability of observing *both* events, $p(E_1 \text{ and } E_2)$. Such computations were based on the mathematical definition of probability and simply involved summing the probabilities of a small subset of results. It is not always convenient to lay out a sample space, however, nor is it always a simple matter to determine the probabilities associated with the individual sample points. In the present section, therefore, we shall develop a method for computing $p(E_1 \text{ and } E_2)$, which involves probabilities at the subset level rather than probabilities at the element level.

But before we can come directly to grips with this problem we must first introduce the concept of conditional probability.

1. Conditional Probability

Suppose that an experimenter has a mini-urn containing three balls: a silver ball (S), a gold ball (G), and a wood ball (W). Suppose, too, that he has a fair coin with the usual assortment of heads (H) and tails (T), one per side. Now define the experiment, draw one ball at random from the urn, and toss the coin. Under these conditions each point in the sample space represents the joint result of the draw and the toss (Figure 1.9).

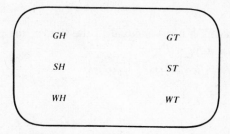

Figure 1.9 *Sample space for toss of one coin and selection of one ball from an urn. See text for explanation.*

Let Event 1 be defined as drawing a metal ball, and Event 2 be defined as obtaining a head on the toss of the coin:

$$E_1 = \text{draw metal ball}$$
$$E_2 = \text{obtain head}$$

As before we indicate these events in the sample space by enclosing the appropriate sample points in rectangles (Figure 1.10).

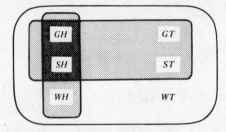

Figure 1.10 *Sample space for toss of one coin and selection of one ball from an urn and two subsets. See text for explanation.*

By either the simple logic and symmetry of the problem or by the mathematical definition of probability, determination of the various probabilities is a very straightforward task.

If the coin is indeed fair, and if one ball is as likely to be drawn as another, then each result in our sample space is as likely as every other result. Therefore, the probability associated with every sample point is equal to $\frac{1}{6}$:

$$p(e \in S) = \frac{1}{6}$$

By the mathematical definition of probability, the probability of drawing a metal ball is equal to the sum of the probabilities of all results in which a metal ball is drawn:

$$P(E_1) = p(GH) + p(SH) + p(GT) + p(ST)$$

$$= \frac{1}{6} + \frac{1}{6} + \frac{1}{6} + \frac{1}{6}$$

$$= \frac{4}{6}$$

$$= \frac{2}{3}$$

Similarly, the probability of observing a head is equal to the sum of the probabilities of all results in which heads appears:

$$p(E_2) = p(GH) + p(SH) + p(WH)$$

$$= \frac{1}{6} + \frac{1}{6} + \frac{1}{6}$$

$$= \frac{3}{6}$$

$$= \frac{1}{2}$$

which should come as a surprise to no one.

Finally, we may determine the probability of observing *both* Event 1 and Event 2. From our discussion of the addition theorem we know that the probability of observing both events is equal to the sum of the probabilities of all sample points common to both subsets. In the present example there are two such results, *GH* and *SH*. Therefore

$$p(E_1 \text{ and } E_2) = p(GH) + p(SH)$$

$$= \frac{1}{6} + \frac{1}{6}$$

$$= \frac{2}{6}$$

$$= \frac{1}{3}$$

Let us now impose the condition that we have *already* drawn our selection from the urn and obtained a metal ball. Under this condition, what is the probability of obtaining a head on the coin toss? That is, what is the probability of E_2 given that E_1 has *already* occurred? This is a problem in *conditional probability*; we are seeking the probability of E_2 *conditional* upon the prior observation of E_1. The notation for the conditional probability is

$$p(E_2 \mid E_1)$$

and is read as the conditional probability of E_2 on E_1 or the probability of E_2 given E_1.

In solving for this probability we first note that imposing the condition that a metal ball has already been (or is certain to be) drawn eliminates from the sample space all results involving observation of a wood ball. Thus, *WH* and *WT* are no longer possible results of the experiment, and our sample space is reduced to that shown in Figure 1.11.

$S_1 =$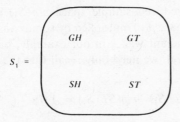

Figure 1.11   Sample space for toss of one coin and selection of one ball from an urn given that ball is metal.

This is identical to the subset of points defining Event 1 in our original set of six possible results. With our sample space now restricted to the results that define Event 1, any result that we may observe is *necessarily* contingent on the occurrence of E_1. To obtain the probability of E_2 given E_1, therefore, we need only compute the probability of E_2 in our reduced sample space S_1. That is,

$$p(E_2 \mid E_1) = p(E_2 \mid S_1)$$

In this regard, we see in Figure 1.11 that the only results in S_1 that satisfy the definition of Event 2, obtain a head, are GH and SH. By the mathematical definition of probability, therefore, $p(E_2 \mid S_1)$ must equal the sum of the probabilities assigned to GH and SH in the reduced sample space. That is,

$$p(E_2 \mid S_1) = p(GH \mid S_1) + p(SH \mid S_1)$$

The reader is cautioned, however, that the probabilities of GH and SH are no longer equal to $\frac{1}{6}$. By definition, the probabilities of all elements in any sample space must total 1.0. Our reduced sample space includes a total of four results, all of which are still equally likely, and the probability of each remaining element must therefore equal $\frac{1}{4}$. Thus

$$
\begin{aligned}
p(E_2 \mid E_1) &= p(E_2 \mid S_1) \\
&= p(GH \mid S_1) + p(SH \mid S_1) \\
&= \frac{1}{4} + \frac{1}{4} \\
&= \frac{1}{2}
\end{aligned}
$$

Once again, however, a manual solution such as this would be extremely tedious if our problem involved very many sample points. Fortunately, we may examine the logic by which we solved our sample problem and abstract a more general solution. Recall first that reduction of our sample space increased the probability of each remaining result. We pointed out, however, that our four remaining results were still equally likely, and each was therefore assigned a new probability of $\frac{1}{4}$. Effectively, then, the original probability of each result $(\frac{1}{6})$ was multiplied by the constant $(\frac{3}{2})$ to obtain its probability in the reduced sample space: $(\frac{1}{6})(\frac{3}{2}) = \frac{1}{4}$. Furthermore, it is true in general that imposing a condition on a sample space will not change the *relative* likelihood of the remaining results. Thus, the probability of any result in a reduced sample space $p(e \mid S_1)$ must equal its probability in the original sample space $p(e)$ multiplied by some constant k. We have seen in retrospect that this constant was $\frac{3}{2}$ in our example, but to see how this constant is determined in general, we need only recall that the sum of probabilities in S_1 must equal 1.0. Thus

$$p(GH \mid S_1) + p(SH \mid S_1) + p(GT \mid S_1) + p(ST \mid S_1) = 1.0$$

We have just established, however, that

$$p(e \mid S_1) = p(e)k$$

and our sum may therefore be rewritten as

$$p(GH)k + p(SH)k + p(GT)k + p(ST)k = 1.0$$

or

$$[p(GH) + p(SH) + p(GT) + p(ST)]k = 1.0$$

Thus

$$p(GH) + p(SH) + p(GT) + p(ST) = \frac{1}{k}$$

However, we know that

$$p(GT) + p(SH) + p(GT) + p(ST) = p(E_1)$$

It must therefore be true that

$$p(E_1) = \frac{1}{k}$$

or

$$k = \frac{1}{p(E_1)}$$

For any result e remaining in the reduced sample space, therefore,

$$p(e \mid S_1) = p(e) \frac{1}{p(E_1)}$$

The final equation in our manual solution

$$p(E_2 \mid E_1) = p(GH \mid S_1) + p(SH \mid S_1)$$

therefore becomes

$$p(E_2 \mid E_1) = \left[p(GH) \frac{1}{p(E_1)} \right] + \left[p(SH) \frac{1}{p(E_1)} \right]$$

$$= [p(GH) + p(SH)] \frac{1}{p(E_1)}$$

$$= \frac{p(GH) + p(SH)}{p(E_1)}$$

It will be recalled, however, that

$$p(GH) + p(SH) = p(E_1 \text{ and } E_2)$$

That is, both $(E_2 \mid E_1)$ and $(E_1 \text{ and } E_2)$ are defined by the *same* subset of results. This is not difficult to understand. The observation of Event 2 *given* Event 1 indicates explicitly that *both* Event 1 *and* Event 2 must occur. Therefore, we may substitute $p(E_1 \text{ and } E_2)$ for $p(GH) + p(SH)$ in the numerator of our last expression:

$$p(E_2 \mid E_1) = \frac{p(E_1 \text{ and } E_2)}{p(E_1)}$$

25

2. The Multiplication Theorem

From the above expression it is a simple algebraic manipulation to obtain

$$p(E_1)p(E_2 \mid E_1) = p(E_1 \text{ and } E_2)$$

or

$$p(E_1 \text{ and } E_2) = p(E_1)p(E_2 \mid E_1)$$

Thus, the probability of observing *both* Event 1 and Event 2 is equal to the probability of Event 1 multiplied by the conditional probability of Event 2 on Event 1. Because this expression involves only the operation of multiplication, it is known as the Multiplication Theorem. Once again this multiplicative property is reflected in the customary mathematical notation for the concept "and," which is expressed as a product. Thus

$$p(E_1 \text{ and } E_2) \quad \text{and} \quad p(E_1 \ E_2)$$

are interchangeable expressions.

a. A Special Case: Independent Events. Two events are said to be independent if the occurrence of nonoccurrence of Event 1 has no effect on the probability of Event 2. This means

$$p(E_2 \mid E_1) = p(E_2)$$

If two events are independent, therefore, we may substitute $p(E_2)$ for $p(E_2 \mid E_1)$ in the multiplication theorem, and

$$p(E_1 \ E_2) = p(E_1)p(E_2 \mid E_1)$$

becomes

$$p(E_1 \ E_2) = p(E_1)p(E_2)$$

For independent events, therefore, the probability of observing both Event 1 and Event 2 is simply equal to the probability of Event 1 multiplied by the probability of Event 2.

b. A Concrete Development of the Multiplication Theorem for Independent Events. Let us suppose that we have a set of 1000 marbles, 500 of which are red and 500 of which are white. Let us further suppose that we have a board, much like a game board for Chinese checkers, with 1000 holes, each of which is deep enough to accommodate only one marble. We divide the board in half and randomly assign one half to red marbles and the other half to white marbles. Then we partition off the two halves and pour all of the red marbles into the half assigned

The Concept of Probability

to red marbles and pour all the white marbles into the other half. If we shake the board around until every hole is filled, it would look like Figure 1.12, with the hatched half representing red and the white portion representing the half containing white marbles. We can therefore define the experiment, select a hole at random, and further define an event, E_1, as obtaining a red marble. With half the holes containing red marbles,

$$p(E_1) = .50$$

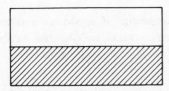

Figure 1.12 Schematic of apparatus described in text. Hatched portion indicates red marbles and white portion indicates white marbles.

Now let us suppose that the floor of each hole is a plunger, or piston, which can be lowered so that each hole, which now contains a marble, can accommodate a second marble. We again divide the board in half, but this time we divide it perpendicular to the first division. As before we randomly assign one half to red marbles and one half to white marbles, and we complete the procedure by pouring another 500 red marbles into their assigned area and pouring another 500 white marbles into the other half. In schematic our board now looks like Figure 1.13.

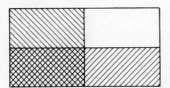

Figure 1.13 Schematic of apparatus described in text. Cross-hatched portion indicates two red marbles, hatched portions indicate one red and one white marble, and white portion indicates two white marbles.

The white portion represents the holes containing two white marbles, the hatched areas represent holes containing one white and one red marble, and the cross-hatched area represents holes containing two red marbles.

Let us once more consider our experiment, select a hole at random. This time, however, let Event 1 (E_1) be defined as obtaining a red marble *from the first* lot of 1000 marbles. In addition, let Event 2 (E_2) be defined as obtaining a red marble from the second lot of 1000 marbles. For both lots half the marbles are red, and therefore

$$p(E_1) = p(E_2) = .50$$

Furthermore, we note that one-fourth of the board is represented with cross-hatching in Figure 1.13, and that one-fourth of the holes must therefore contain two red marbles. Therefore, the probability of selecting a hole containing two red marbles is .25, or

$$p(E_1E_2) = .25$$

Let us now conduct our same experiment in a slightly different fashion. Instead of dividing the board into halves, let us divide it into fourths and randomly assign two-fourths to red marbles and two-fourths to white marbles. If we pour in our first lot of 1000 marbles according to these assignments, our board might look like Figure 1.14. Once again half the holes are filled with red marbles,

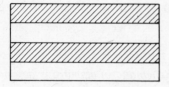

Figure 1.14   Schematic of apparatus described in text. Hatched portions indicate red marbles and white portions indicate white marbles.

and half are filled with white marbles. Now we lower our plungers, again making it possible to fit a second marble into each hole, and we again divide the board into fourths. As before we randomly assign two-fourths to red marbles and two-fourths to white marbles and pour in our second lot of marbles accordingly. Our board, which is now taking on a decidedly checkered appearance, should look something like Figure 1.15. If we define our experiment and two events as before, the probabilities are seen to remain the same. Red marbles comprise half of the first lot and half of the second lot, and thus

$$p(E_1) = p(E_2) = .50$$

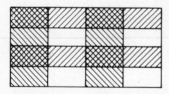

Figure 1.15   Schematic of apparatus described in text. Cross-hatched portions indicate two red marbles, hatched portions indicate one red and one white marble. and white portions indicate two white marbles.

Moreover, we see that 4 of the 16 subdivided areas are cross-hatched, indicating that these areas contain two red marbles. Thus

$$p(E_1E_2) = .25$$

It should be apparent by now that the board can be divided into as many parts as we wish, and the probabilities will remain the same. The probability of finding a red marble in the bottom layer (Event 1) is equal to the probability of finding a red marble in the top layer (Event 2) is equal to .50. And, the probability of randomly selecting a hole containing two red marbles (E_1E_2) remains equal to .25.

Let us examine the mechanics of this conclusion in terms of the first example where the board was simply divided in half for the first lot and divided in half again (crosswise to the first division) for the second lot. After we poured in the first lot, half the board contained red marbles. When we assigned areas for the second pour half of the holes already containing red marbles were assigned to white marbles and half were again assigned to red marbles. Thus, a total of one-half of one-half of the entire board was assigned red marbles on both pours. Therefore, the proportion of holes containing two red marbles must be

$$\left(\frac{1}{2}\right)\left(\frac{1}{2}\right) = \frac{1}{4}$$

In terms of the probability of randomly selecting a hole containing two red marbles, this becomes

$$(.50)\,(.50) = .25$$

In the first expression the $(\frac{1}{2})$ factors correspond to the proportion of red marbles in each lot and therefore correspond respectively to the probability of Event 1 (obtaining a red marble from the first lot) and the probability of Event 2 (obtaining a red marble from the second lot). In terms of these probabilities our result becomes

$$p(E_1)p(E_2) = .25$$

Let us now consider the problem in its limit, where the board is divided up into thousandths. This means, of course, that each part contains only one hole. Therefore, we randomly assign 500 holes to white marbles and 500 holes to red marbles. Continuing with our usual procedure we lower the plungers and make our random assignments for the second lot of marbles. Once again, on the basis of the mechanics described in the previous section, we should expect one fourth of the holes to contain two red marbles, or

$$p(E_1E_2) = .25 = p(E_1)p(E_2)$$

It should be apparent to the reader, however, that randomly assigning 500 holes to red marbles and 500 to white marbles is no different from simply pouring in the entire lot of 1000 and letting them fill holes at random. Moreover, if we then pour in our second lot at random, it should be apparent that the results of the second pour are completely independent of the results of the first pour. That is, whether or not a red marble from the first lot winds up in a given hole has absolutely no effect on whether or not a red marble from the second lot subsequently falls into that same hole. This is, of course, in contrast to the previous experiments where areas for the second lot were assigned to contrive an even distribution of red and white marbles. Therefore, we are now able to generalize our conclusions and say that

$$p(E_1E_2) = p(E_1)p(E_2)$$

when E_1 and E_2 are independent events.

F. Problems, Review Questions, and Exercises

1. An urn contains four white balls, two red balls, and two green balls.
   a. What is the probability of obtaining a single red ball on one draw from the urn? Solve the problem by using the classical definition of probability.
   b. What is f?
   c. What is m?
   d. What is the probability of obtaining a single black ball on one draw from the urn?
   e. If five balls are drawn without replacement, what is the probability that at least one will be white?

2. "Hollywood Roulette" is a game invented by southern California grocers. When a customer purchases a six-pack of diet soft drinks, one can is beer. Given the purchase of one such six-pack and assuming that the cans are identical:
   a. What is the probability of drawing the beer on the first try?
   b. Can this problem be solved by using the classical definition of probability?
   c. If your answer to 2b was yes, draw the sample space for the experiment and assign appropriate probability values to each possible result.

3. Define the sample space for problem 1 in terms of three exhaustive and mutually exclusive events.
   a. What probability do you assign to each event?
   b. If you knew only these three probabilities could you solve 1a using the classical definition of probability?
   c. If you knew only these three probabilities could you solve 1a using the mathematical definition of probability?

4. Define the sample space for problem 2 in terms of two exhaustive and mutually exclusive events.
   a. What probability do you assign to each event?
   b. If you knew only these two probabilities could you solve 2a using the classical definition of probability?
   c. If you knew only these two probabilities could you solve 2a using the mathematical definition of probability?

5. Fred the Fed has obtained a warrant from Judge Roy Bean permitting him to search a certain apartment building for marijuana. Because "The Law West of the Pecos" has not yet incorporated the Bill of Rights, Bean simply specified that Fred was to search the apartment of the first person encountered leaving the elevator. The building in question has three stories of apartments above the foyer. On the top floor, seven of the apartments contain the contraband and three do not. On the second floor, two do and eight do not, and on the first floor, six do and four do not. Assume that all of the apartments house the same number of residents.
   a. If Fred waits in the foyer, what is the probability that the first person to leave the elevator will be a resident of the third floor?

b. Using the mathematical definition of probability, what is the probability that the first person to leave the elevator in the foyer will live in one of the third floor apartments that has marijuana?

c. Using the answers to **5a** and **5b** use the formula for conditional probability to determine Fred's probability of success if he waits in front of the elevator on the top floor instead of the foyer.

6. Suppose you roll a fair die once.
   a. What is the probability of obtaining a 5?
   b. What is the probability of obtaining a prime number (a number divisible only by itself and 1)?
   c. What is the probability of obtaining a 5 *given that* the number you obtain is a prime number?
   d. Are these two events independent? Why or why not?
   e. Use the multiplication theorem to find the probability of obtaining both a prime number and a 5.

chapter two

functions, random variables and probability distributions

A. Functions

One of the most important and powerful concepts in mathematics is the function. Intuitively, a function is simply a mathematical operation or group of mathematical operations by which a number x is paired with, or transformed into, another number y. Thus, it is sometimes useful to think of a function as a simple machine. One puts an X-value into the hopper, turns the crank, and out pops a Y-value (Figure 2.1).

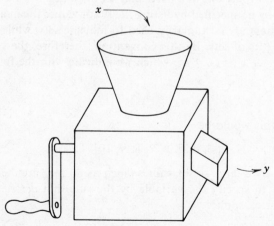

Figure 2.1   A "function machine."

Customarily, however, functions are expressed as mathematical equations, such as

$$Y = X^2$$
$$Y = X + 1$$

or

$$Y = \frac{1}{X} + \sqrt{X}$$

Such an equation is called the *function rule* or *rule of correspondence*, because it specifies the quantitative operations that must be performed on x in order to yield y. To develop some of the formal properties of functions, consider the function rule

$$Y = X + 1$$

where x can be any integer from 1 to 10. This rule tells us that for any permissible value of X, the corresponding value of Y is found by adding 1 to the X-value. Thus, if we let x equal 4,

$$y = 4 + 1$$
$$= 5$$

Similarly, if we let $x = 10$,

$$y = 10 + 1$$
$$= 11$$

If, however, we set x equal to 15, our rule yields no Y-value, because 15 does not lie within the function's *domain of definition*, the integers from 1 to 10.

From this one example several things are apparent. First, we note that the symbol X may be replaced by any one of a particular set of values. Such a symbol is called a *variable* and may be specified by listing the set of values it can assume. In this regard, capital letters are customarily used to indicate sets, while lower-case letters indicate elements of sets. By this convention, therefore, the variable X is defined by the *value set* $x_1, x_2 \ldots x_n$, which we indicate with the following notation:

$$X = \{x_1, x_2 \ldots x_n\}$$

In our present example this would be

$$X = \{1, 2, 3, 4, 5, 6, 7, 8, 9, 10\}$$

If, however X is defined by a large set of values (such as all real numbers) it is usually more convenient to specify the variable by the rule that *determines* its value set. Again from our example,

$$X = \{x \,|\, x \text{ is an integer from 1 to 10}\}$$

where the vertical line | is read "given that" and is thus defined as in conditional probability.

In the present example, it should also be apparent that y will assume several values. While this may not always be true, it is nonetheless customary to assume that Y is a variable. Like X, it may therefore be specified either by its value set, which in our example would be

$$Y = \{2, 3, 4, 5, 6, 7, 8, 9, 10, 11\}$$

or by the rule defining its value set

$$Y = \{y \mid y = x + 1\}$$

Consider now the relationship established by our function rule between the sets X and Y. For any particular value x_i it defines one and only one corresponding Y-value, y_i. Thus, if $x = 1$, $y = 2$. If $x = 7$, $y = 8$, and so forth. Like X and Y, this collection may be represented as a set:

$$\{x_1, y_1; x_2, y_2 \ldots x_n, y_n\}$$

Finally, it will be observed that our pairs of corresponding values are *ordered*. That is, an X-value must always be defined *before* we can determine the corresponding value of Y. Thus, X is often called the *independent* variable and Y the *dependent* variable.

These properties are comprehended in the formal definition of a function:

A function is a set of ordered pairs (x,y) such that for every value $x \, \epsilon \, X$ there exists one and only one value $y \, \epsilon \, Y$.

Our formal definition notwithstanding, the reader is advised that the mathematical relationship between X and Y will often be of more immediate concern than the pairs of values (x, y) generated by this relationship. Frequently, therefore, the term "function" is used when "function rule" would be more formally acceptable. Furthermore, it must be noted that our present discussion of functions has been based on one particular function rule:

$$Y = X + 1$$

In general, however, mathematicians use the notation

$$Y = f(X)$$

to indicate functional dependence of Y on X. This expression is read "Y equals f of X," or "Y is a function of X," which seems to imply that the dependent variable "is a function" of the independent variable. In some contexts, therefore, the term "function" appears virtually synonymous with "dependent variable."

1. Functions as Curves

Thinking of a function as a set of pairs of values leads to a useful application of geometry. If we represent our two variables X and Y as axes, we may consider every pair of corresponding values (x_i, y_i) to represent a point in the Cartesian plane defined by these axes. Our set of points

$$\{x_1, y_1; x_2, y_2 \ldots x_n, y_n\}$$

therefore describes a line that graphically represents the equation defining our function. Let us take a simple function,

$$Y = 2X + 1$$

We may arbitrarily select X-values and compute the corresponding Y-values by performing the mathematical operations specified in the function. We then plot the X-values on the horizontal axis and the corresponding values of Y on the vertical axis. It is seen from Figure 2.2 that our function describes a straight line. Moreover, it will always be true that when the independent variable is not raised to any power greater than 1 (e.g., it is not squared or cubed), the function will be rectilinear. Such functions are therefore called *linear* functions.

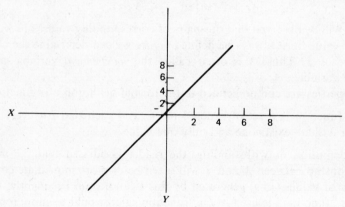

Figure 2.2   Graphic representation of function $Y = 2X + 1$.

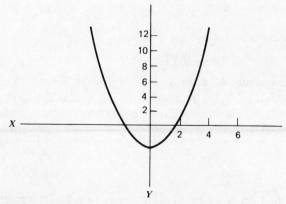

Figure 2.3   Graphic representation of function $Y = X^2 - 3$.

When X is raised to any power greater than 1, the function describes a curve and is thus said to be *curvilinear*. As an example of a curvilinear function, we may consider $Y = X^2 - 3$ (Figure 2.3). This parabolic curve is typical of quadratic functions. That is, we can expect this symmetrical, curvilinear form whenever we graph a function in which X is the *only* independent variable and 2 is the highest power to which it is raised.

Cubic functions are those in which at least one term contains the factor X^3, and quartic functions are those in which X^4 appears at least once. Such functions have equally characteristic forms.

B. Random Variables as Functions of Random Experiments

Thus far in our consideration of probability we have illustrated a number of principles with various experiments. We have, for example, tossed coins, rolled dice, and drawn cards. All of these experiments have had one common property: the sample space for the experiment includes more than one result. This property is the defining characteristic of a *random experiment*. The reader will note, however, that our examples have generally involved more than simply conducting a random experiment and reporting its result. In general we assign *numerical values* to the results of random experiments. Were we to toss a coin 10 times, for example, we would ordinarily count the number of heads (or tails) and indicate the result of our experiment with this value. Such quantification of a random experiment is the defining property of a *random variable*.

In this regard, we know by definition that a random experiment can yield a number of different results, the set of which defines the sample space. It follows, therefore, that a random variable is not a *single* value, but like any variable is a *set* of values. In 10 tosses of a fair coin, for instance, we might obtain any of the following numbers of heads:

$$\{0, 1, 2, 3, 4, 5, 6, 7, 8, 9, 10\}$$

Whatever the result of our experiment, moreover, it will be assigned only *one* number from this value set. The operations by which each result e_i yields a numerical value x_i therefore defines a *functional* relationship between the sample space and the random variable. Thus

A random variable X is a function of the elements defining a sample space S.

To illustrate, let us define a random variable X as the number of heads observed in 10 tosses of a fair coin. To obtain the numerical value x_i associated with a particular result e_i we simply conduct the experiment and count the number of heads. Suppose now that our experiment produces the following result

H H H T H T H H T H

By counting the number of heads we find that the corresponding X-value is 7. When children are first taught to count, it is typically imposed as an exercise in simple recitation. Thus, they learn to repeat a series of words, "one, two, three . . . ," long before they have acquired any comprehension of numerosity. Consequently, few of us ever consider the mathematical operations implicit in enumeration. To obtain the result $x_i = 7$, for example, we implicitly performed two quantitative operations. First, we assigned a value of 1.0 to each head and a value of 0 to each tail. Then, we let x_i equal the sum of these 10 values. That is, we have taken the result

$$e_i = H\ H\ H\ T\ H\ T\ H\ H\ T\ H$$

and transformed it into

$$x_i = 1 + 1 + 1 + 0 + 1 + 0 + 1 + 1 + 0 + 1 = 7$$

This transformation may be formally expressed in functional notation:

$$x_i = X(e_i)$$

where the function X is defined by the mathematical operations involved in counting, and $X(e_i)$ is read "X of e_i" not "X times e_i." In the present example

$$X(e_i) = 7$$

and therefore

$$x_i = 7$$

And, in general,

$$x_1 = X(e_1)$$
$$x_2 = X(e_2)$$
$$\cdots\cdots$$
$$x_m = X(e_m)$$

We see, therefore, that quantification of the m results defining our sample space produces m numbers, but it should be apparent that several of our m possible results may assume the *same* numerical value. Consider, for example, two possible results of our coin toss experiment:

$$H\ H\ H\ T\ H\ T\ H\ H\ T\ H$$

$$T\ H\ T\ H\ H\ H\ T\ H\ H\ H$$

Although these are distinguishable results, and therefore distinct elements of the sample space, they have precisely the same number of heads and would thus be assigned the same value, 7. Hence, while we may have a large number of elements m in our sample space, we may have considerably fewer elements in the value set of our random variable. Consequently, we indicate the number of values, in contrast to the number of results, by the small letter n.

When considering random variables, we remind the reader that the use of capital and lowercase letters reflects an important distinction. In general, a capital letter denotes the variable under consideration, while a lowercase letter represents a specific value of the random variable. In our coin toss experiment, therefore, we represented the random variable, number of heads, by the letter X, and a particular number of heads, 7, by the small letter x.

1. The Quantitative Nature of Events

In Chapter One we defined an event as a subset of elements that share some defined property. In general, the properties with which we are concerned in the study of probability are quantitative. Thus, an event E_i becomes the subset of all results e that assume the same numerical value x_i. That is,

$$E_i = \{e \mid X(e) = x_i\}$$

which is read "Event i is the set of all results e given that X of e is equal to x_i." Recall, too, that the probability of any event E_i is equal to the sum of the probabilities of the elements that comprise the event:

$$p(E_i) = \sum p(e\epsilon\, E_i)$$

If we define membership in the subset E_i by the numerical value assigned to its elements e, this becomes

$$p(E_i) = \sum p[e \mid X(e) = x_i]$$

We see, therefore, that the probability of an event E_i is really the probability of observing some particular value of a random variable, x_i:

$$p(E_i) = p(x_i)$$

We also know from our mathematical definition of probability that the sum of the probabilities of all m elements in the sample space is equal to 1.0. Therefore, if we combine all of the points in a sample space into a set of n exhaustive and mutually exclusive events, the sum of the probabilities of the events must equal the sum of the probabilities of the individual points, that is, 1.0:

$$p(E_1) + p(E_2) + \ldots + p(E_n) = 1.0$$

It must therefore follow that the sum of the probabilities of all values of the random variable must also equal 1.0:

$$p(x_1) + p(x_2) + \ldots + p(x_n) = 1.0$$

C. Probability Distributions

In the last section we concluded that the sum of the probabilities of all values of a random variable must always equal 1.0. This does not, however, tell us anything about the manner in which this total probability of 1.0 is distributed among the elements of the value set. And, indeed, this is the principal concern of probability theory. This does not mean that we are primarily concerned with the probabilities of individual X-values. In Chapter One we developed a number of techniques for determining the probability of an event, and in the present chapter we have seen that computing the probability of an event E_i is tantamount to computing the probability of some value of a random variable, x_i. Thus, were we concerned entirely with individual probabilities, our study of probability would be complete. In point of fact, however, probability theory is concerned not so much with individual probabilities as with general relationships between probability and random variables. That is, the probability theorist is interested in the *distribution* of probability across *all* of the values that the random variable can assume. To address such questions, we must therefore turn from probabilities per se to probability *distributions*.

A probability distribution is a representation of all the values a random variable can take $\{x_1, x_2 \ldots x_n\}$ together with the probability associated with each value $\{p(x_1), p(x_2) \ldots p(x_n)\}$.

1. Mathematical Representation: The Probability Function

As defined above, a probability distribution expresses a relationship between two sets, a random variable and probability, which pairs one value of the random variable, x_i, with precisely one probability, $p(x_i)$. This relationship satisfies our definition of function, and it therefore follows that a probability distribution may be represented as a mathematical function:

A probability function is defined as a function $Y = f(X)$ such that X is a random variable and Y is the set of probabilities associated with X.

For example, if X is the number of heads obtained in N tosses of a coin, we shall soon be able to prove that the probability that $x = r$ is given by the following function:

$$p(x = r) = \frac{N!}{r!(N - r)!} \, p^r q^{N-r}$$

where p equals the probability of a head on any toss, and q equals the probability of a tail $= 1 - p$.

The reader should not be confused by our use of r instead of x_i. This is simply to keep our notation consistent with some well-established mathematical conventions that will be introduced in the next chapter when we discuss permutations and combinations.

2. Tabular Representation: The Probability Table

From our discussion of functions it should be apparent that a function may be used to compute any given Y-value, y_i, from its corresponding X-value, x_i.

If one knows the appropriate probability function, therefore, he may select specific values of the random variable and compute their corresponding probabilities. He would thus obtain several pairs of values

$$\{x_1, p(x_1); x_2, p(x_2) \ldots x_n, p(x_n)\}$$

which could be displayed in a table or chart. In statistical work, this sort of probability table is the most frequent display of probability distributions that one is likely to encounter. It is simply a pair of columns or rows in which one row (or column) lists the selected values of the random variable, and the other row lists the corresponding probabilities. Thus, if we wished to represent the probability distribution of the number of heads in four tosses of a fair coin, the probability table would look something like Table 2.1.

Table 2.1 Probability Distribution for Number of Heads Obtained in Four Tosses of a Fair Coin

x_i	0	1	2	3	4
$p(x_i)$	.06	.25	.38	.25	.06

The probability function is not, of course, essential to the production of a probability table. The desired probabilities can often be computed manually, and if the number of X-values is small, as in our example, this may require less work than using mathematical formulas.

3. Graphic Representation: The Histogram

Recall from earlier discussion that functions may be represented graphically by plotting points that correspond to selected pairs of values (x_i, y_i). In plotting probability functions we let the horizontal axis (X-axis) represent the random variable and the vertical axis (Y-axis) represent probability. By way of example let us consider the pairs of values given in Table 2.1. If we let the X-axis represent the number of heads and the Y-axis represent probability, our probability function is expressed as five points, (x_i, y_i), where $y_i = p(x_i)$ (Figure 2.4).

When many points are involved, it is sometimes difficult to identify each point with its appropriate X-value, so we generally drop perpendicular lines from each point to the X-axis. Each point therefore becomes a vertical line, the length of which represents the probability associated with the X-value from which it originates (Figure 2.5).

While this type of graphic representation is mathematically accurate, it is still not as easily interpretable as we might wish. To create a more readable visual

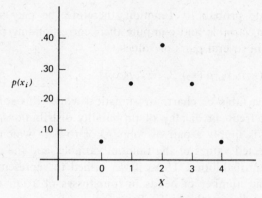

Figure 2.4   Point graph representing probability distribution in Table 2.1.

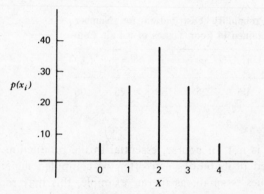

Figure 2.5   Line graph representing probability distribution in Table 2.1.

display, we generally substitute solid rectangles for our vertical lines. This familiar "bar graph" is known as a histogram and is probably the most popular method of representing probability distributions. The probability distribution that we have graphed above is reproduced histographically in Figure 2.6.

In this display, probability is represented as the height of the rectangle whose base is centered on the X-value under consideration. By the nature of our experiment, we know tht X can only assume the values 0, 1, 2, 3, and 4. Nonetheless, in order to exploit the graphic convenience of the histogram, each bar must necessarily span an entire set of values. Thus, the base of the bar representing the probability of $x = 0$ extends from -0.5 to $+0.5$; the bar whose height represents the probability of $x = 1$ spans the interval $+0.5$ to $+1.5$, and so forth. The reader is reminded, however, that the only possible X-values are those which lie at the *midpoints* of these intervals. We shall engage in this bit of graphic fiction periodically throughout the text, and the reader is therefore cautioned that all of the points spanned by the bar *do not necessarily* represent possible values of the random variable.

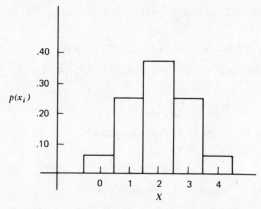

Figure 2.6   Histogram representing probability distribution in Table 2.1.

D. Cumulative Probability Distributions

In games of chance the events under consideration are generally quantitative in nature. In craps, for instance, the results of tossing two presumably fair dice are converted into numerical values by counting the spots on the uppermost faces of the dice. Frequently, a player must gamble on the probability that this number will not exceed some maximum value. In the game of blackjack, for example, each player tries to accumulate cards whose numerical total is equal to 21. The player whose total is closest to 21 wins, but anyone whose sum exceeds 21 is automatically eliminated from play. Suppose, then, that a player has been dealt two cards totalling 15. He must now decide whether to refuse another card, and risk losing to someone with a higher total, or to accept another card and risk pushing his total over 21. If he is to keep his total under 21, he cannot draw a card that exceeds six. The probability that an additional card will improve his hand is therefore equal to the probability of drawing a card whose value is equal to *or less than* six:

$$p(x \leq 6)$$

This is defined as the *cumulative probability* of six. Cumulative probability is symbolized by the capital letter F, thus

$$p(x \leq 6) = F(6)$$

and more generally,

$$p(x \leq x_j) = F(x_j)$$

In the present example we can see that the cumulative probability of six is equal to the probability of the event defined by the set of results {Ace, two, three, four, five, six}. By the mathematical definition of probability we know that this probability is equal to

$$p(\text{Ace}) + p(2) + p(3) + p(4) + p(5) + p(6)$$

Or, if we define a random variable X as the numbers appearing on the cards in our deck,

$$F(6) = \sum p(x_i)$$

for all values of X less than or equal to 6. And, in general,

$$F(x_j) = \sum_i p(x_i)$$

For all values of X less than or equal to x_j. Formally, then

The cumulative probability $F(x_j)$ is defined as $p(x \leq x_j)$ and is equal to $\sum\limits_i p(x_i)$ for all values $x_i \leq x_j$.

To illustrate cumulative probability, consider again the distribution of heads in four tosses of a fair coin. In the table given we see that

$$p(0) = .06$$
$$p(1) = .25$$

By our definition of cumulative probability, therefore,

$$F(1) = p(0) + p(1)$$
$$= .06 + .25$$
$$= .31$$

Furthermore, we know that

$$p(2) = .38$$

Thus

$$F(2) = p(0) + p(1) + p(2)$$
$$= .06 + .25 + .38$$
$$= .69$$

If we complete our computations for all values of the random variable, we obtain the cumulative probability distribution shown in Table 2.2.

Table 2.2 Cumulative Probability Distribution for Number of Heads Obtained in Four Tosses of a Fair Coin

x_i	0	1	2	3	4
$F(x_i)$	.06	.31	.69	.94	1.0

The reader will note that $F(4) = 1.0$. Furthermore, it will always be true that $F(x_n) = 1.0$, simply because $F(x_n)$ is equal to the sum of the probabilities of *all* values in the value set. It may also be observed that knowing the cumulative probability of every value in the distribution allows us to compute the probability of any particular value. In computing the cumulative probability of $x = 2$, for example, we took the following sum:

$$F(2) = p(0) + p(1) + p(2)$$

Recall, however, that

$$p(0) + p(1) = F(1)$$

Thus

$$F(2) = F(1) + p(2)$$

and

$$p(2) = F(2) - F(1)$$

This technique is of considerable practical importance to the statistician, because most probability distributions are tabled in cumulative form. Thus, to obtain the probability of, say, x_j, one must take the *cumulative* probability of x_j and subtract from it the *cumulative* probability of the next smaller value, x_{j-1}. That is,

$$p(x_j) = F(x_j) - F(x_{j-1})$$

By way of example, let us use the cumulative probabilities tabled above to determine the probability of $x = 2$.

$$p(2) = F(2) - F(1)$$
$$= .69 - .31$$
$$= .38$$

which we may confirm by checking the probability distribution from which we computed our cumulative probabilities.

E. Problems, Review Questions, and Exercises

Consider the following experiment. Throw two fair dice and count the total number of spots on the uppermost faces of the dice.
1. How many results can this experiment produce. Draw the sample space for the experiment.
2. What is the random variable in this experiment?
3. What is function relating each result to its corresponding value of the random variable?

4. How many values can the random variable assume? Give the value set for the random variable.
5. What the probability associated with each value of the random variable? Express the relationship between the random variable and its corresponding probability values in both tabular and histographic form.
6. Tabulate the cumulative probability distribution for the random variable.

(*Note:* Save your answer to problem **5** for use in later problem sets.)

chapter three

binomial probability

A. Permutations and Combinations

1. Basic Rules of Counting

When we introduced the notion of conditional probability we discussed
a problem in which an experimenter drew a ball from an urn and tossed
a coin. It will be recalled that the coin could come up either heads (*H*)
or tails (*T*) and that he could select from the urn a gold ball (*G*), a silver
ball (*S*), or a wood ball (*W*). We then defined each element in the sample
space as a joint result of the draw and the toss. That is, each point
represents the simultaneous occurrence of either a head or a tail and one
of the three balls (Figure 3.1).

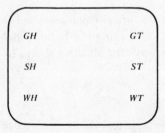

GH	*GT*
SH	*ST*
WH	*WT*

*Figure 3.1 Sample space for toss of
one coin and selection of one ball from
an urn. See text for explanation.*

As before we count the points in the space and find that we have six possible results. If the sample space were very large, however, it would be more convenient to be able to compute the number of elements than to solve for the number manually. Without recourse to the sample space we can see in the above experiment that the coin toss has two possible results, H and T. For each of the two possible results of the coin toss, the draw can have three results, G, S, or W. This gives a grand total of $2 \times 3 = 6$ ways in which our coin toss and our selection from the urn can occur together. This illustrates a basic rule that is inherent in the operation of counting.

> If something can happen in N_1 ways, after which something else can happen in N_2 ways, then the total number of possible ways for both things to happen together is $N_1 N_2$.

Let us now complicate matters a bit by adding the condition that the experimenter draw a card from an ordinary deck of playing cards, noting only the suit of the card; clubs, diamonds, hearts, or spades. Our sample space of joint results thus becomes considerably larger (Figure 3.2). Once again, for each of

$G, H,$ club	$G, H,$ diamond	$G, H,$ heart	$G, H,$ spade
$S, H,$ club	$S, H,$ diamond	$S, H,$ heart	$S, H,$ spade
$W, H,$ club	$W, H,$ diamond	$W, H,$ heart	$W, H,$ spade
$G, T,$ club	$G, T,$ diamond	$G, T,$ heart	$G, T,$ spade
$S, T,$ club	$S, T,$ diamond	$S, T,$ heart	$S, T,$ spade
$W, T,$ club	$W, T,$ diamond	$W, T,$ heart	$W, T,$ spade

Figure 3.2 Sample space for toss of one coin, selection of one ball from an urn, and draw of one card from ordinary playing deck. See text for explanation.

the two possible results of the coin toss, the selection of a ball from the urn can have three possible results, giving us $2 \times 3 = 6$ ways in which the two can occur together. For each of these six, four possible suits can be drawn from the deck of playing cards, thus giving us $2 \times 3 \times 4 = 24$ ways in which the coin toss, the selection of a ball from the urn, and the draw of a card from the deck can occur together. We see, then, that our counting rule can be extended.

> If something can happen in N_1 ways, after which a second thing can happen in N_2 ways, after which a third thing can happen in N_3 ways, then the total number of possible ways for all three things to happen together is $N_1 N_2 N_3$.

And, in general

> If something can happen in N_1 ways, after which a second thing can happen in N_2 ways, . . . after which an rth thing can happen in N_r

ways, then the total number of possible ways for all r things to happen together is $N_1N_2 \ldots N_r$.

Our demonstrations of the rules of counting are in no way offered as proof of the rules. Indeed, one can no more prove the counting rules than one can prove that the sum of 2 plus 2 is always equal to 4. By definition of the symbols involved $2 + 2 = 4$, and the rules we have discussed above are simply inherent in the operations of counting.

2. Permutations

A permutation is an arrangement or configuration of a subset of objects. It is therefore characterized by the identity of the elements it includes *and* by the order in which they appear. Thus

$$ABC, BCA, \text{ and } BAC$$

are three different permutations of the letters A, B, and C.

Many probability problems require that we determine the *number* of possible permutations of several objects, and in order to solve a permutation problem two values must be specified. First, it is necessary to know the number of objects in the set under consideration. In the example given above the number of objects (letters) is three. Second, one must specify how many elements of the set are going to be included in each permutation. In our example all three letters were included in each arrangement. We might, however, have taken the letters two at a time, in which case some of our permutations would have been

$$AB, AC, \text{ and } CB$$

We therefore speak of the number of permutations of N objects taken r at a time, or

$$_NP_r$$

where N is the total number of objects in the set and r is the number of objects included in each permutation. Thus, if we consider the possible arrangements of the three letters A, B, and C where each permutation includes all three letters, the number of such permutations is symbolized by

$$_3P_3$$

If, on the other hand we were concerned with permutations involving only two letters at a time, the number of permutations would be symbolized as

$$_3P_2$$

Let us take an example. How many ways can one arrange four people at a counter with four seats? That is, what is $_4P_4$? to illustrate the general strategy used in computing a solution to this sort of problem, let us diagram the situation and solve the problem manually. First, it should be apparent that any of the four individuals may occupy the first seat:

SEAT 1
1
2
3
4

Thus we see that person 1, 2, 3, or 4 may occupy the first seat. Once that seat is assigned, however, there are only three eligible candidates remaining for the second seat. Thus, if person 1 takes the first seat, the second seat may be occupied by either person 2, person 3, or person 4. If person 2 occupies the first seat, then persons 1, 3, and 4 are eligible for the second seat, and so on, as illustrated in Figure 3.3. Once the second seat is filled, there remain only two persons as yet

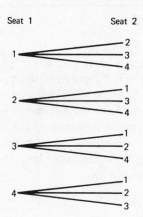

Figure 3.3 Seating arrangements for four people in two seats.

unseated, and either of them may occupy the third seat. Thus, we see in Figure 3.4 that if person 1 occupies the first seat and person 2 occupies the second seat, then the third seat may be filled by either person 3 or person 4. Moving down to the next block, we see that if person 2 occupies the first seat and person 4, say, occupies the second seat, then either person 1 or person 3 may occupy the third seat.

Finally, when the first three seats have been filled, there is only one person left to occupy the last seat. As indicated in Figure 3.5, therefore, the first seat may be occupied four ways, after which the second may be occupied three ways. after which the third may be filled two ways, after which the last seat may be filled in only one way. By the general rule of counting discussed above, this means that there must be

$$4 \times 3 \times 2 \times 1 = 24$$

possible arrangements of four people in four seats. Or

$$_4P_4 = 4 \times 3 \times 2 \times 1 = 24$$

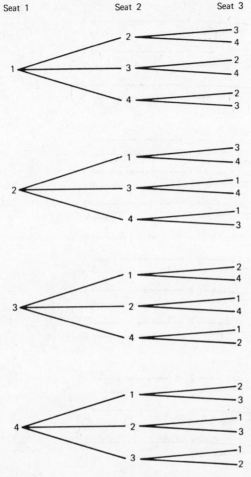

Seat 1   Seat 2   Seat 3

Figure 3.4   Seating arrangements for four people in three seats.

This may be confirmed by inspecting the branch diagram (Figure 3.5). If one counts the number of separate "pathways" from seat 1 to seat 4 (or, more simply, count the number of entries in the last column) he finds that there are indeed 24.

Another example. What if one of the seats were defective, leaving just three seats to be occupied by four people? Once again we have a set of four objects, but only three of them will be included in any given arrangement. Thus, the total number of seating arrangements is equal to the number of permuations of four things taken three at a time, that is,

$$_4P_3$$

It does not take a great deal of imagination to see that this problem may be solved by examining the *first three* columns of our branch diagram. (or, we may simply look at the earlier diagram, which considered only the first three seats.)

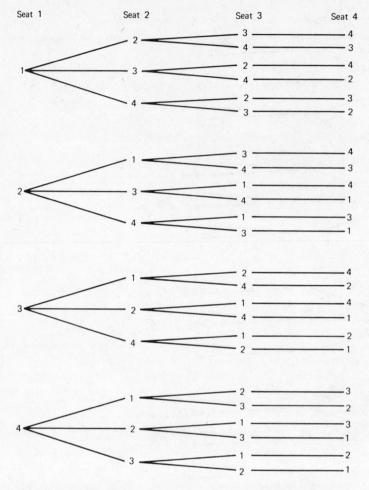

Figure 3.5 Seating arrangements for four people in four seats.

In either case we can see that the first seat may be occupied by any of the four persons, after which the second seat may be occupied by any of the remaining three persons, after which the last seat may be filled by either of the two persons yet unseated. Once again our general counting rule tells us that the total possible number of seating arrangements must be

$$4 \times 3 \times 2 = 24$$

And, therefore

$$_4P_3 = 4 \times 3 \times 2 = 24$$

Still another example. Let us now imagine that for some perverse reason two of the four seats are reserved or otherwise unusable. Thus, we will be seating our four persons just two at a time. The total number of possible seating arrangements is now given by

$$_4P_2$$

By now the reader should be able to anticipate the solution. As before the first seat may be occupied by any of the four persons (probably the fleetest of foot) after which the second seat may be occupied by any one of the other three persons. Thus, the total number of seating arrangement must be

$$4 \times 3 = 12 = {_4P_2}$$

For the examples above, we may summarize the solutions we have obtained as follows:

$$_4P_4 = 4 \times 3 \times 2 \times 1$$

$$_4P_3 = 4 \times 3 \times 2$$

$$_4P_2 = 4 \times 3$$

These solutions embody a number of general principles that we may abstract to establish a general formula for solving permutations problems. First, we note that the solution is always given by the product of several numbers. This, of course, is a consequence of the general rule of counting.

Second, we see from the solutions that the first number is always equal to 4, the number of persons to be seated. This is, of course, the total number of objects in our set, and in general the first value in the product will always be equal to N, the number of objects in the set.

Third, we can see that the numbers form a series, with each successive number one less than the preceding number, that is, 4 followed by 3 followed by 2, and so on. Formally, then our numbers constitute an arithmetic progression with the constant (-1) added to each successive value. In general, therefore, our solution will be a product that looks like

$$N(N - 1)(N - 2)(N - 3)\ldots$$

where, once again, N is equal to the total number of objects in the set.

Finally we note that there are as many values in each series as there are seats to be occupied. Thus, in the example with four seats the series has four values; when there were only three seats, we find only three numbers; and, with only two seats to fill, there are also two values. And, in general, the number of factors in the product (i.e., the number of values in the series) is equal to r, the number of objects included in each permutation. Thus, the general solution for the number of permutations of N objects taken r at a time is

$$_NP_r = \underbrace{N(N - 1)(N - 2)(N - 2)\ldots}_{r \text{ factors}}$$

a. Factorial Notation: $N!$ Before proceeding further with our consideration of permutations, it is necessary to introduce some new notation. The expression $N!$, which is read "N factorial," is a shorthand abbreviation for the product of the N values

$$1, 2, 3, \ldots N$$

Thus

$$5! \text{ (or 5 factorial)} = 1 \times 2 \times 3 \times 4 \times 5$$
$$3! = 1 \times 2 \times 3$$

and

$$83! = 1 \times 2 \times 3 \ldots 81 \times 82 \times 83$$

More generally, however, $N!$ is represented as the product of the N values, 1 through N inclusive, in descending order. For the examples given above, therefore,

$$5! = 5 \times 4 \times 3 \times 2 \times 1$$
$$3! = 3 \times 3 \times 1$$

and

$$83! = 83 \times 82 \times 81 \ldots 3 \times 2 \times 1$$

Thus, N factorial is given by the general definition

$$N! = N(N - 1)(N - 2)(N - 3) \ldots 1$$

A special case is $0!$, which is defined by convention as equal to 1.

b. Partitioning $N!$ Recalling the permutations problems that we just solved, it can be seen that the solution for the number of N objects taken r at a time is *always some portion of $N!$.* More particularly, $_N P_r$ is always the first r factors of $N!$. Thus, for the number of permutations of four things taken two at a time,

$$_N P_r = {_4 P_2} = 4 \times 3 = \text{the first 2 factors of 4!}$$

For $_4 P_3$

$$_N P_r = 4 \times 3 \times 2 = \text{the first 3 factors of 4!}$$

and for $_4 P_4$

$$_N P_r = 4 \times 3 \times 2 \times 1 = \text{the first 4 factors of 4!}$$

$$= 4!$$

Therefore, we may think of the entire expression

$$N! = N(N - 1)(N - 2) \ldots 1$$

56

as having two components. One component is the product of the first r factors, which is equal to the number of permutations of N things taken r at a time:

$$N! = \underbrace{N(N - 1)(N - 2) \ldots 1}_{\text{first } r \text{ factors}}$$

$$= {}_NP_r$$

The second component, however, is of equal importance and warrants some discussion. The entire series is composed of the values 1 through N (in descending order) and must have a total of N factors. If we consider the first r factors to be a single component (equal to ${}_NP_r$), this leaves a residual of some $N - r$ factors remaining in the progression. Because they are the *last* $N - r$ factors in the series, they must necessarily comprise the values

$$1, 2, 3 \ldots (N - r - 2), (N - r - 1), (N - r)$$

More particularly, the residual factors are the product of these $N - r$ factors arranged in descending order:

$$(N - r)(N - r - 1)(N - r - 2) \ldots 1$$

which is by definition equal to

$$(N - r)!$$

Therefore, $N!$ can be partitioned into two components:

$$N! = \underbrace{\underbrace{N(N - 1)(N - 2) \ldots (N - r + 1)}_{\text{first } r \text{ factors } = {}_NP_r} \underbrace{(N - r)(N - r - 1) \ldots 1}_{\text{last } N - r \text{ factors } = (N - r)!}}_{\text{Total of } N \text{ factors}}$$

Therefore

$$N! = ({}_NP_r)(N - r)!$$

and

$$\frac{N!}{(N - r)!} = {}_NP_r$$

The number of permutations of N objects taken r at a time is therefore given by the formula

$$_NP_r = \frac{N!}{(N - r)!}$$

3. Combinations

A combination is a subset of r objects drawn from a set of N objects. From this definition we see that a combination, unlike a permutation, is specified only by the identity of its elements and not their order. Suppose, for instance, that we have a set of three objects $\{A, B, C\}$. Suppose, too, that we arrange them into subsets of two objects per group. Some of our subsets would be

$$AB, BA, CB, \text{ and } BC$$

These four configurations constitute four different permutations of our three letters taken two at a time, but we have represented only two different combinations. AB and BA include precisely the same elements, A and B, and are therefore alternative arrangements of the *same* combination. The same is true for CB and BC.

Probability problems often require that we determine the number of ways to combine N objects into distinguishable subsets of r objects each, $_NC_r$. In this regard, it should be apparent from the preceding paragraph that any one combination can be permuted, or ordered, many ways. Indeed, for any subset of r objects, the number of permutations must be

$$_rP_k = \frac{r!}{(r-k)!}$$

where k is the number of objects in each permutation. Thus, if we stipulate that every permutation includes all r of the elements defining the combination, the number of ways to permute the combination becomes

$$_rP_r = \frac{r!}{(r-r)!}$$

$$= \frac{r!}{1}$$

$$= r!$$

It should be further apparent that for fixed values of N and r, exhaustively permuting every possible combination is no different from simply taking every possible permutation. Because every combination generates $r!$ permutations, therefore, the number of combinations multiplied by $r!$ must equal the number of permutations. That is

$$_NC_r\,(r!) = {_NP_r}$$

Thus

$$_NC_r = \frac{_NP_r}{r!}$$

$$= \frac{\dfrac{N!}{(N-r)!}}{r!}$$

$$= \frac{N!}{r!\,(N-r)!}$$

B. The Binomial Distribution

In the last chapter we discussed the notion of a probability distribution and explained that such a distribution can be represented in the form of a mathematical function where

$$Y = f(X) = p(X)$$

and

$$y_i = p(x_i)$$

Probability functions are extremely useful, of course, because they permit us to calculate the probability associated with any value of a random variable without recourse to the sorts of manual solution we have employed so far. That is, if

$$f(X) = p(X)$$

we need only select our value, x_i, perform the operations specified in the function, and thereby compute the probability associated with x_i.

It should be apparent, however, that random variables generated by different random experiments are likely to have very different probability distributions. Two such variables would, therefore, be related to their respective probability sets by different functions. Thus, if we are to compute probabilities on the basis of probability functions (or, as is actually the case, if we are to use values that are already computed and available in statistical tables), we must first determine the probability distribution that is appropriate to our experiment.

The most widely applicable probability distributions are the binomial distribution, the normal distribution, the Student's t distribution, the chi-square distribution, and the F-distribution. Each distribution is applicable to a distinctive type of statistical problem. That is, the experimental procedures and the operations used to quantify the experimental results determine the appropriate probability distribution to use. Therefore, these different distributions will be developed as we introduce the various types of statistical problems to which they are appropriate.

In this section we shall introduce the binomial distribution. While this distribution itself has little application in statistics, it is important for a number of reasons. First, the binomial probability function can be derived directly from the basic probability concepts with which we are already familiar. Thus, the student can follow its development without recourse to mathematics beyond algebra. Second, the development of the function recapitulates most of the concepts and principles we have discussed to this point and places them in a coherent framework. Finally, the binomial distribution is a springboard from which other probability distributions may be approached with some continuity.

1. The Binomial Experiment

We indicated above that the probability distribution of a random variable is determined by the experimental procedures by which the variable is generated.

Thus, a binomial random variable is produced by a binomial experiment, which is defined by the following characteristics:

i. *The results of the experiment can be classified into one of two mutually exclusive categories, success or failure.* Introductory students frequently have difficulty with the terms "success" and "failure." These terms derive from games of chance where one outcome resulted in monetary gain and the other in monetary loss. In contemporary statistical usage, however, the designation of success and failure is an arbitrary matter and simply serves to distinguish the two classes of results

ii. *The experiment is repeated a fixed number of times, N.* There may be any finite number of such repetitions, or trials, but it must be established before the experiment commences.

iii. *Each trial is independent of every other trial.* By the definition of independence given in Chapter One this implies that the probability of a success *p* remains constant from trial to trial.

2. A Typical Binomial Probability Problem

In a binomial experiment the random variable under consideration, X, is the number of successes. Thus, given a binomial experiment with a fixed number of trials, N, and a known probability of success, p, the task is to determine the probability of obtaining $x = r$ successes. The student may legitimately inquire why we set x equal to r instead of x_i as is the usual convention. First, the mathematical expressions become quite complex, and the addition of subscripts can only serve to confuse. Second, it is an effort to maintain consistency with the conventional notation of permutations and combinations. The necessity for this will become apparent later.

Let us begin with the following experiment: Toss a biased coin, $p(head) = \frac{2}{3}$, three times. It can be seen that this simple experiment satisfies all of the conditions that define a binomial experiment.

i. Each toss can result in either a head or a tail. Because the probability of obtaining a head ($\frac{2}{3}$) is given, we shall arbitrarily call heads a success and tails a failure.

ii. The experiment is repeated three times. $N = 3$.

iii. The results of one toss have absolutely no effect on the probability of a success on the next toss, and so the trials are independent.

Given this experiment, a typical binomial probability problem would ask the question, what is the probability of obtaining exactly two heads? To solve the problem let us represent all possible results of the experiment in a sample space. Furthermore, let us define the event E_1 as obtaining two heads. If we enclose the elements of Event 1 in a rectangle, our sample space looks like Figure 3.6. Our task, then, is to determine the probability of Event 1. By the mathematical

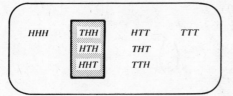

Figure 3.6 Sample space for three tosses of a coin and event defined by observation of exactly two heads.

definition of probability, the probability of E_1 is equal to the sum of the probabilities of all sample points that are elements of E_1:

$$p(E_1) = \sum p(e \in E_1)$$

$$= p(THH) + p(HTH) + p(HHT)$$

To compute the probability of Event 1, therefore, we must determine the probabilities associated with the three sample points, *THH, HTH*, and *HHT*. Every element is comprised of the results of three tosses, each of which represents one trial. By definition the trials of binomial experiment are independent. By the multiplication rule for independent events, therefore, the probability of any sample point e_i must equal the product of the probabilities associated with the results of its three tosses.

Therefore

$$p(THH) = p(T)p(H)p(H)$$
$$p(HTH) = p(H)p(T)p(H)$$

and

$$p(HHT) = p(H)p(H)p(T)$$

Thus

$$p(E_1) = p(THH) + p(HTH) + p(HHT)$$

$$= p(T)p(H)p(H) + p(H)p(T)p(H) + p(H)p(H)p(T)$$

Our only task remaining, therefore, is to determine the probability of a tail. That is, we must determine $p(\text{failure}) = q$. Intuitively it is obvious that $p(T) = \frac{1}{3}$. Nonetheless, it is useful to articulate the formal considerations that lead to this conclusion. We know from the mathematical definition of probability that the sum of all elements in a sample space must equal unity. For *any* binomial trial, therefore,

$$p(\text{success}) + p(\text{failure}) = 1.0$$

Thus

$$p(\text{failure}) = 1 - p(\text{success})$$

or

$$q = 1 - p$$

$$= 1 - \left(\frac{2}{3}\right)$$

$$= \frac{1}{3}$$

Moreover, we know from the definition of a binomial experiment that p is constant from trial to trial. Therefore, q must also be constant from trial to trial. We are now in a position to compute the probability of Event 1:

$$p(E_1) = \left(\frac{1}{3}\right)\left(\frac{2}{3}\right)\left(\frac{2}{3}\right) + \left(\frac{2}{3}\right)\left(\frac{1}{3}\right)\left(\frac{2}{3}\right) + \left(\frac{2}{3}\right)\left(\frac{2}{3}\right)\left(\frac{1}{3}\right)$$

$$= \frac{4}{27} + \frac{4}{27} + \frac{4}{27}$$

$$= \frac{12}{27}$$

$$= \frac{4}{9}$$

3. The Binomial Probability Function

It should be apparent that if our number of trials and the number of successes defining our event were very large, this sort of manual solution would be extremely cumbersome. Therefore, let us examine the various stages of the solution and try to abstract some participles that will permit us to derive a general formula for calculating $p(E)$ where E is defined as obtaining r successes in N trials of a binomial experiment.

Note that each of the three elements of our subset E_1 has the same probability as every other element:

$$p(e_i \in E_1) = p(THH) = p(HTH) = p(HHT) = \frac{4}{27}$$

Therefore

$$p(E_1) = \frac{4}{27} + \frac{4}{27} + \frac{4}{27}$$

can be more easily expressed as

$$p(E_1) = 3\left(\frac{4}{27}\right)$$

where

$$3 = \text{the number of elements in } E_1$$

and

$$\frac{4}{27} = \text{the probability of each element}$$

$$= p(e_i \in E_1)$$

In the general case, therefore,

$$p(E) = (\text{the number of elements in } E)\, p(e_i \in E)$$

Therefore, if we can find general solutions for computing the number of elements in a binomial event E and for the probability of each such element ($e_i \in E$), we shall have generated a general formula for computing $p(E)$.

a. To Calculate the Probability of Each Element of a Binomial Event: $p(e_i \in E)$. In our example recall that

$$p(e_i \in E) = \left(\frac{2}{3}\right)\left(\frac{2}{3}\right)\left(\frac{1}{3}\right)$$

We note that the product consists of three factors, and it should be obvious that each factor corresponds to one of the $N = 3$ trials comprising the event.

Second, each factor is equal to either

$$\frac{2}{3} = p(H) = p$$

or

$$\frac{1}{3} = p(T) = q$$

Finally, the number of times p and q appear corresponds to the number of successes and failures specified in the event. That is, E_1 is defined as obtaining two heads and one tail, and thus $p(H) = \frac{2}{3}$ appears twice, and $p(T) = \frac{1}{3}$ appears only once.

Let us, then, consider the general case where E is defined as obtaining r successes in N trials. On the basis of the considerations given above, we should expect $p(e)$ to be given by a product of N factors, each of which is equal to either p or q. Moreover, r of these factors should equal p, and the remaining $N - r$ factors should equal q.

The entire expression may be schematized as follows:

$$p(e_i \in E) = \underbrace{\underbrace{(p)(p) \ldots (p)}_{r \text{ factors}}\underbrace{(q)(q) \ldots (q)}_{N - r \text{ factors}}}_{N \text{ factors}}$$

Or, more compactly

$$p(e_i \in E) = p^r q^{N-r}$$

b. To Calculate the Number of Elements in E.

It should be kept in mind that the number of elements in our event E is the number of results with precisely r successes in N repetitions of a binomial experiment. That is, we are deriving a formula for computing the number of possible ways to obtain r successes in N trials. To get a feel for the nature of the problem, however, let us first examine some problems that may seem entirely unrelated to binomial probability.

While on a vacation trip through the western United States I once had an extended delay in Wendover, Nevada, waiting for a radiator hose to be replaced on my neurotic Italian automobile. Having exhausted the available reading material in the local garage (one 4-year-old copy of *Playboy*), I turned my attention to the four slot machines arrayed against one wall At that moment the mechanic walked in and nodded at the machines, saying, "Four *more* ways the customer can lose money here." To which the station manager added drily, "*If* he only plays them one at a time."

Problem. Assuming that the slot machines never pay off, how many ways can one lose money if he plays them two at a time? That is, how many ways can one place money in two machines from a group of four? First, he might put a coin in machine 1 and machine 2. Or, he could put coins in machine 1 and machine 3. Or, he might put coins in machine 1 and machine 4, and so forth. In short, he can put money in every conceivable pair of machines.

$$M_1M_2$$
$$M_1M_3$$
$$M_1M_4$$
$$M_2M_3$$
$$M_2M_4$$
$$M_3M_4$$

Thus, we see that there are six ways of placing coins in two machines from a group of four machines. Or, alternatively, there are six ways of simultaneously putting two coins in four machines.

Problem. How many ways can the customer lose money if he plays three machines at a time? Again this asks how many ways a man can place coins in three machines from a group of four. If we list every possible group of three machines we obtain the following:

$$M_1M_2M_3$$
$$M_1M_2M_4$$
$$M_1M_3M_4$$
$$M_2M_3M_4$$

We find, then, that there are four ways of simultaneously placing three coins in four machines.

Note that the strategy we used to solve these problems was very simple. In the first problem we simply listed every combination of four machines taken two at a time. Thus, the solution to the first problem is given by

$$_4C_2 = 6$$

For the second problem we listed every combination of four machines taken three at a time, and the solution is thus given by

$$_4C_3 = 4$$

In general, then, the number of ways of placing r coins in N slot machines must be

$$_NC_r$$

It should be recalled that our original intent was to derive a formula for computing the number of elements in a binomial event E. That is, how many results of a binomial experiment yield precisely r successes in N trials? In our coin toss problem Event 1 was defined by two successes (heads) in three tosses. With the values of r and N thus defined our problem is to determine the number of ways of obtaining two successes in three trials. This question may be slightly reworded without altering its meaning. How many ways can one place two successes in a group of three trials? It then becomes analogous to the problem of placing two coins simultaneously in three slot machines. The experimenter could obtain successes on trial 1 and trial 2 or on trial 1 and trial 3 and so forth. If we methodically list the possibilities, we obtain

$$T_1T_2$$
$$T_1T_3$$
$$T_2T_3$$

Thus, if we define E_1 as obtaining two successes (heads) in three tosses of a coin' the number of results which satisfy this condition is given by

$$_3C_2 = 3$$

And in general, the number of ways of obtaining r successes in N trials of a binomial experiment is simply the number of combinations of N objects taken r at a time:

$$_NC_r = \frac{N!}{r! \, (N - r)!}$$

This final expression, therefore, gives a general formula for computing the number of elements in a binomial event E.

To review our development thus far, we have established that the probability of a binomial event E, defined as obtaining r successes in N trials, is equal to the number of results that satisfy the conditions defining the event multiplied by the probability of each such result. Or

$$p(E) = (\text{number of elements in } E) \, p(e_i \, \epsilon \, E)$$

Moreover, we have determined that

$$p(e_i \in E) = p^r q^{N-r}$$

and that

$$\text{the number of elements in } E = \frac{N!}{r!\,(N-r)!}$$

Therefore, the binomial probability function is given by the expression

$$p(E) = \frac{N!}{r!\,(N-r)!}\, p^r q^{N-r}$$

c. The Binomial Probability Function and the Binomial Expansion.

For many students, this treatment of binomial probability is the first encounter with the word "binomial." Others will recall the binomial expansion, or expansion of the binomial series, from high school algebra. In this connection, many of you probably learned that

$$(a + b)^2 = a^2 + 2ab + b^2$$

Others may still remember that

$$(a + b)^3 = a^3 + 3a^2b + 3ab^2 + b^2$$

Some may even recall a rather ritualistic set of arithmetic manipulations for expanding the general binomial $(a + b)^N$. I remember that the solution I faithfully committed to memory in Algebra III was given in five steps:

1. The number of terms in the expansion is always equal to one more than the exponent, N.
2. Each of the $N + 1$ terms in the expansion is a product of a, b, and a coefficient, K. Moreover, the binomial constants a and b are raised to some power in each term:

$$\underbrace{Ka^?b^? + Ka^?b^? + \ldots + Ka^?b^?}_{N + 1 \text{ terms}}$$

3. The exponent of a in the first term is always equal to N. The exponent *decreases* by one in each successive term until the last term where the exponent is equal to 0:

$$Ka^N b^? + Ka^{N-1}b^? + \ldots + Ka^{N-r}b^? + \ldots + Ka^0 b^?$$

4. The exponent of b in the first term is always equal to 0. The exponent *increases* by one in each successive term until the last term where the exponent is equal to N:

$$Ka^N b^0 + Ka^{N-1}b^1 + \ldots + Ka^{N-r}b^r + \ldots + Ka^0 b^N$$

Recall, however, that any value raised to the 0th power is equal to 1 and any value raised to the 1st power is equal to itself. Therefore, our series reduces to

$$Ka^N + Ka^{N-1}b + \ldots + Ka^{N-r}b^r + \ldots + Kb^N$$

5. Finally, we were told that the coefficient of the first term will always be equal to 1. Thereafter, the coefficient of each successive term is computed by multiplying the coefficient of the immediately preceding term by the exponent of a in the immediately preceding term divided by the ordinal position of that term.

Following this computational outline we can generate the expansion for a binomial $(a + b)$ raised to any power, N. In the general case, the expansion is given by the following expressions:

First term: $(1)\, a^N$ $\qquad\qquad = a^N$

Second term: $(1)\dfrac{N}{1}\, a^{N-1}b$ $\qquad\qquad = Na^{N-1}b$

Third term: $\left[1 \cdot \dfrac{N}{1}\right]\dfrac{(N-1)}{2}\, a^{N-2}b^2 \qquad = \dfrac{N(N-1)}{1 \cdot 2}\, a^{N-2}b^2$

$\qquad\qquad \ldots \qquad\qquad\qquad\qquad\qquad \ldots$

$(r+1)^{\text{th}}$ term: $\left[1 \cdot \dfrac{(N)}{1} \cdot \dfrac{(N-1)}{2} \cdots \dfrac{(N-r+2)}{r-1}\right]\dfrac{(N-r+1)}{r}\, a^{N-r}b^r$

$$= \dfrac{N(N-1)\cdots(N-r+1)}{r!}\, a^{N-r}b^r$$

$\qquad\qquad \ldots \qquad\qquad\qquad\qquad\qquad \ldots$

$(N+1)^{\text{th}}$ term: $\left[1 \cdot \dfrac{(N)}{1} \cdot \dfrac{(N-1)}{2} \cdots \dfrac{2}{(N-1)}\right]\dfrac{(N-N+1)}{N}\, b^N = b^N$

If the reader finds this to be a confusing exercise in rote manipulation of mathematical symbols, we are in complete agreement! The structure of the binomial expansion becomes much more comprehensible, however, when we examine the probability distribution of a binomial random variable. If we set the number of trials of a binomial experiment equal to N, it is seen that X, the number of possible successes, can take any value from 0 to N. That is,

$$X = \{0, 1, \ldots r, \ldots N\}$$

If we apply our general formula for the probability of r successes in N trials to each possible value, we obtain a complete tabulation of the probability distribution of X. When $N = 2$, for example, the distribution of X is as shown in Table 3.1.

The reader will immediately see that the set of probabilities is structurally identical to the expansion of $(a + b)^2$.

And, if $N = 3$, the distribution we obtain is shown in Table 3.2.

Table 3.1 Probability Distribution in Terms of p and q for $X{:}B\,(2, p)$

x_i	0	1	2
$p(x_i)$	q^2	$2pq$	p^2

Table 3.2 Probability Distribution in Terms of p and q for $X{:}B(3, p)$

x_i	0	1	2	3
$p(x_i)$	q^3	$3pq^2$	$3p^2q$	p^3

Once again the series of values $p(x_i)$ corresponds exactly to the terms of the expansion of $(a + b)^3$. The apparent relationship of the binomial expansion to the probability distribution of a binomial random variable becomes even more explicit if we perform a few simple transformations on the formula for binomial probability.

$$p(x = r) = \frac{N!}{r!\,(N - r)!}\,p^r q^{N-r}$$

If we divide both the numerator and the denominator by the same value $(N - r)!$, the expression becomes

$$p(x = r) = \frac{N(N - 1)\ldots(N - r + 1)}{r!}\,p^r q^{N-r}$$

Then, if we reverse the positions of the q and p factors, it becomes

$$p(x = r) = \frac{N(N - 1)\ldots(N - r + 1)}{r!}\,q^{N-r}p^r$$

which is seen to be precisely equal to the $(r + 1)^{\text{th}}$ term in the general expression for the binomial expansion given above. Thus, we see that each value in a binomial expansion corresponds to the probability of some value of a binomial random variable. And, specifically, the $(r + 1)^{\text{th}}$ term of the expansion is equal to $p(x = r)$.

C. Binomial Probability as a Review of Probability Theory

When we introduced the binomial distribution we indicated that the development of the binomial probability function recapitulates a good deal of the probability theory we have discussed. An explication of the principles and concepts involved in the development may serve as a useful review of the student.

1. The Manual Solution

Our development began with the manual solution of a specific problem. What is the probability of obtaining two heads in three tosses of a coin which is biased with $p(\text{head}) = \frac{2}{3}$? Our first step was to represent all possible results of the experiment in a sample space. Then we formally defined an event, E_1, as obtaining two heads in three tosses and represented E_1 as the subset of all such results. These operations required an understanding of sample spaces and sets.

Next we set $p(E_1)$ equal to

$$p(THH) + p(HTH) + p(HHT)$$

This was based on the mathematical definition of probability and thus required a knowledge of that definition.

In computing the individual probabilities of each of the three elements, THH, HTH, and HHT, we stated that the probability of each was equal to the product of the probabilities associated with its constituent trials. Thus

$$p(THH) = p(T)p(H)p(H)$$

This was an application of the multiplication theorem for independent events and was justified by the definitional requirement that the trials of a binomial experiment be independent. This step, then, required an understanding of the development of the multiplication theorem for independent events, which involves the concepts of conditional probability, the general form of the multiplication theorem, and independence of events.

Finally, we had to solve for $p(T) = q$. This was a trivial problem but required the knowledge that the sum of the probabilities of all possible results of an experiment must equal unity.

2. The General Solution for $p(e_i \, \epsilon \, E)$

In moving from a specific problem to the general case, where an event E is defined as obtaining r successes in N trials, we first abstracted a general formula for the probability associated with each element in the event. This required only that we generalize our manual solution and no probability concepts were involved in the procedure.

3. The General Solution for Computing the Number of Elements in E

We demonstrated that this factor was equal to the number of combinations of N objects taken r at a time, and this part of the development thus required that the student understand the concepts and solutions for permutations and combinations.

D. Problems, Review Questions, and Exercises

1. In how many ways can a five-card hand be constructed so that it contains exactly two aces and exactly two kings? (For the remaining card, assume that the different suits are not distinguishable from one another.)
2. In how many ways (orders) can five persons be seated in a row?
3. If three of the five people are men, in how many of these seating arrangements do men occupy the two end seats?
4. Many years ago, *Look* magazine ran a contest in which participants had to identify pictures of 20 prominent persons and match each picture with one of 24 quotations. A handbook of reference sources used this contest as an example and how to conduct reference research. The author indicated that the problem had 480 (that is, 20×24) possible solutions. Was he correct? If not, how many possible solutions are there?
5. A missile manufacturer claims that his missiles are 90 percent effective. The Lower Slobovian Air Force checks the claim by firing 10 missiles at Upper Slobovia. It obtains 6 hits. What is the probability of obtaining 6 or fewer successes if indeed $p = .90$. Work the answer to the fourth decimal place. (*Note.* Save your answer to problem **5** for use in later problem sets.)

part two
properties
of distributions

In Part Two we leave the study of probability theory per se and move directly into statistics. This does not mean that we may dismiss the probability concepts we have studied thus far. Indeed, the first three chapters have provided the reader with the basic conceptual foundation upon which the remainder of the text is erected. In this regard, probability stands in the same relationship to statistics as does physics to engineering. Engineering is a body of techniques based on the principles of physics, which the contractor uses in the construction of bridges and dams. Similarly, statistics is a body of techniques based on the principles of probability, which the scientist uses to build scientific generalizations. Thus, beginning with Chapter Four we become less concerned about what the theorist *says* and relatively more concerned about what the scientist *does*.

In the last several years the popular conception of the scientist has undergone considerable change. Even the entertainment media no longer present him in a white laboratory smock surrounded by improbably complex arrangements of glass tubing or equally improbable electronic gadgetry. For better or for worse, the public more often encounters the scientist as a witty and urbane philosopher exchanging quips and profound observations with television talk-show hosts. To some extent this change is unfortunate, because the old image carried some obscure grains of truth. Invariably the television scientist of the 1950s was found either peering intently into a microscope (or telescope or test tube) or standing in front of a blackboard covered with incomprehensible mathematical symbols (or writing same in a leatherbound notebook). That is, he was generally engaged in either *making observations* or *describing his observations*. And, indeed, observation and description are two of the most fundamental activities of science.

Before conducting his observations, however, be they with a telescope or an intelligence test, the scientist must first define his *units of observation*. That is, he must decide whether he is going to investigate molecules of DNA, persons with a particular genetic anomaly, university students, or electromagnetic particles. Whatever the objects of investigation, it is likely that they will differ from one another in many ways. DNA molecules will differ in their number and configuration of nucleic acid bases. University students will differ in hair color, liquor preference, and number of siblings, and electromagnetic radiations will differ in wavelength and amplitude. Generally, however, the scientist is concerned with only one or a few of those attributes that his objects may exhibit in varying degrees. Many such attributes are inherently qualitative and must be described naturalistically. Thus, an ethologist might observe a troop of baboons and record detailed descriptions of expressive gestures, dominance behaviors, grooming activity, and the like. If, however, the scientist wishes to treat his data statistically, the attributes he observes must be represented numerically. That is, he must define a set of quantitative operations that permit the assignment of a number to every object he observes. Hence, a scientist might observe the *number of guanine bases* in a crystal of DNA, the *I.Q.'s* of persons of a particular genotype, or the *heights* of university students.

From these examples we may infer that the various units of observation can assume different numerical values. Thus, the operations that yield these values constitute a function as defined in Chapter Two, and scientific observations therefore become values of a random variable. Similarly, scientific description of observations becomes statistical description of a random variable. And, in general, such descriptions concern the manner in which the variable is distributed. The reader may recall from the introduction that description is one of the major emphases of statistics. Furthermore, we briefly mentioned three properties that can describe distributions of random variables: location, dispersion, and covariation. Introductory statistics texts customarily devote considerable attention to statistical description, and, indeed, the descriptive aspects of location, dispersion, and covariation are thoroughly developed in Part Two. It is felt, however, that these properties have theoretical importance that often transcends their simple descriptive value. In addition to descriptive properties, moreover, distributions have various mathematical properties, such as continuity and location of maxima, which are also of theoretical interest and which are also discussed in this unit. We shall therefore depart a bit from tradition and subsume statistical description under the more general rubric of distribution properties.

chapter four

populations and samples

A. Representations of Data

In the introduction to Part Two we concluded that scientific description involves making a descriptive statement about the distribution of a random variable. When data are first collected, however, they usually defy *any* sort of coherent description. The scientist is generally confronted with an immense list of numbers, and his first task is to organize his data into some comprehensible arrangement or display. Before we begin the business of description, we shall therefore consider some conventional statistical techniques for data management and representation.

1. Frequency Distributions

Let us suppose that 25 people each toss four coins. If we define our random variable X as the number of heads obtained by each experimenter, we might obtain the following results:

$$2\ 2\ 0\ 1\ 1\ 1\ 2\ 3\ 1\ 2\ 2\ 3\ 3\ 4\ 1\ 3\ 2\ 3\ 3\ 1\ 2\ 2\ 1\ 2\ 2$$

It is obvious that even with a sample size of only 25 we must organize our observations into some more manageable display. One very simple expedient would be to rearrange them so that they appear in numerical order:

$$0\ 1\ 1\ 1\ 1\ 1\ 1\ 1\ 2\ 2\ 2\ 2\ 2\ 2\ 2\ 2\ 2\ 2\ 3\ 3\ 3\ 3\ 3\ 3\ 4$$

An aggregate of numerical observations arranged in order of magnitude is called an *array*, and even a cursory glance at our array reveals an interesting characteristic of our data. Although we have 25 observations, these observations are comprised of only five *different*, or distinguishable, values of the random variable. Each of our 25 observations is equal to either 0, 1, 2, 3, or 4. These five numbers are an exhaustive list of the *observed* values assumed by our random variable and may be represented using the same convention established in Chapter Two to represent the set of all *possible* values of a random variable. Thus, our observed value set is

$$\{0,\ 1,\ 2,\ 3,\ 4\}$$

We customarily let the capital letter N represent the total number of observations, and as before, the number of elements in the value set is indicated by the small letter n. In the present example, therefore, $N = 25$ and $n = 5$. This sort of representation condenses our data considerably, but unfortunately it indicates nothing about the number of times each value x_i is observed. It does not, for example, tell us that only one person obtained 0 heads while 10 people obtained 2 heads. If we are to retain this information, therefore, we must also include a listing of the frequencies f_i associated with our X-values (Table 4.1).

Table 4.1 Frequency Distribution for Number of Heads Obtained in Toss of Four Coins with Experiment Repeated 25 Times

x_i	0	1	2	3	4
f_i	1	7	10	6	1

This sort of tabulation is known as a *frequency distribution* and contains all of the information contained in the original collection of raw data.

Another useful arrangement of data is the *cumulative frequency distribution*. The cumulative frequency is very similar to the cumulative probability and is computed in much the same way. The cumulative frequency of x_j is equal to the sum of the frequencies for all observed values x_i less than or equal to x_j. Selecting convenient notation for the cumulative frequency is always an arbitrary matter, and we shall symbolize the cumulative frequency of x_j as Cf_j. Formally, then,

The cumulative frequency Cf_j is equal to $\sum_i f_i$ for all values $x_i \leq x_j$.

As with cumulative probabilities, we can tabulate our cumulative frequencies and represent them in a cumulative frequency distribution:

A cumulative frequency distribution is defined as a representation of all observed values of a random variable $\{x_1, x_2, \ldots x_n\}$ together with the cumulative frequency associated with each value $\{Cf_1, Cf_2, \ldots Cf_n\}$.

It should be apparent that the cumulative frequency for x_n must necessarily be equal to N, the number of observations, just as the cumulative probability of x_n is always equal to 1.0.

Finally, we may divide each of our n frequencies f_i by the total number of observations N and obtain the relative frequency associated with each value x_i.

$$\frac{f_1}{N}, \frac{f_2}{N}, \ldots \frac{f_n}{N}$$

In our present example each of our 25 persons observed one repetition of the experiment, so if we divide each frequency f_i by the total number of observations $N = 25$, we obtain the values shown in Table 4.2.

Table 4.2 Relative Frequency Distribution for Number of Heads Obtained in Toss of Four Coins with Experiment Repeated 25 Times

x_i	0	1	2	3	4
$\dfrac{f_i}{N}$	.04	.28	.40	.24	.04

This sort of tabulation is the most generally used representation of data and is known as a *relative frequency distribution*:

A relative frequency distribution is defined as a representation of all observed values of a random variable $\{x_1, x_2, \ldots x_n\}$ together with the relative frequency associated with each value $\{f_1/N, f_2/N, \ldots f_n/N\}$ where N is the total number of observations, n is the number of elements in the value set, and f_i is the number of times x_i is observed.

2. Types of Data: Discrete and Continuous Random Variables

Thus far we have largely restricted our examples to experiments in which the m results in the sample space could be placed in a one-to-one correspondence with the real integers, $1, 2, \ldots n$. Consider once more, for example, the experiment in Chapter Two in which the experimenter tossed 10 coins and recorded the number of heads. In this example the result

$$\text{H H T T T T T T T T}$$

corresponds to the integer 2. The results

$$\text{H T H T H T T T T T}$$

and

$$\text{T H T T T H H T T T}$$

both correspond to the integer 3, and so forth. Furthermore, it will be recalled that this correspondence is accomplished by assigning the value 1 to each head, assigning 0 to each tail, and summing the values thus assigned to each possible result. That is, the numerical value of the experimental result is determined by counting. It should be apparent that any random variable whose values are generated by counting must have a countable number of values. More particularly, it must be possible to count the number of possible values in any interval of the value set. Such variables are called *discrete random variables* and are formally defined as follows:

> X is a discrete random variable if it can assume only a countable number of values in the interval x_a to x_b.

One implication of this definition is that a discrete variable may not assume *all* possible values in the interval x_a to x_b. If, for instance, we are told that the experimenter in our coin-toss example observed some value between 6 and 8, inclusive, we know that he must have observed 6, 7, or 8 heads. He could *not* have observed 6.3 heads or 6.791 heads or 7.34 heads. This restriction is not, however, true of variables such as weight and length whose values are obtained by *measuring* rather than counting. Consider elapsed time. Suppose we know that a sprinter's time in the 50-yard dash is between 6 and 8 seconds. Unlike the number of heads in 10 tosses of a coin, the number of seconds required to run 50 yards might very conceivably assume a value of 6.3 or 6.791 or 7.34. Indeed, the runner's time might have taken *any* value between 6 and 8. This is the defining property of a *continuous random variable*.

> X is a continuous random variable if it can assume any value in the interval x_a to x_b.

This does not mean, of course, that all possible values of a continuous random variable are actually measurable. If our sprinter were being timed with an ordinary wristwatch, for example, his time would be recorded to the nearest second and could therefore take only three values, 6, 7, or 8. Even a precision

78

stopwatch is only accurate to the nearest tenth of a second, and any time between 6 and 8 seconds could thus assume but 21 values,

$$\{6.0, 6.1, \ldots 8.0\}$$

In world class competition his time would be recorded electronically to the nearest hundredth of a second and might therefore assume any of 210 values,

$$\{6.00, 6.01, \ldots 8.00\}$$

Even this, however, is a far cry from the infinite number of values that any continuous variable such as time might theoretically assume.

a. Grouped Distributions of Discrete and Continuous Data: Real and Apparent Class Limits. Let us suppose that a psychologist is recording the number of errors made by each of 100 laboratory rats in a maze. Every animal remains in the maze until he finds the goal box, and every time he enters a blind alley an error is added to his score. This experiment will generate an array of 100 scores that the investigator may wish to represent in the form of a frequency or relative frequency distribution. Suppose, however, that the "brightest" animal makes only 116 errors, while the "dullest" rat makes 195 errors. This means that the random variable could take 80 possible values (116, 117, . . . 195), and a complete tabulation of the (relative) frequencies would not really achieve any economy of representation. Moreover, it is likely that many of the possible values will not even be observed. Ordinarily, therefore, the experimenter would group his array into *classes*, or intervals, and compute the frequency or relative frequency associated with each interval. Thus, instead of representing observations of 135, 116, 128, and 119 errors as four separate values in a distribution, he might put them all in a single class 116 to 135, along with any other scores that fall in this interval. Data from this fictitious experiment might therefore look like Table 4.3.

While this is the most common sort of grouped representation, it may be more meaningful in some cases to leave the first or last interval "open ended." If in the present example a few atypical rats had made, say 30 or 40 errors, we would not ordinarily include all of the class intervals required to accommodate them. Instead, we would simply identify our first interval as -135, meaning "135 or fewer."

The pair of numbers that define the upper and lower extremes of each interval are called the *class limits*, and the range of values included between them defines

Table 4.3   **Grouped Frequency Distribution for Maze-Running Errors**

Number of Errors	116–135	136–155	156–175	176–195
f	9	28	41	22

the width of the class interval, or the *class width*. Except for the occasional departure mentioned above, data are generally grouped in intervals of equal width. The reader should be careful to note that the class limits define an *inclusive* list, and the intervals given above therefore have a class width of 20. If, however, we simply subtract the lower limit from the upper limit of any class, the difference we obtain is 19. To resolve this inconsistency, we must introduce the distinction between *real* and *apparent* class limits.

In this regard, consider the weight distribution of senior class competitors in the Flint Judo Club (Table 4.4).

Table 4.4 Grouped Frequency Distribution for Weights of Judo Competitors

Weight	116–135	136–155	156–175	176–195
f	3	9	13	7

As before, each interval is represented by its class limits. Thus, the lightweight division includes all players weighing from 116 to 135 pounds. Unlike maze-running errors, however, weight is a continuous random variable, and if we begin the next weight class at 136 pounds, we omit the infinite number of possible values between 135 and 136 pounds. As indicated above, however, the number of places to which a continuous observation is carried is limited by the precision of the measuring instrument. An ordinary gymnasium scale is only accurate to the nearest pound, and therefore a weight of, say, 116 pounds will be indicated for any individual between 115.5 and 116.5 pounds. In point of fact, therefore, the first interval includes all players between 115.5 and 135.5 pounds. Thus, while the *apparent* class limits of the first interval are 116 and 135 pounds, the *real* class limits are 115.5 and 135.5 pounds. The real class limits of the second interval are 135.5 and 155.5 pounds, and so forth. That is, the real class limits extend *one half of a measurement unit* (in this case 0.5 pounds) beyond the apparent class limits. Thus, if we represent the distribution given above using the real instead of apparent limits, it looks like Table 4.5.

The reader will observe that each upper class limit is precisely equal to the lower class limit of the next interval. While this overlap guarantees that every possible value of the random variable is represented in the distribution, it also

Table 4.5 Grouped Frequency Distribution for Weights of Judo Competitiors with Intervals Expressed in Terms of Real Class Limits

Weight	115.5–135.5	135.5–155.5	155.5–175.5	175.5–195.5
f	3	9	13	7

means that some values (e.g., 135.5 and 155.5) appear in *two* classes. It must be understood, however, that by defining our real limits to the nearest half pound, we have necessarily carried them to one decimal place further than the actual observations, which, as indicated above, are measured to the nearest pound. thus, the problem of classifying, say, the 135.5-pound player can never arise.

The distinction between real and apparent class limits not only permits us to represent all values of a continuous variable, but it provides the means of computing the class width of any interval. In our maze-running problem, we discovered that subtracting the lower class limit from the upper class limit of any interval yields a difference of 19 even though the class width was 20. Observe what happens, however, when we perform the same calculation on the limits of an interval in our weight distribution:

$$155.5 - 135.5 = 20$$

The difference is, of course, that the data in our discrete example were classed in terms of apparent limits, while the continuous data were classed in terms of real limits. To avoid the computational difficulty we encountered in the maze-running example, therefore, we simply adopt the pretense that *all* variables are continuous and thus express discrete intervals, like continuous intervals, in terms of their real class limits. To generalize these considerations we shall introduce the expression ΔX (read *delta X*). In mathematics the *delta* notation represents the difference between two values x_{j-1} and x_j where $x_1 < x_2 < \ldots < x_n$. That is,

$$x_j - x_{j-1} = \Delta X$$

If in general, therefore, we define x_a and x_b, respectively, as the lower and upper real limits of the ith interval, then the class width of the interval is given by

$$x_b - x_a = \Delta X$$

Even if all intervals are of equal width ΔX, it is sometimes necessary to distinguish them by subscripts that indicate their ordinal positions in the distribution. We therefore identify the ith class as Δx_i. In the example given above, therefore,

$$\Delta x_1 = 115.5 \quad \text{to} \quad 135.5$$

$$\Delta x_2 = 135.5 \quad \text{to} \quad 155.5$$

and so forth. In general notation, then, our frequency distribution of maze-running errors would be represented as shown in Table 4.6

Table 4.6 Grouped Frequency Distribution for Maze-Running Errors Illustrating Use of Δ-Notation to Identify Intervals

Δx_i	115.5–135.5	135.5–155.5	155.5–175.5	175.5–195.5
f_i	9	28	41	22

In Table 4.6, f_i is the frequency of observations falling in the interval Δx_i.

In some contexts, each class interval is represented by its *midpoint* rather than by its limits. The midpoint of the ith interval is the X-value that lies halfway between the upper and lower real limits of the interval and is designated x_i:

$$x_i = x_a + \frac{(x_b - x_a)}{2}$$

$$= x_a + 0.5\Delta X$$

Conversely, the real limits of the interval may be expressed in terms of the midpoint x_i and the class width ΔX:

$$x_a = x_i - 0.5\Delta X$$

and

$$x_b = x_i + 0.5\Delta X$$

In the distribution presented above, we have defined our real class limits to the nearest "half an error," even though a random variable such as number of errors can assume only integer values. Similarly, in grouping our weight data, we defined our real class limits to the nearest half pound, even though our scale could measure only to the nearest pound. Whether a variable is discrete or continuous, therefore, we see that real class limits are a mathematical fiction. Indeed, we established this fiction in Chapter Two. Recall, in this regard, that we based our presentation of graphic display on the probability distribution of the number of heads observed in four tosses of a fair coin (Table 4.7).

Table 4.7 Probability Distribution for Number of Heads Obtained in Four Tosses of a Fair Coin

x_i	0	1	2	3	4
f_i	.06	.25	.38	.25	.06

Recall, too, that the most nearly mathematically correct graphic representation of this distribution is a series of five vertical lines, the heights, of which correspond to the probabilities of the five possible values of X. We also indicated, however, that to enjoy the greater readability of a histogram, we would hold formal mathematical concerns in abeyance and treat each possible X-value as an interval rather than a single point. We noted, for example, that the bar representing the probability of $x = 1$ was erected over the interval $+.05$ to $+1.5$. In the light of our present discussion, it is seen that this convention simply represents each discrete value as though it were a continuous interval of unit width ($\Delta X = 1.0$). The apparent limits of this fictitious interval are both equal to x_i, the midpoint, but the *real* class limits extend one-half a measurement unit beyond the apparent limits and are therefore equal to

$$x_i - 0.5 \quad \text{and} \quad x_i + 0.5$$

It is obvious, of course, that grouping integer data into classes of unit width achieves no economy of representation. In general, therefore, data are grouped into larger intervals. While this produces a more compact display of data, this manageability is achieved only at the expense of information. In our maze-running example, for instance, we can see that 29 observations fell in the second interval 135.5 to 155.5, but we do not know the frequency of any specific score, for example, $x = 42$. Depending on the application for which the data are intended, therefore, it is customarily assumed *either* that the observations in a given class are uniformly distributed across all possible values in the interval *or* that they are all concentrated at the midpoint of the interval.

B. Populations and Parameters

In Part One the presentation of probability theory was facilitated by the use of set notation. We frequently found it convenient, for example, to represent the set of all possible results as a sample space and to represent various subsets as events. Statistics also uses set notions, but as we move away from probability theory, we find that the nomenclature is sometimes a bit different. Instead of trials, for instance, we generally speak of *observations*, the elements of our sets become *samples* rather than results, and instead of sample spaces in general, we are likely to be concerned with one particular sample space, the *population*.

To explain the distinction between sample spaces and populations, let us suppose that a certain university has 25,000 students with each student's name filed on a separate IBM card. Now let us define the random experiment: *draw five IBM cards*. The sample space for this experiment would include every subset of five cards drawn from our set of 25,000 and would therefore contain

$$m = {}_{25,000}C_5$$

elements, or samples. If our experiment called for only three observations, our sample space would be defined by

$$m = {}_{25,000}C_3$$

samples. Suppose, however, that we make only one observation. For this experiment our sample space would include

$$m = {}_{25,000}C_1$$
$$= 25,000$$

samples. This particular sample space is defined as the population. Thus, the population is the set of all possible samples that may result from a single observation. More simply, we may define the population as the collection of all possible or potential units of observation. Thus, we might speak of the population of students in a particular university, the population of individuals with a particular genetic anomaly, or the population of bolts which a particular factory has produced or ever will produce.

These considerations notwithstanding, we noted in the introduction to Part Two that the scientist is seldom interested in all of the various attributes that the objects in his population may manifest. Instead, he is generally concerned about one or a few attributes that can be represented as the values of a random variable. Thus, as mentioned before, he might observe the *heights* of university students, the *I.Q.*'s of persons of a particular genotype, or the *diameters* of bolts forged in a particular plant. In this context, therefore, the population is conceptualized not so much as a collection of objects, but as a set of values. Furthermore, if the scientist observes every object in his population, this set necessarily includes every possible value that the random variable can assume within the definition of his experiment

$$\{x_1, x_2, \ldots x_n\}$$

When every object in the population has been observed, we say that the experiment has *exhausted* population. In this circumstance it will be recalled from Chapter One that the relative frequency f_i/N of each value x_i is precisely equal to the probability $p(x_i)$ associated with that value. If the entire population could be observed, therefore, the investigator would not only have the value set of his random variable, but the set of corresponding probabilities

$$\{p(x_i), p(x_2), \ldots p(x_n)\}$$

By the definition given in Chapter Two, therefore, he would have the probability distribution of his random variable X. Thus, we customarily treat the population distribution of a random variable as though it were a theoretical probability distribution, and the conventions by which we describe probability distributions are therefore applicable to description of population distributions.

Let us consider, therefore, what is involved in the description of a probability distribution. It makes intuitive sense that if a probability distribution is thoroughly described, we should be able to determine the probability of any value of the random variable. Recall from Chapter Two that such computations require the appropriate mathematical function. Thus, if a random variable X is distributed binomially, we know that the appropriate formula for computing the probability of any X-value is given by the binomial probability function

$$p(x = r) = \frac{N!}{r! \, (N - r)!} \, p^r q^{N-r}$$

It should be apparent, however, that even knowing the proper function we would be unable to compute the probability for any value of $X = r$ unless we also knew

84

the number of trials N and the probability of success p. These two values are the *parameters* of the binomial probability distribution.

> A parameter is a numerical index that describes or summarizes some characteristic of a probability distribution and that, together with the appropriate function, determines the probability of every value of the random variable.

Thus, when the description of a probability distribution includes both its probability function and the values of its parameters, it is possible to compute every probability associated with the random variable, and the distribution is therefore said to be completely *specified*. While it may seem superfluous, it should also be mentioned that different probability functions have different parameters. The number of trials N and the probability of success p are the parameters for the binomial distribution, but they are not the parameters for the normal distribution, Student's t distribution, the χ^2 distribution, or the F distribution.

We customarily specify the probability function and parameter values of a random variable using notation of the following form:

$$X : B(N, p)$$

which is read, X [is] distributed binomially with number of trials equal to N and probability of success equal to p. Thus, the character just to the right of the colon indicates the appropriate probability function, and the characters in parentheses give the values of its parameters. Other examples are

$$X : N(\mu, \sigma^2)$$

which is read, X [is] distributed normally with expectation *mu* and variance *sigma* squared, and

$$X : \chi^2 (\nu)$$

which is read X [is] distributed as chi-square with *nu* degrees of freedom.

Finally, it is important to note that parameter values remain constant within the context of any particular problem. In this unit we shall develop several parametric concepts, and it shall become apparent that we simplify much of our algebra by treating parameters as arithmetic constants. Barring coincidence, however, parameter values can be expected to differ from problem to problem. These two properties have given rise to the somwhat misleading characterization of parameters as "variable constants."

C. Samples and Statistics

Frequently a scientist is unable to observe every member of the population he is investigating. If, for example, a geneticist were studying all persons carrying a particular recessive gene, he would not be able to identify all of the members of his population, because it is unlikely that heterozygous carriers would be diagnosed. Alternatively, his population might defy exhaustive observation because it is purely conceptual. An industrial metallurgist, for example, might be concerned with all ingots of steel, *past*, *present*, and *future*, produced by a particular mill. Or, as is more often the case, the population might just be too large to treat exhaustively. Such would be the case for the behavioral scientist investigating some attribute of undergraduate university students. In cases such as these, the scientist must rely on observations taken on some subset of the population. Such a subset is called a *sample*. Thus, 25 diagnosed diabetics, 50 ingots of tool steel, and 25 students selected from an entire student body are samples of their respective populations. It should be understood that the individual observations that comprise the sample are of little intrinsic importance to the scientist. Rather, the observed values are of interest only inasmuch as their relative frequency distribution faithfully reflects the distribution properties of the population from which they are drawn. A sample is of little use, for example, if it includes a high frequency of values that have low probabilities in the parent population.

To illustrate how this might happen, let us suppose that a developmental physiologist is concerned with the distribution of heights in the population of male freshmen at some university. It is not feasible for him to measure every incoming student, so he decides to construct a relative frequency distribution based on the sample of 25 freshmen. He knows that during the week preceding registration all freshmen undergo physical examinations at the student health center. To obtain his sample, therefore, he goes to the health center and records the heights of the first 25 men to complete their examinations. Suppose, however, that the basketball coach has arranged with the health center to schedule all freshmen basketball players for the first day of physical examinations. Given this circumstance, the experimenter's sample would probably include a disproportionate representation of basketball players, and it is therefore likely that the relative frequency of heights exceeding, say, 74 inches would be considerably greater in his sample than $p(x > 74)$ in the freshman population. To avoid this sort of sampling error, we generally specify that our observations be selected by *random sampling*.

> Random sampling is defined as a method of selection that guarantees that all elements in the sample space have the same probability.

It should be apparent that this condition is not met in our example. Of all possible samples (i.e., of all possible combinations of 25 male freshmen), those which include basketball players are more probable than those which do not. Fortunately, most scientists are keenly aware of the danger of sampling error and have developed highly sophisticated methods of random sampling. There are, for example, computer programs that could take the entire roster of a freshman class and list every combination of 25 men. The experimenter could then blindly draw one such list and thus have his random sample. This procedure is tantamount to putting each man's name on a slip of paper and drawing 25 slips.

The reader should assure himself, incidentally, that the constraints of random sampling are satisfied whether the slips are drawn simultaneously or consecutively.

When an experimenter requires only one sample, the type of simple random sampling just described assures that his observations are as representative as possible of the parent population. Frequently, however, an experimenter requires several samples, and in such instances it is generally desirable that these samples be independent of one another. That is, for any two samples e_1 and e_2 drawn from the same population, it is often necessary that

$$p(e_2|e_1) = p(e_2)$$

Suppose, in this regard, that the investigator in the first example required two independent samples of 25 freshmen each. To obtain his second sample, let us further suppose that he simply choose the first 25 men to complete their physicals on the *second* day of examinations. It should be obvious that none of the students included in his first sample could possible appear in his second sample. Conversely, any combination e_2 that includes individuals observed in his first sample could not be chosen. That is,

$$p(e_2|e_1) = 0$$

This is, of course, because the first 25 individuals have been removed from the population, and all samples that include any of these persons are no longer included in the sample space. When an experiment requires repeated sampling, therefore, we generally stipulate that each sample be returned to the population before the next sample is drawn. This is known as *sampling with replacement* and assures the independence of all samples.

Once his data have been collected, however he selects his observations, the scientist must order them into some sort of comprehensible arrangement. Generally this will take the form of a relative frequency distribution. Whether his data be continuous or discrete, and whether his distribution be grouped or not, he now faces the task of describing the distribution. Just as population distributions are described by their parameters, so do we have have a corresponding body of conventional indices for describing relative frequency distributions of sample data. These indices are called *descriptive statistics*.

> A descriptive statistic is a numerical index that describes or summarizes some characteristic of a relative frequency distribution.

In summary, then, the probability distribution of a random variable in a population is described by various parameters, and the relative frequency distribution of a random variable in a sample is described by various statistics. It follows, therefore, that the exact values of population parameters are likely to remain unknown unless the experimenter can sample his population exhaustively. In contrast, the values of descriptive statistics can always be computed directly from the data that the experimenter observes.

D. Development of Statistics and Parameters in Part Two: A Preview

As the reader might imagine, one can generally conceive of a number of alternative mathematical expressions to describe any particular distribution property. Some of these alternatives, however, incorporate more information or are easier to handle than others. The descriptive statistics that are most frequently used have largely evolved by the process of elimination or modification to meet the following criteria:

i. *A descriptive statistic should be single-valued.* That is, a descriptive property should be represented by a single number.

ii. *A descriptive statistic should be algebraically tractable.* One should be able to calculate, transform, or otherwise manipulate a descriptive statistic using ordinary algebraic operations.

iii. *A descriptive statistic should consider every observed value of the random variable.* That is, every observed X-value should be included in the computation of the statistic.

iv. *A descriptive statistic should consider the frequency of every observed X-value.* The observed values of the random variable should contribute to the statstic in proportion to the frequency with which they appear in the sample.

As we consider various distribution properties we shall use these criteria to formulate appropriate descriptive statistics. The presentations will often parallel the historical evolution of the statistics and will permit us to develop the required formulas step by step in much the same way as we generated the binomial probability function in Chapter Three. We feel that it is easier to follow the development of a mathematical expression than to analyze it in retrospect. Once we have established a suitable statistic for describing a particular distribution property, we can use it to generate the formula for the corresponding population parameter. Recall from Chapter One that the relative frequency f_i/N of any value x_i approaches the probability of that value $p(x_i)$ as the number of observations N approaches infinity. A probability distribution, therefore, may be conceptualized as a relative frequency distribution based on an "infinite" number of observations. Thus, if we modify the mathematical expression of a descriptive statistic to accommodate the "infinitely large" sample, we obtain a parameter whose formula is structurally analogous to the statistic from which it is derived.

Statistical notation customarily distinguishes parameters from statistics by using Greek letters to represent parameters and Roman letters to represent statistics. Unfortunately, this convention was established after the development of the binomial distribution. The parameters of the binomial distribution are symbolized by the Roman letters N and p, and the reader has therefore been introduced to a major exception before being advised of the rule.

E. Problems, Review Questions, And Exercises

1. Toss 10 coins and record the number of heads. Repeat this experiment 10 times and tabulate the frequency distribution for the number of heads.
   a. Is this a discrete random variable or a continuous random variable?
   b. What are the apparent limits of your intervals?
   c. What are the real limits of your intervals?
2. Watch 3 hours of television, including 1 hour of some late-night talk show. Using a watch with a sweep hand, time to the nearest second every single commercial you see in this 3 hours. Tabulate the frequency of commercials as a function of length (in seconds).
   a. Is this a discrete random variable or a continuous random variable?
   b. What are the apparent limits of your intervals?
   c. What are the real limits of your intervals?
3. Conduct the following experiment. Write the name of each member of your immediate family (spouse, children, parents, and siblings) on a separate piece of paper. Select a pair of names by drawing two slips simultaneously.
   a. What is the sample space for this experiment?
   b. How many elements are in your sample space?
   c. What is the population in this experiment?
   d. Define the experiment for which the sample space would be identical with population.
4. If problem 3 had required that your two slips be drawn consecutively without replacement, would the sample space for the experiment have been any different?

chapter five

the concept of location

A disgruntled congressman is credited with the recent comment that adequate federal funding for social programs would be assured if ABC replaced its "Wide World of Sports" with the "Wide World of Hunger." While this story is probably apocryphal, it is grounded in the indisputable truth that news media devote more coverage to sports than to any other single category of news events. Consequently, most of us have grown up in an environment saturated with the statistics that describe various aspects of competitive athletics. Although we may not know how they are computed, for example, most of realize that batting averages, golf handicaps, and bowling averages tell us something about an individual athlete's *typical* performance.

These measures summarize many observations collected over many trials and therefore comprehend considerable information about an athlete's overall skill. Any summary, however, sacrifices a certain amount of information for the sake of brevity. Roger Maris's lifetime batting average of .260 simply tells us that over his entire major league career 26 percent of his turns at bat resulted in base hits. It does not tell us that he had a hot streak with the Yankees in 1961 and, while breaking Babe Ruth's record for home runs hit in a single season, averaged .269. Nor does it tell us that in 1966 he hit the ball only 81 times in 348 times at bat and was ignominiously traded to St. Louis. These statistics, therefore, tell us nothing about the extreme values of a random variable, and they are thus called measures of *central tendency*. In purely quantitative terms, these statistics tell us where the bulk of observations is located on the real line. Hence, measures of central tendency are also called measures of *location*. Because empirical data generally have a rather restricted range of potential values, applied statisticians prefer the term central tendency. On the other hand, many probability distributions are defined across the entire real line from $-\infty$ to $+\infty$, and probability theorists thus prefer to speak of location. Our own emphasis on concepts and principles inclines us toward the latter convention.

Location is an important property of a distribution, regardless of its shape or form. If a teacher wishes to gear his presentation to the capability of his students, he may assume that I.Q.'s will follow a symmetrical distribution with a few students in the extremes and most of the students in the middle. But it is important for him to know whether this popular middle range clusters around 100, 150 or 65. We have therefore devoted an entire chapter to this concept, and we begin with a discussion of the most common statistics for describing location: the mode (Mo.), the median ($\tilde{x}$), and the sample mean ($\bar{x}$).

A. The Mode: Mo.

In common parlance, something is in the mode if it is fashionable or popular. In statistics "popularity" refers to frequency of observation, and therefore the most frequently observed value of a random variable is called the *mode*. In our discussion of relative frequency distributions, we simulated the following distribution of heads observed in 25 tosses of four coins (Table 5.1).

Table 5.1 Frequency Distribution for Number of Heads Obtained in Toss of Four Coins with Experiment Repeated 25 Times

x_i	0	1	2	3	4
f_i	1	7	10	6	1

In this distribution two heads is the most frequently observed value of the random variable, and thus

$$Mo. = 2$$

The mode is easily determined and simple to interpret, but it has several undesirable properties. First, it is not necessarily single-valued. Two or more values may be observed with equal frequency. If half of our coins were biased in favor of heads and half were biased in favor of tails, we might obtain the *biomodal* distribution shown in Table 5.2.

Table 5.2 Bimodal Frequency Distribution for Number of Heads Obtained in Toss of Four Coins with Experiment Repeated 25 Times

x_i	0	1	2	3	4
f_i	3	7	6	7	2

Moreover, the mode (or modes) is determined entirely by inspection; it cannot be calculated arithmetically. Finally, the mode tells us only about the most popular value(s) in the distribution and considers neither the other values nor their relative frequencies.

B. The Median: $\tilde{x}$

A more commonly used statistic is the *median*, which, as the name suggests, is the value that falls in the middle of our array of observations. If N *ungrouped* observations are arranged in numerical order and if N is odd, the median is therefore equal to the value of the middle element. If N is even the median is equal to the average of the middle two elements. Thus, given the observations

$$x_1, x_2, \ldots x_N$$

where N is odd, the median value is given by the element in the $(N + 1)/2$ position. If N is even, the median is computed by taking the average of elements $(N/2)$ and $(N/2) + 1$. In either case the median is the point that separates the lower half of the array from the upper half and must therefore be single-valued.

For *grouped* data the median is defined as the value of X at or below which exactly 50 percent of the observations fall. This means that the sum of the frequencies for all X-values less than or equal to the median must equal $.50N$, where N is the number of observations:

$$\sum_i^n f_i = .50N$$

for all values $x_i \leq \tilde{x}$. Or, in terms of cumulative frequency,

$$Cf_{\tilde{x}} = .50N$$

This is generally considered to be the defining property of the median, and we therefore find the median frequently defined in terms of the frequencies of other values in the distribution. This was reflected in the statement above that the median is the value of X *at or below which exactly 50 percent of the observations fall.* The most obvious implication of this statement is that 50 percent of the observations also fall *above* the median, and it is therefore apparent that the median considers the frequencies of all values in the distribution. While it does indeed consider all frequencies, it does not consider the other values per se. That is, X-values are considered to be either greater than or less than the median, but their relative magnitudes are not distinguished. Thus, an extreme X-value and a moderate X-value make the same contribution to the median if both are observed with equal frequency. This is illustrated in the two distributions given in Table 5.3. Even though the second distribution includes an unusually small value (14), both distributions have the same median $\tilde{x} = 100$.

Table 5.3 Two Frequency Distributions with the Same Median

x_i	85	90	92	100	107	111	116
f_i	1	2	2	1	2	2	1
x_i	14	89	94	100	108	110	114
f_i	1	1	1	1	1	1	1

The median's most serious defect, however, is that it is nonalgebraic. This is sometimes difficult to see, because the median may be expressed in an algebraic formula. For example, if N is odd,

$$\tilde{x} = x_{(N+1)/2}$$

This expression involves two operations. First, the subscript $(N + 1)/2$ tells us to count the elements in our array until we reach the value in the $(N + 1)/2$ position. The equals sign tells us to set this value equal to the median $\tilde{x}$. Both of these operations, enumeration and equation, are indeed algebraic. However, the formula presupposes that the observations have been arrayed in numerical order, and numerical ordering is not an algebraic operation. To arrange a set of observations in numerical order, each value must be compared with every other value to determine its position. Such comparisons are considered to be logical rather than algebraic.

C. The Sample Mean: $\bar{x}$

Sometime in the course of your education you were taught to compute an average. The average, you were told, is simply the sum of several values divided by the number of values comprising the sum. In the intervening years you may have used this simple formula to compute such things as your average monthly income, the average cost of books per class, or perhaps your average highway speed maintained on a long trip. Certainly very few of you have escaped the omnipresent gradepoint average. This childhood friend is the most widely used of all measures of location and is about to renew your acquaintance as the arithmetic *mean* $\bar{x}$.

1. The Summation Symbol: Σ

Before we can discuss the sample mean, however, we must first introduce some new notation. Statistical work often requires that we represent the sum of an array of numerical values. These values might be the n elements of a value set, the N observations in a collection of data, or any other ordered arrangement of numbers. The sum of N observations, for example, could be represented in either of the following ways:

$$\begin{aligned} & x_1 \\ & x_2 \\ + & \ldots \qquad x_1 + x_2 + \ldots + x_N \\ & \underline{x_N} \end{aligned}$$

Both of these expressions are rather cumbersome, however, so statisticians use the symbol Σ (capital Greek *sigma*) to indicate a summation. In this notation the sum given above may be expressed more compactly as

$$\sum_{i=1}^{N} x_i$$

which is read, the sum of x_i with i running from 1 to N. The values 1 and N are called the *limits of summation* and tell us that our sum includes every observation in the array from x_1 to x_N. If it is understood that the sum begins with the first number in our array x_1, we may eliminate the lower limit of summation from our notation. Thus

$$\sum_{i}^{N} x_i$$

The Concept of Location

is identical to

$$\sum_{i=1}^{N} x_i$$

The upper limit of summation, however, will usually identify the particular group of numbers to be summed and will therefore always be expressed if there is *any* possibility of confusion. In this regard the reader must be alert to such important distinctions as

$$\sum_{i}^{N} x_i$$

the sum of the N observations, as opposed to

$$\sum_{i}^{n} x_i$$

the sum of the n elements of the value set.

To facilitate later reference to these notation conventions, they are presented again in Appendix I, where we also derive the rules governing algebraic use of the summation sign. The rules that will be encountered most frequently, however, are summarized below and should be thoroughly understood before proceeding.

Rule 1 If c is a constant,

$$\sum_{i}^{N} c = Nc$$

Rule 2 If c is a constant and X is any set of N real numbers,

$$\sum_{i}^{N} cx_i = c \sum_{i}^{N} x_i$$

Rule 3 If X is a set of N real numbers and Y is a set of N real numbers,

$$\sum_{i}^{N} (x_i y_i) = x_1 y_1 + x_2 y_2 + \ldots + x_N y_N$$

Rule 4 If X is a set of N real numbers and Y is a set of N real numbers,

$$\sum_{i}^{N} (x_i + y_i) = \sum_{i}^{N} x_i + \sum_{i}^{N} y_i$$

To continue with our discussion of the arithmetic mean, let us suppose that we have made the following observations on a random variable X:

5 4 2

As indicated above, we compute the average, or mean, by taking the sum of these values and dividing by the number of observations:

$$\bar{x} = \frac{5 + 4 + 2}{3} = \frac{11}{3} = 3.67$$

In the general case of N observations, this may be expressed by the definitional formula

$$\bar{x} = \frac{\sum\limits_{i}^{N} x_i}{N}$$

Statistical problems rarely deal with so few observations, however, so let us now suppose that our sample is comprised of the following nine observations:

$$5\ 5\ 4\ 4\ 4\ 4\ 2\ 2\ 2$$

Our sample mean is therefore given by

$$\frac{\sum\limits_{i}^{N} x_i}{N} = \frac{5 + 5 + 4 + 4 + 4 + 4 + 2 + 2 + 2}{9}$$

$$= \frac{32}{9} = 3.56$$

The reader will note, however, that our data consist of repeated observations of the same three values: 5, 4, and 2. That is, the observed value set of X is

$$\{5, 4, 2\}$$

Therefore, our observations can be represented more conveniently in the form of a frequency distribution.

x_i	5	4	2
f_i	2	4	3

In a similar fashion the sum of our nine observations can also be represented in terms of our three X-values and their respective frequencies:

$$5 + 5 + 4 + 4 + 4 + 4 + 2 + 2 + 2 =$$

$$(5 + 5) + (4 + 4 + 4 + 4) + (2 + 2 + 2) = 32$$

Each of the three sums enclosed in parentheses can, however, be expressed as a product. The sum of three 2's, for example, is equal to (2×3). Therefore the sum of our nine observations may be expressed as

$$(5 \times 2) + (4 \times 4) + (2 \times 3) = 10 + 16 + 6 = 32$$

It may be seen from the frequency distribution given above that each term in this expression is equal to the product of an element of the value set multiplied by its frequency. And in general any sum of N observations may be expressed as the sum of the n elements of the value set x_i each multiplied by its frequency f_i:

$$x_1 + x_2 + \ldots + x_N = x_1 f_1 + x_2 f_2 + \ldots + x_n f_n$$

or

$$\sum_i^N x_i = \sum_i^n x_i f_i$$

Recall that the numerator in the definitional formula for the arithmetic mean is simply the sum of the N observations comprising the sample

$$\sum_i^N x_i$$

If we substitute for this the sum given in frequency notation as derived above we obtain an alternative expression for the sample mean:

$$\bar{x} = \frac{\sum_i^N x_i}{N} = \frac{\sum_i^n x_i f_i}{N}$$

In this form it is readily apparent that the arithmetic mean satisfies all four of our criteria for a good statistic. It is single-valued, and it is algebraically tractable. In this regard, the above expressions indicates that computation of the mean involves only the operations of addition, multiplication, and division. Moreover, our summation runs from x_1 to x_n and every value of the random variable is therefore included in the calculation. Finally, we see that the frequency of every X-value f_i is also included in the computation. These last two criteria are violated only when the sample mean is computed from grouped data. In such instances, f_i becomes the frequency of the interval Δx_i, and x_i is the midpoint of the interval.

2. Interpretations of the Sample Mean

a. The Mean as the Center of Mass of the Distribution. Let us consider a random variable that for the sake of simplicity can take only positive values. We could represent the frequency distribution of this variable in a number of ways. We might, for example, simply tabulate the X-values and their frequencies. Or, we might graphically represent the distribution in the form of a histogram such as we discussed in connection with probability distributions. If we wished to be more concrete we might even construct a physical model of a frequency histogram. We could, for example, use a 12-inch ruler or a yardstick to represent the horizontal axis of our model, and the magnitude of each *X*-value would thus be given in inches. At each position on the ruler corresponding to an observed value of *X* we could place a wood block or other weight proportional to the frequency with which that value appears. These weights would then be analogous to the bars of a histogram. If we now locate a fulcrum at the point corresponding to the value of the mean $\bar{x}$, our weighted yardstick would be perfectly balanced. Using the data from our previous example, our model would look like Figure 5.1.

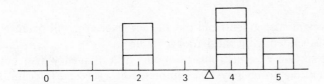

Figure 5.1 Physical model of frequency distribution illustrating mean as center of mass.

In this sort of physical representation, therefore, the mean is the exact center of mass of the distribution.

b. The Mean as the Point about which the Sum of Deviations is Zero. To understand why our physical model is precisely balanced at the mean, we must review some basic high school physics. In Figure 5.2 we have a lever, a fulcrum, and a weight of 100 pounds suspended from one end of the lever at a distance of 1 foot from the fulcrum. If the other end of the lever is 10 feet from the fulcrum, how much weight must we apply to balance the 100-pound weight?

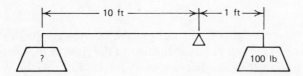

Figure 5.2 Physical model of frequency distribution representing sum of deviations about mean as total rotational force about fulcrum.

We all know from personal experience that when we apply weight to one end of a lever it causes the other end to move in the opposite direction. Moreover, the end of the lever does not move in a straight line, but pivots about the fulcrum with a certain amount of rotational force known as torque. Some 2000 years ago a waterlogged Greek named Archimedes informed the world that this rotational force is equal to the weight applied to the lever multiplied by its distance from the fulcrum. That is,

$$\text{Force} = \text{weight} \times \text{distance}$$

Therefore, the force exerted by the 100-pound weight must be

$$100 \text{ pounds} \times 1 \text{ foot} = 100 \text{ foot-pounds}$$

If we balance this weight, therefore, we need 100 foot-pounds at the other end of the lever. Recall that the longer lever arm is 10 feet long, and the required weight is therefore

$$\frac{100 \text{ foot-pounds}}{10 \text{ feet}} = 10 \text{ pounds}$$

Thus, a 10-pound weight suspended from the 10-foot lever arm will balance the 100-pound weight suspended from the 1-foot lever arm.

In the preceding section we constructed a model of a frequency distribution that we may now consider to be a system comprised of a lever, a fulcrum, and various weights. Because it is balanced, we know that the rotational forces on both sides of the fulcrum must be equal. However, the weights to the right of the fulcrum exert their force in a clockwise direction, while the weights to the left of the fulcrum exert an anticlockwise rotational force. The two forces therefore operate in opposite directions and therefore take opposite algebraic signs. Thus

$$(\text{weight} \times \text{distance})_{\text{right}} = -(\text{weight} \times \text{distance})_{\text{left}}$$

Recall that each "weight" represents the frequency f_i associated with a particular value x_i and that the position of the fulcrum is given by the mean $\bar{x}$. It is necessarily true, therefore, that the distance of any weight f_i from the fulcrum $\bar{x}$ must equal the difference

$$x_i - \bar{x}$$

Thus, the rotational force exerted by any weight f_i is equal to the product

$$f_i(x_i - \bar{x})$$

and the *total* force exerted by weights to the right of the fulcrum is equal to the *sum* of the products $f_i(x_i - \bar{x})$ for all values x_i located to the right of the mean $\bar{x}$. Similarly, the weights to the left of the fulcrum exert a total force equal to the sum of the products $f_i(x_i - \bar{x})$ for all X-values to the left of the mean. Once again, these two forces, or sums, are of *equal* magnitude but of *opposite* sign, and we can therefore say

$$\sum_i^L f_i(x_i - \bar{x}) = -\sum_i^R f_i(x_i - \bar{x})$$

where the limits of summation L and R indicate that the first sum is taken across all X-values to the left of $\bar{x}$, and the second sum is taken across all values to the right of $\bar{x}$. If we transpose the equation we obtain

$$\sum_i^R f_i(x_i - \bar{x}) + \sum_i^L f_i(x_i - \bar{x}) = 0$$

This means, of course, that the sum of the products $f_i(x_i - \bar{x})$ across *all n* values of X must equal zero:

$$\sum_{i}^{n} f_i(x_i - \bar{x}) = 0$$

Furthermore, we may convert our expression from frequency notation by removing the factor f_i and allowing our summation to run from 1 to N, the number of observations. Thus

$$\sum_{i}^{N} (x_i - \bar{x}) = 0$$

The difference $x_i - \bar{x}$ is defined as the *deviation* of x_i from the mean, and it is therefore seen that the mean is that value about which the sum of deviations is equal to zero.

D. The Population Mean: μ

In the preceding section we discovered that the sample mean is an index for describing central tendency that has all of the properties considered desirable in a descriptive statistic. If we consider the expression for the sample mean that we formulated using frequency notation, we may derive the corresponding parameter, the population mean μ. As given above,

$$\bar{x} = \frac{\sum_{i}^{n} x_i f_i}{N}$$

We may bring the denominator N into the numerator with some trivial algebraic manipulation:

$$\bar{x} = \frac{\sum_{i}^{n} x_i f_i}{N}$$

$$= \frac{1}{N} \sum_{i}^{n} x_i f_i$$

$$= \sum_{i}^{n} x_i \frac{f_i}{N}$$

Hence, the sample mean is equal to the sum of the n elements of the value set, each multiplied by its relative frequency f_i/N. If, however, our sample includes every member of the population, we know from Chapter One that the relative frequency f_i/N is precisely equal to the probability $p(x_i)$. For the entire population, therefore, the mean becomes

$$\mu = \sum_i^n x_i\, p(x_i)$$

Conceptualizing μ as the average of an exhaustive sample has considerable intuitive appeal and provides a comfortable rationale for generalizing many properties from the sample mean to the population mean. Given this interpretation, for example, it is easily seen that μ is the center of mass and, likewise, the point about which the sum of deviations is zero in the population distribution. In point of fact, however, conceptualizing μ as a "population average" is strictly appropriate under special conditions, and a more general development of μ is therefore presented below.

1. The Population Mean as the Expected Value of a Random Variable: $E(X)$

Suppose that an individual is approached by a service organization and is asked to buy a raffle ticket for $1. Unless he simply considers it to be charitable donation, he would probably try to determine the value of such an investment. That is, what is the ticket actually worth in terms of its potential returns compared with the certain forfeiture of $1? This assessment depends on two factors. First he must know the value of the prize. If the prize were worth less than $1 there is no way that the ticket could be considered a good investment. If, however, the prize were an $80 wristwatch the purchaser could conceivably make a $79 profit on his investment.

Even with a potential gain of $79, however, the value of the investment would also depend on the total number of tickets sold. If, for example, only one ticket is sold, he is certain to win, and his purchase is worth $79: $80 worth of merchandise less the $1 cost of the ticket. If, on the other hand, 100 tickets are sold he has a .99 probability of simply losing his $1 and only a .01 chance of making his $79 gain. He could therefore compute the worth of his ticket by multiplying the probability of winning by the net value of the win and subtracting from this the probability of losing multiplied by the cost of the loss:

$$(.01)\,(\$79.00) - (.99)\,(\$1.00) = \$.79 - \$.99 = -\$.20$$

This means that if an individual purchased tickets in many such raffles, he could expect a long-run loss of $.20 per ticket. Thus we say that the *expected value* of a ticket is $-\$.20$.

As the reader might have guessed the concept of expected value comes to us from games of chance. In this context a gambler may use "odds" to compute the expected gain or loss of a given wager over repeated play. We may therefore cast our question in probability terms and consider a game of chance to be an experiment with two possible outcomes, success and failure. In these terms the

102

monetary gain or loss associated with each outcome constitute the numerical values of a random variable. In our present example, then

$$X = \{79, -1\}$$

and the expected value of the ticket

$$E(X) = .01(79) + .99(-1)$$

In general, therefore, the expected value of a random variable is given by the sum of the values of the variable, each multiplied by its probability:

$$E(X) = \sum_i^n x_i\, p(x_i)$$

which is precisely the same formula we derived for the population mean μ. Recall, however, that our earlier development of μ as a "population average" was based on the notion of exhaustive sampling. Implicitly, therefore, it assumed a finite population. The derivation of μ as an expected value does not require this assumption and therefore provides a more widely applicable conception of the population mean.

E. The Algebra of Expectations

In the chapters that follow we shall have occasion to compute a number of expectations, and it is therefore imperative that the student know the algebraic conventions that will be employed. Rules for computing and manipulating mathematical expectations are given below, and they are derived in detail in Appendix II.

Rule 1   If X is a random variable with expectation $E(X)$ and Y is a random variable with expectation $E(Y)$

$$E(X + Y) = E(X) + E(Y)$$

Rule 2   If c is a constant,

$$E(c) = c$$

Rule 3   If c is a constant and X is a random variable,

$$E(X + c) = E(X) + c$$

Rule 4   If c is a constant and X is a random variable,

$$E(cX) = c\, E(X)$$

F. The Expectation of a Binomial Random Variable: $E(X)$ when $X:B(N, p)$

We shall conclude our treatment of location by using our newly acquired algebra of expectations to derive the expected value of a binomial random variable. Recall that a binomial random variable is generated by an experiment with the following characteristics:

i. The results of the experiment can be classified into one of two mutually exclusive categories, success or failure.
ii. The experiment is repeated a fixed number of times, N.
iii. Each trial is independent of every other trial, which implies that the probability of success p is constant across trials.

If we quantify the outcome of any binomial trial by assigning the value 1 to a success or the value 0 to a failure, we obtain what is known as an *indicator variable*. That is, a random variable I that can assume only the values 0 and 1:

$$I = \{0, 1\}$$

From our earlier discussions of binomial probability, we know that the probabilities associated with success and failure are p and $(1 - p) = q$, respectively, for every trial. Thus

$$p(I = 1) = p$$

$$p(I = 0) = q$$

We thus know the probability of every value that our indicator variable can take, and we can therefore compute its expectation:

$$E(I) = \sum_{i}^{n} I_i \, p(I_i)$$

$$= 0 \, (q) + 1 \, (p)$$

$$= p$$

Recall from Chapter Three that a binomial random variable is the total number of successes in N binomial trials. If we think of our binomial random variable as a function of a binomial experiment, it is therefore apparent that the variable is generated by assigning a value of 1 to each success, a value of 0 to each failure, and taking the sum across N trials. Therefore our random variable $X : B(N, p)$ must constitute the sum of N indicator variables:

$$X = I_1 + I_2 + \ldots + I_N$$

Therefore

$$E(X) = E(I_1 + I_2 + \ldots + I_N)$$

Recall from the algebra of expectations that the expectation of a sum of random variables $(X + Y)$ is equal to the sum of their individual expectations $E(X) + E(Y)$. Therefore

$$E(X) = E(I_1) + E(I_2) + \ldots + E(I_N)$$

By definition all indicator variables have the same value set $\{0, 1\}$. Because binomial trials are independent, moreover, we know that our N indicators also have the same set of probabilities $\{q, p\}$. This means that they are identically distributed and must therefore have identical expectations:

$$E(I_1) = E(I_2) = \ldots = E(I_N) = E(I) = p$$

Therefore

$$E(X) = \underbrace{p + p + \ldots + p}_{N \text{ times}}$$

$$= Np$$

The expectation of a binomial random variable is therefore given by the number of trials N multiplied by the probability of success p.

G. Problems, Review Questions, and Exercises

1. Draw a sample of 20 one-digit numbers from Table **B** of Appendix V.
   a. Identify the mode(s) and median of your distribution.
   b. Draw additional samples of 10, 5, and 2 one-digit numbers. Compute the sample mean $\bar{x}$ for each of the four samples you have drawn.
   c. The numbers in Table **B** are uniformly distributed. That is,

   $$p(0) = p(1) = \ldots = p(9) = .10$$

   If Table **B** is considered a population of one-digit numbers, what is $E(X)$?
   d. Compare the sample means computed in **1b** with $E(X)$ computed in **1c**. What happens to $\bar{x}$ as N increases?
   (*Note.* Save your data from problem **1** for use in later problem sets.)
2. Compute the expected value of the random variable discussed in problem **4** of Chapter Two.
3. Suppose a person invests 50¢ in a state lottery ticket. Suppose further that for every million tickets sold, the state will draw 17 prizes of $10,000 each.
   a. Tabulate the probability distribution of the random variable in question.
   b. Compute $E(X)$ for the random variable.

A. The Range
B. The Mean Deviation from the Mean
C. The Mean Absolute Deviation from the Mean
D. The Sample Variance: s^2
E. The Population Variance: σ^2
  1. The Population Variance as an Expected Squared Deviation:
     $V(X) = E[X - E(X)]^2$
  2. The Computational Formula for the Population Variance:
     $V(X) = E(X^2) - [E(X)]^2$
F. The Algebra of Variances
G. The Variance of a Binomial Random Variable: $V(X)$ when $X:B(N, p)$
H. Problems, Review Questions, and Exercises

the concept of variability

Let us suppose that an economics instructor has one introductory class that meets during regular day session and one that meets at night. Both classes receive the same lectures and use the same text, and he therefore gives the same midterm examination to both. He knows, however, that his day class is composed primarily of full-time students and that his night class includes many business people and working students who attend school on a limited basis. He suspects that this difference in class composition may influence test performance, and he therefore compares the examination scores of his two classes with considerable care. The examination has 100 possible points, and the distributions of scores for the two classes are represented in Figure 6.1.

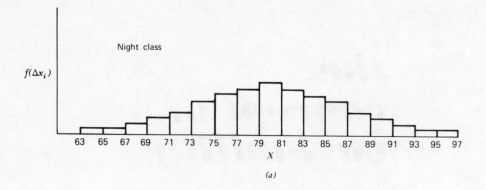

(a)

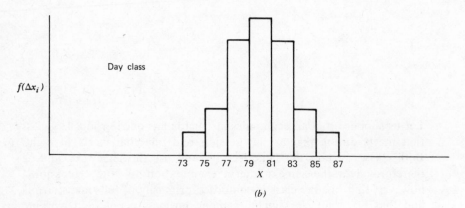

(b)

Figure 6.1   Frequency distributions of examination scores for two classes.

As indicated in Figure 6.1, both groups averaged approximately 80 points on the examination, but it is fairly evident that the scores of the night class were spread across a much wider range than the scores of the day class. On this basis the instructor might well surmise that some of his night students find the material exceptionally difficult while others find it too elementary. If possible, therefore, he might divide the class into one accelerated section and one elementary section.

Note, however, that neither the form nor the location of either distribution provides the basis for such a decision. Both distributions are symmetrical and unimodal with means, medians, and modes in the interval 79 to 81. Instead the decision would be prompted by the degree of spread in the two groups of scores. This spread indicates a considerable degree of *variability* in the performance of the night class, and we can therefore see that such variability, or *dispersion*, is an important characteristic of distributions.

In the present chapter we examine the concept of dispersion, or variability, in much the same fashion as we examined location in Chapter Five. We therefore begin our presentation with a discussion of the most frequently suggested indices

The Concept of Variability

for describing variability. As before, our evaluation of such statistics will be based on their conformity to the criteria presented in Chapter Four.

A. The Range

Perhaps the most obvious measure of dispersion is simply the difference between the largest observation and the smallest observation. This statistic is known as the *range* and is given by the formula

$$\text{Range} = x_{max} - x_{min}$$

Thus, in the following array, the range is equal to $90 - 70 = 20$.

$$70 \quad 73 \quad 76 \quad 80 \quad 85 \quad 87 \quad 90$$

Like the mode, the range is easy to interpret and simple to compute, but unfortunately it satisfies only one of our criteria for a good statistic: It is single-valued. This value, however, is obtained entirely by inspection. Determination of the range requires that we first identify the minimum and maximum values of the random variable. This means that each observation must be compared with every other observation, and the reader will recall that such comparisons are logical rather than algebraic. Second, the range utilizes only two values. The other $n - 2$ values make no contribution to its computation. Finally, the frequencies of even the maximum and minimum X-values are not considered in the calculation of the range. If we examine the two ungrouped frequency distributions in Table 6.1 it becomes evident that these deficiencies can give a misleading assessment of variability.

Table 6.1   Two Frequency Distributions With The Same Range

x_i	50	51	52	53	54	55	56	57	58	59	60	61	62	63	64	65	66	67	68	69	70
f_i	2	2	2	2	2	2	2	2	2	2	2	2	2	2	2	2	2	2	2	2	2

x_i	50	51	52	53	54	55	56	57	58	59	60	61	62	63	64	65	66	67	68	69	70
$!_i$	1	0	0	0	0	0	0	3	6	8	10	6	4	2	0	0	0	0	0	0	2

In the first distribution the observations are uniformly distributed from 50 to 70, and the 20-point range therefore provides a meaningful index of dispersion. In the second distribution, however, all but three of the observations lie between 57 and 63. Nonetheless, the three atypical, or outlying, values give this distribution also a range of 20 even though we have no observations between 51 and 56 or between 64 and 69.

B. The Mean Deviation from the Mean

As indicated above, the range depends on the difference between just two values, x_{max} and x_{min}, which may or may not be typical of the entire sample. We could obtain a more representative index, and thereby avoid the difficulties illustrated in the last example, by taking the average difference among all observations. To compute this statistic we subtract each of the N observations x_j from each of the other $N - 1$ observations x_i, giving us a total of $N(N - 1)$ differences to average.

$$\frac{\sum (x_i - x_j)}{N(N - 1)}$$

By stipulating that we include every difference, however, we include both $(x_i - x_j)$ and $(x_j - x_i)$. These two values will, of course, be of equal magnitude but of opposite sign, giving us a net sum of zero. We could avoid this problem by considering only one difference for every pair of values, thus giving us a total of $N(N - 1)/2$ terms. But even if we only considered half of the differences, this could still involve a prohibitively large number of calculations. With 100 observations, for example, we would have 4950 differences to compute and sum together. As an alternative, therefore, we might consider the average difference, or deviation, of all observations from some constant. This would give us only N differences to compute, but it begs the question of selecting an appropriate constant. It makes intuitive sense that the value of our constant should be typical of the values observed in our data, and we might therefore select the most frequently observed value, the mode. However, if we want our index of variability to reflect the average difference among all observations, our constant should be some function of *all* N observations. On these bases the most appropriate constant is the sample mean $\bar{x}$.

Our reasoning has therefore led us to suggest that the average, or mean, deviation from the sample mean might be an appropriate and convenient measure of dispersion:

$$\frac{\sum_{i}^{N} (x_i - \bar{x})}{N}$$

However, it will be recalled from Chapter Five that the mean is that point about which the sum of the deviations is always zero. Therefore, the numerator in our expression is zero, and our mean deviation from the mean must likewise equal zero irrespective of the variability of the distribution. A singularly unattractive property in a statistic designed to measure variability!

110

C. The Mean Absolute Deviation from the Mean

Recall from Chapter Five that the sum of the positive deviations from the mean is always equal in magnitude to the sum of the negative deviations from the mean. This is why the sum of *all* deviations must always equal zero. If, however, all deviations were treated as *positive* values, this property would disappear, and the mean deviation would become a meaningful index of variability. When we consider only the magnitude of a number and disregard its sign, we speak of the *absolute* value of the number. We signify absolute value by enclosing the number between vertical lines. Thus, in absolute terms,

$$+10 = -10 = |10|$$

We might therefore define our index of dispersion as the average absolute deviation from the mean:

$$\frac{\sum_{i}^{N} (|x_i - \bar{x}|)}{N}$$

This is a meaningful dispersion measure and is frequently used in nonstatistical contexts. Unfortunately, absolute quantities are not algebraically tractable. One cannot, for example, perform simple algebraic addition with absolutes. We know that $5 - (+3) = 2$, and $5 - (-3) = 8$, but what does $5 - |3|$ equal?

When one considers the intuitive appeal of this statistic, its sole defect may seem trivial. And indeed statisticians could probably establish a convention governing algebraic manipulation of absolutes. Such a convention would be superfluous, however, as there already exists an established method for converting negative values to positive values.

D. The Sample Variance: s^2

As the reader will recall from elementary algebra, the product of two negative numbers is always positive. We therefore remove all negative signs by the simple expedient of squaring our deviations. Thus, if $y_i = -y_j$

$$(y_i)^2 = (y_j)^2$$

If we let $y_i = (x_i - \bar{x})$ and $y_j = (x_j - \bar{x})$, then

$$(x_i - \bar{x})^2 = (x_j - \bar{x})^2$$

Squaring is simply an application of multiplication and is therefore entirely algebraic. We may therefore define our index of dispersion as

$$\frac{\sum_{i}^{N} (x_i - \bar{x})^2}{N}$$

This is the average *squared* deviation from the mean and is defined as the sample variance, s^2.

In a large sample we may expect any value x_i to appear several times, and as before we may therefore represent our N observations more economically as a set of n values $\{x_1, x_2, \ldots x_n\}$ and n corresponding frequencies $\{f_1, f_2, \ldots f_n\}$. In this connection, recall from Chapter Five that the sum of N observations may be expressed as the sum of the n elements, each multiplied by its frequency.

$$\sum_i^N x_i = \sum_i^n x_i f_i$$

Similarly,

$$\sum_i^N (x_i - \bar{x})^2 = \sum_i^n (x_i - \bar{x})^2 f_i$$

and

$$s^2 = \frac{\sum_i^N (x_i - \bar{x})^2}{N} = \frac{\sum_i^n (x_i - \bar{x})^2 f_i}{N}$$

When expressed in this form it may be seen that the variance satisfies all four criteria for a good statistic. It is single-valued, and as indicated above it involves only algebraic operations. Moreover one can see from the limits of summation that the variance incorporates all n observed values x_i. Finally, each squared deviation $(x_i - \bar{x})^2$ enters the sum with the same frequency f_i as the X-value from which it is computed.

The only drawback to the sample variance is that it is given in units that are difficult to interpret. The other dispersion statistics that we discussed are in the same metric as the values from which they are computed. If our data were heights of incoming freshmen men, for example, a range of 24 would indicate that there was a 24-inch difference between the tallest man and the shortest man. When we compute the variance, however, each observation (or, more precisely, the deviation of each observation) is squared, and the variance is therefore given in units squared. In the present example, a variance $s^2 = 16$ would represent 16 *square* inches, which has very little meaning in the measurement of height.

For purposes of description, therefore, we frequently take the square root of the variance, which is defined as the *sample standard deviation:*

$$s = \sqrt{\frac{\sum_i^N (x_i - \bar{x}^2)}{N}} \quad or \quad s = \sqrt{\frac{\sum_i^n (x_i - \bar{x})^2 f_i}{N}}$$

The standard deviation is always assumed to be the *positive* root of the variance and therefore gives us an index in the same units as the random variable.

The procedures for computing s^2 or s from grouped data are no different, but as with computation of the sample mean, f_i becomes the frequency associated with Δx_i, x_i is the midpoint of Δx_i, and n is the number of intervals in the distribution.

The Concept of Variability

E. The Population Variance: σ^2

In Chapter Five we used the formula for the sample mean $\bar{x}$ to generate the definitional expression for the population mean μ. In the same fashion we may generate the formula for the population variance σ^2 from the sample variance s^2.

$$s^2 = \frac{\sum_i^n (x_i - \bar{x})^2 f_i}{N}$$

If we bring the denominator N into the numerator and into the summation sign, this becomes

$$s^2 = \sum_i^n (x_i - \bar{x})^2 \frac{f_i}{N}$$

As in Chapter Five, let us now assume that our sample exhausts the parent population. Under this circumstance

$$\frac{f_i}{N} = p(x_i)$$

$$\bar{x} = \mu$$

and the formula for s^2 given above thus becomes

$$\sum_i^n (x_i - \mu)^2 \, p(x_i)$$

which defines the population variance σ^2.

As with the sample variance s^2, the population variance σ^2 is expressed in units that are difficult to interpret. To describe the dispersion of a population distribution in the same metric as the observations, we therefore use the population standard deviation:

$$\sigma = \sqrt{\sigma^2}$$

$$= \sqrt{\sum_i^n (x_i - \mu)^2 \, p(x_i)}$$

1. The Population Variance as an Expected Squared Deviation: $V(X) = E[X - E(X)]^2$

Recall from our development of s^2 that the sample variance is by definition the *average* squared deviation from the sample mean. Any such sample average, however, requires division by the sample size N, and it should therefore be apparent that our development of

$$\sigma^2 = \sum_i^n (x_i - \mu)^2 \, p(x_i)$$

as the variance of an exhaustive sample is not applicable to infinite populations. This same restriction was, of course, also encountered in our development of μ in Chapter Five, but it will be recalled that conceptualizing the population mean as an *expected* value rather than as an *average* value permitted a more general derivation. If, in a similar fashion, we consider the population variance to be the *expected* squared deviation from the population mean $E[X - E(X)]^2$, we may likewise derive our definitional formula for σ^2 without assuming a finite population. When expressed as an expectation, the population variance is customarily symbolized as $V(X)$. Thus

$$V(X) = E[X - E(X)]^2$$

For the sake of simplicity, however, let us represent our deviation $[X - E(X)]$ by the letter D. Thus

$$V(X) = E(D^2)$$

By the definition of expected value, then

$$E(D^2) = \sum_{i}^{n} d_i^2\, p(d_i^2)$$

By the substitution we stipulated above, however, d_i must equal $[x_i - E(X)]$, which means that

$$d_i^2 = [x_i - E(X)]^2$$

Thus, d_i^2 occurs when and only when $x = x_i$, and it therefore follows that

$$p(d_i^2) = p(x_i)$$

Thus

$$E(D^2) = \sum_{i}^{n} d_i^2\, p(x_i)$$

We know, however, that $d_i^2 = [x_i - E(X)]^2$, and making this substitution our expression becomes

$$E(D^2) = \sum_{i}^{n} [x_i - E(X)]^2 p(x_i)$$

or

$$V(X) = \sum_{i}^{n} [x_i - E(X)]^2\, p(x_i)$$

This is the definitional formula traditionally given for the population variance. It may be seen that it is completely parallel to the expression developed above.

$$\sigma^2 = \sum_{i}^{n} (x_i - \mu)^2\, p(x_i)$$

with $V(X)$ substituted for σ^2 and $E(X)$ substituted for μ.

2. The Computational Formula for the Population Variance:
$$V(X) = E(X^2) - [E(X)]^2$$

Both of the formulas discussed above are somewhat tedious to use in actual computation. We can, however, use the conceptual formula to derive an expression that considerably facilitates manual or machine calculation of the population variance.

$$V(X) = E[X - E(X)]^2$$
$$= E\{X^2 - 2X E(X) + [E(X)]^2\}$$
$$= E(X^2) - E[2X E(X)] + E\{[E(X)]^2\}$$

In the second term, $E(X)$ and 2 are both constants, and it will be recalled from the algebra of expectations that $E(cX) = c E(X)$. In the third term, $[E(X)]^2$ is a constant, and $E(c) = c$. Therefore, our expression becomes

$$V(X) = E(X^2) - 2 E(X) E(X) + [E(X)]^2$$
$$= E(X^2) - 2 [E(X)]^2 + [E(X)]^2$$
$$= E(X^2) - [E(X)]^2$$

F. The Algebra of Variances

We shall compute a number of variances in the chapters that follow, and it is therefore important that the student understand the conventions governing their manipulation. The rules are derived in detail in Appendix II and are summarized below.

Rule 5 If X and Y are independent random variables with variances $V(X)$ and $V(Y)$, respectively,

$$V(X + Y) = V(X) + V(Y)$$

Rule 6 If c is a constant

$$V(c) = 0$$

Rule 7 If c is a constant and X is a random variable

$$V(X + c) = V(X)$$

Rule 8 If c is a constant and X is a random variable

$$V(cX) = c^2 V(X)$$

G. The Variance of a Binomial Random Variable: $V(X)$ when $X:B(N, p)$

At the end of the last chapter we derived this expected value of a binomial random variable, and we shall conclude this chapter with a similar derivation of the variance of a binomial random variable.

As before we define the indicator variable I to represent the possible outcomes of any single binomial trial:

$$I = \{0, 1\}$$

where 0 corresponds a failure and 1 corresponds to a success. Hence

$$p(I = 1) = p$$
$$p(I = 0) = q$$

From the computational formula for variance

$$V(I) = E(I^2) - [E(I)]^2$$

We first solve for the second term $[E(I)]^2$. From our derivation of the expectation of a binomial random variable we know know that

$$E(I) = p$$

Therefore

$$[E(I)]^2 = p^2$$

We now solve for the first term, $E(I^2)$. By the definition of expected value we know that

$$E(I^2) = \sum_i^n I_i^2\, p(I_i^2)$$

where

$$I^2 = \{0^2, 1^2\} = \{0, 1\}$$

Moreover, since I cannot be negative, $I^2 = 1$ only when $I = 1$. Thus

$$p(I^2 = 1) = p(I = 1) = p$$

Furthermore, since I^2 can assume only the two values 1 and 0,

$$p(I^2 = 0) = 1 - p = q$$

Thus

$$E(I^2) = \sum_i^n I_i^2\, p(I_i^2)$$
$$= (1^2)p + (0^2)q$$
$$= p$$

116

Therefore

$$V(I) = E(I^2) - [E(I)]^2$$
$$= p - p^2$$
$$= p(1 - p)$$
$$= pq$$

Recall again that a binomial random variable is the total number of successes in N binomial trials. Therefore our random variable $X : B(N, p)$ may be represented as the sum of N indicator variables:

$$X = I_1 + I_2 + \ldots + I_N$$

and

$$V(X) = V(I_1 + I_2 + \ldots + I_N)$$

By definition all trials of a binomial random experiment are independent, and from our algebra of variances the variance of a sum of independent random variables $(X + Y)$ is equal to the sum of their individual variances $V(X) + V(Y)$. Therefore

$$V(X) = V(I_1) + V(I_2) + \ldots + V(I_N)$$

By definition all indicator random variables must be identically distributed with the same value set $\{0, 1\}$ and the same set of probabilities $\{q, p\}$. They must therefore have identical variances.

$$V(I_1) = V(I_2) = \ldots = V(I_N) = V(I) = pq$$

Therefore

$$V(X) = \underbrace{pq + pq + \ldots + pq}_{N \text{ times}}$$

$$= Npq$$

The variance of a binomial random variable is given by the product of the number of trials N multiplied by the probability of success p multiplied by the probability of failure q.

H. Problems, Review Questions, and Exercises

1. Compute the sample variance s^2 for each sample drawn in problem **1** of Chapter Five.
2. Recall that the numbers in Table **B** of Appendix II are uniformly distributed.
   a. If Table **B** is considered a population of one-digit numbers, what is $V(X)$?

b. Compare the sample variances computed in problem **1** with $V(X)$ computed in **2a**. What happens to s^2 as N increases?

3. Compute the variance of the random variable discussed in problem **4** of Chapter Two.

4. Compute the variance $V(X)$ of the random variable defined by the experiment in problem **1** of Chapter Four.

the concept of standardization

A. Measurement in the Sciences

It is generally accepted as a truism in the scientific community that basic knowledge develops only as fast as our ability to measure precisely the phenomena that we observe. Thus, classical Newtonian physics was firmly grounded in the empirical experiments of Galileo, who determined the acceleration of gravity at the earth's surface:

$$g = 32 \text{ ft/sec}^2$$

Similarly, modern physics could not have developed without the earlier experiments of Michelson, who precisely measured the speed of light. And more recently, the molecular structure of DNA was established only after X-ray diffraction techniques provided Watson and Crick with precise estimates of intramolecular distances in the DNA crystal. The importance of measurement in the sciences is aptly expressed in the following quotation by the nineteenth-century British physicist, Lord Kelvin:

I often say that when you can measure what you are speaking about, and express it in numbers, you know something about it; but when you cannot express it in numbers, your knowledge is of a meager and unsatisfactory kind.

Certainly not even Lord Kelvin would claim that scientific knowledge proceeds *directly* from measurement. What is crucial to the establishment and confirmation of scientific hypotheses is *comparisons* among measurements. Thus, Galileo completely revised the seventeenth century conception of the physical universe when he compared the times required by light and heavy objects to fall the same distance and found them to be identical.

In order to compare two measurements, however, it is necessary that both be expressed in the *same scale of measurement*. This requirement has two conditions. First, both measurements must be taken from the *same reference point*. When measuring physical properties, such as length or volume, it is generally assumed that the reference point is absolute zero. Thus, a length of 15 inches is usually assumed to mean 15 inches *greater than* 0 *inches*. This is not always the case, however. Consider the measurement of temperature. Which is warmer, 25° C or 293° K? Although 293 is a larger number than 25, the Kelvin and Centigrade scales do not have the same zero point; 0° C is equal to 273° K and 25° C is therefore 5 degrees warmer than 293° K.

The second requirement is that our measurements be in *units of equal magnitude*. If object A is 15 units long and object B is only 11 units long, we cannot conclude that A is longer than B unless we know that both are expressed in units of the same length. If A is 15 *inches* long, for example, and B is 11 *yards* long, then B is decidedly longer than A. In this regard, suppose that an American tourist discovers that gasoline averages 50¢ per gallon in Canada as compared with 40¢ per gallon in the home town. On the basis of this comparison he concludes that Canadian gasoline is 10¢ per gallon more expensive than the domestic product. This is based on the implicit assumption that Canadian gallons and United States gallons are of equal volume. In point of fact, however, Canadian petroleum products are measured in *Imperial* gallons, which are approximately equal to 1.2 United States (or *wine*) gallons, and our fictitious tourist therefore paid precisely the same price in Canada as he would have paid at home.

1. Comparison of Physical Measurements

To illustrate the basic problem of comparing incompatible measurements, let us fabricate a familiar example. Jim and Joe have planned a hunting trip and are debating which is the shorter of two possible routes. They know that their maps are outdated by new highway construction, and they therefore elect to take both cars, one by each route, and compare the distances recorded on their odometers. They leave from Jim's house the next morning and when they arrive at the trailhead in the afternoon, Jim's odometer reads 16,225 and Joe's reads 5050. Before they can compare their measured distances, however, they must first subtract the mileage that was *already* on the two cars when they departed that morning. They are, after all, concerned only with the *net* mileage logged on this particular trip and not the total mileage accumulated since the cars were purchased.

	Jim	Joe
Odometer readings at trailhead	16,225	5,050
Odometer readings at Jim's house	16,000	5,000
Net distance traveled	225	50

Note that by subtracting the mileage recorded at the beginning of the trip they have satisfied the first requirement for a meaningful comparison of measurements. They have transformed the two measurements so that both originate from the same reference point, Jim's house. At this juncture it certainly appears that Jim took the long way around. But, this assumes that both odometers were recording distance in the same units of measurements. In fact, however, Joe's car is a rare model manufactured in the Black Forest (by elves), and his odometer is calibrated in *leagues*. Jim's car, on the other hand, is a somewhat better-known model manufactured in Milan (in hysterics), and his odometer is calibrated in *kilometers*. Before they can conclude that Joe's route is shorter, therefore, they must first express their readings in some common unit of measurement and thus satisfy the second requirement for meaningful comparison.

In a spirit of international compromise they agree to convert their measurements to English miles. For Joe, this conversion requires that he divide his net distance, given in leagues, by the number of leagues per mile:

$$\frac{\text{Distance in leagues}}{\text{Number of leagues per mile}} = \text{Distance in miles}$$

Similarly, Jim must divide his recorded distance in kilometers by the number of kilometers per mile:

$$\frac{\text{Distance in kilometers}}{\text{Number of kilometers per mile}} = \text{Distance in miles}$$

Being well-equipped outdoorsmen they carry the most recent edition of the *Handbook of Chemistry and Physics* (in paperback), and a quick glance at the appropriate conversion table tells them that a mile equals .333 leagues and that a mile equals 1.61 kilometers. Expressed in miles, therefore, the league of Joe's route is

$$\frac{50 \text{ leagues}}{.333 \text{ leagues per mile}} = 150 \text{ miles}$$

and to the nearest mile, the length of Jim's route is

$$\frac{225 \text{ kilometers}}{1.61 \text{ kilometers per mile}} = 140 \text{ miles}$$

With the two measurements thus converted to a common scale of measurement we can now see that Joe's route was 10 miles longer than Jim's.

2. Comparison of Behavioral Measurements: Standardization

Although we chose an example involving measurement of a physical property, distance, the importance of measurement is certainly not restricted to the physical sciences. Indeed, measurement also finds manifold application in the social and behavioral sciences. Economists conduct market surveys to measure consumer demand. Teachers administer examinations to measure acquisition of course material. Politicians hire pollsters to measure public opinion. Psychologists devise tests to measure personality traits. And, as we all know, colleges use entrance examinations to measure scholastic aptitude. And as in the physical sciences, these measurements are used to make comparisons. Thus, college entrance examinations are used to select the most promising applicants, market surveys are used to identify new territories with high sales potential, and examinations are used to determine the grade that each student in a class will receive.

In making these comparisons the behavioral scientist is generally confronted with precisely the same problems as our two hunters. The quantities he wishes to compare must be expressed in common units of measurement relative to a fixed reference point of origin. The nature of behavioral properties, however, makes this problem considerably more complex than in the physical sciences. Comparison of physical measurements is facilitated by the establishment of conventional units whose meaning is defined *independently* of the particular measurements under consideration. Thus, when we say that Jim traveled 225 kilometers, his distance may be interpreted in terms of an *external* reference, the standard meter. On the other hand, when we say that a student received a score of 600 on the Scholastic Aptitude Test, the measurement assumes meaning *only* in comparison to the scores of other individuals who complete the *same* examination. That is, there exists no external standard in terms of which this measurement may be meaninfgully interpreted. To convert his measurements into a scale that permits meaningful comparisons, therefore, the behavioral scientist relies on statistics rather than extrinsic standards. This procedure is called *standardization* and is illustrated on the following example.

Suppose that a college instructor is especially concerned about the progress of one particular student. He has the student's test results from the first two midterm examinations, and he wishes to determine whether his performance improved, declined, or remained constant. The two scores are 63 and 75, and in absolute terms it appears that the student performed better on the second test. However, the instructor has no reason to assume that his two tests were scored in the same scale of measurement. The first examination, for example, might have been more difficult than the second, in which case the student's relative performance might not have improved at all. These raw test scores are therefore analogous to the final odometer readings in the last example, and, like the hunters, our instructor must express his scores in reference to some common point and convert them to some common unit of measurement.

Recall, in this regard, that the two principals in our last example established

Jim's house as their point of reference. That is, they were concerned only with the distance to the trailhead *in relation* to Jim's house. Each driver therefore subtracted from his final odometer reading a value (i.e., his initial reading) that *described the location* of Jim's house. In the present example, meaningful interpretation requires that the test scores be expressed *in relation* to the scores of other students who took the same examinations. Our instructor therefore subtracts from each score a value that *describes the location* of its distribution. From Chapter Five we know that the location of a distribution is best described by its mean, μ, and the instructor therefore defines the class mean as his reference point. The class mean for the first examination was 50, and the mean for the second examination was 70. Subtracting these two values from the corresponding test scores, we obtain the following *deviation scores*:

$$x_1 - \mu_1 = 63 - 50 = 13$$

$$x_2 - \mu_2 = 75 - 70 = 5$$

If he is to interpret these deviation scores as "distances" from a common reference point, the instructor must necessarily assume that the two means, μ_1 and μ_2, represent the *same* point in much the same way as the two initial odometer readings, 16,000 and 5000, represented a common point of departure in the earlier problem. In the present context this is not an unreasonable assumption. Although the distribution of scores on any test depends on the metric properties of the *test*, the distribution of individual differences that underlie test performance is a property of the test *population*. The distribution of test results may thus change from test to test, but the distribution of abilities will not. Hence, while the test mean may take different numerical values from test to test, it will always reflect the average ability of the test population, which we assume remains fixed. On the first midterm, therefore, the student scored 13 points above *the* class mean but he only exceeded *the* class mean by 5 points on the second midterm.

Once again, however, our instructor has no guarantee that the two tests were scored in the same units of measurement. If, for example, the first midterm had 150 possible points and the second had only 100 possible points, then the points on the second examination might be "larger" than the points on the first, much as leagues are larger than kilometers. Thus, a deviation score of 5 points on the second test might actually represent *better* performance than a deviation score of 13 on the first, just as 50 leagues is longer than 225 kilometers. The two deviation scores must therefore be converted to some common unit of measurement.

Although the instructor has no external standard to which he can refer his measurements, this is not an insurmountable difficulty. A standard is, after all, nothing more than a fixed interval that is expressed in the same units as the various measurements one wishes to compare. In the travel problem the drivers were able to convert their measurements to miles, because the mile is an interval of fixed magnitude that their conversion table expressed in both leagues and kilometers. In the present example, therefore, our instructor requires some fixed interval that is expressed in the "points" for which both examinations were scored.

If he once again assumes that the distribution of student performance remains constant from test to test, he must conclude that the class should exhibit precisely the same amount of *variation* on the two examinations. This implies that the standard deviations of his two tests, which appear below, represent intervals of equal length.

$$\sigma_1 = 15$$

$$\sigma_2 = 5$$

Although we assume the two standard deviations to be equal, we note that they take different numerical values. In this regard, recall from Chapter Six that the standard deviation is always expressed in *the same units of measurement as the variable it describes*. We may therefore change the above expressions to read

$$\sigma = 15 \text{ test points on the first examination}$$

$$\sigma = 5 \text{ test points on the second examination}$$

Thus, we see that the standard deviation satisfies our definition of a standard. It is a fixed interval that is expressed in the units (test points) of the two scores we wish to compare. The σ-values given are therefore analogous to the conversion constants used in the last problem:

$$mile = .333 \text{ leagues}$$

$$mile = 1.61 \text{ kilometers}$$

It follows, therefore, that the instructor may adopt σ as his common unit of measurement and convert his two incompatible test scores to σ-*units*, just as Jim and Joe converted their leagues and kilometers to *miles* in the last problem. To make these conversions, each driver divided his net distance (expressed in either leagues or kilometers) by the appropriate conversion constant. Thus, for Joe

$$\frac{50 \text{ leagues from Jim's house}}{.333 \text{ leagues per mile}} = 150 \text{ miles from Jim's house}$$

Similarly, our instructor divides each deviation score by the appropriate σ-value and thereby converts both scores to σ-units, or *standard units*. Thus, for the first examination

$$\frac{13 \text{ test points from the mean}}{15 \text{ test points per } \sigma} = .87 \; \sigma\text{-units from the mean}$$

and for the second examination

$$\frac{5 \text{ test points from the mean}}{5 \text{ test points per } \sigma} = 1.0 \; \sigma\text{-units from the mean}$$

Deviation scores expressed in standard units are called *standard scores*, or Z-scores, and are defined by the general expression

$$z_i = \frac{x_i - \mu}{\sigma}$$

B. The Expectation and Variance of a Standardized Random Variable: $E(Z)$ and $V(Z)$

To establish a common reference point for the measurements in our last problem, we expressed the two raw test scores as deviations from their respective population means. If we standardize an entire distribution, therefore, we subtract the population mean from every score, and it should be apparent that such transformations will affect the location of the distribution. At an intuitive level, it is easily seen that the mean of such a distribution becomes zero. Recall that every score x_i assumes the standard form

$$z_i = \frac{x_i - \mu}{\sigma}$$

Consider now the score, which we call x_M, that is precisely equal to the population mean, μ. That is,

$$x_M = \mu$$

If we convert x_M to standard form, we obtain

$$z_M = \frac{x_M - \mu}{\sigma}$$

$$= \frac{\mu - \mu}{\sigma}$$

$$= 0$$

Thus, the standard score for any value that is precisely equal to the population mean is zero. This implies, of course, that the standardized value of the mean itself must equal zero. This may be proved more formally by the algebra of expectations.

Let us define X as a random variable with a population mean equal to μ and a variance equal to σ^2. In addition, let us define a second random variable Z such that

$$Z = \frac{X - \mu}{\sigma}$$

This provides us with a formal definition of a standardized variable, and we can now prove that the expected value of Z, $E(Z)$, is equal to zero.

$$E(Z) = E\left[\frac{X - \mu}{\sigma}\right]$$

$$= E\left[\frac{1}{\sigma}(X - \mu)\right]$$

However, σ is a parameter, and $1/\sigma$ must therefore be a constant. By the algebra of expectations, we know that a constant can be factored out of the expectation. Thus

$$E(Z) = \frac{1}{\sigma} E(X - \mu)$$

which by the algebra of expectations is equal to

$$\frac{1}{\sigma} [E(X) - E(\mu)]$$

We know that $E(X) = \mu$. Moreover, μ is a parameter and must therefore be a constant. Thus, $E(\mu) = \mu$, and

$$E(Z) = \frac{1}{\sigma}(\mu - \mu)$$

$$\frac{1}{\sigma}(0)$$

$$= 0$$

As the reader may have anticipated, standardization also affects the variance of a distribution. Both of the problems discussed above required that incompatible measurements be converted to some common unit of measurement. In the travel problem the two drivers converted their measurements from leagues and kilometers to miles. Thus, the mile became the common unit of measurement. In this new scale of measurement, therefore,

mile = 1.0 unit of measurement

Quite simply, if the unit of measurement is the mile, then a mile must be equal to 1.0 units of measurement. We have an analogous situation in the second problem where the instructor converted his two test scores to standard units. When this conversion is performed, the unit of measurement becomes an interval that is equal to the standard deviation, σ. If the unit of measurement is σ, it is conversely true that

σ = 1.0 unit of measurement

Thus, for any distribution of standardized scores, the standard deviation must always equal 1.0. Once again this may be proved by the algebra of expectations.

The Concept of Standardization

As before we define X as a random variable with expectation μ and variance σ^2, and we define Z as the standard variable

$$\frac{X - \mu}{\sigma}$$

By the definition of variance, therefore,

$$V(Z) = V\left[\frac{X - \mu}{\sigma}\right]$$

$$= V\left[\frac{1}{\sigma}(X - \mu)\right]$$

However, $1/\sigma$ is a constant, and we know from the algebra of expectations that a constant can be factored out of the variance. However, when it is factored out it is squared. Hence

$$V(Z) = \frac{1}{\sigma^2} V(X - \mu)$$

Once more, however, μ is a parameter, and $-\mu$ must therefore be a constant. By the algebra of expectations we know that

$$V(X + c) = V(X)$$

Therefore

$$V(X - \mu) = V(X)$$

and

$$V(Z) = \frac{1}{\sigma^2}[V(X)]$$

$V(X)$ was given as σ^2, and our expression thus becomes

$$V(Z) = \frac{1}{\sigma^2}(\sigma^2)$$

$$= 1.0$$

Standardization of any random variable involves two arithmetic operations. First, we subtract the population mean, μ, from each observation. We saw in the last section that this locates the distribution at zero. Second, we divide these deviation scores by the population standard deviation, σ. We have just demonstrated that this sets the standard deviation equal to one. Thus, *any* distribution of standardized values has a mean of 0 and a standard deviation of 1.0.

Frequently, of course, the population parameters μ and σ are unknown, and it is therefore impossible to compute

$$z_i = \frac{x_i - \mu}{\sigma}$$

Generally, however, scores are standardized to facilitate comparisons, and if such comparisons are restricted to the individuals at hand, one may legitimately substitute sample statistics for the population values. Suppose, for instance, that the college instructor in our second example gave the same midterm examinations every year. His test population would thus include all persons who ever took or ever will take his two tests, and the present class would represent only a sample of this population. Recall, however, that he was only concerned with the progress of one particular student. Because student populations change from year to year (even though tests may not), the student's performance would be most meaningfully assessed in relation to the performance of his own classmates. The comparison was therefore confined to students in the present class, and his "sample" thus assumed the status of a population.

When the set of observations thus defines the entire population, it is just as if the experimenter had drawn an exhaustive sample. That is, the sample includes every element in the population, and for every value x_i it must necessarily be true that

$$\frac{f_i}{N} = p(x_i)$$

We know from Chapter Five that

$$\bar{x} = \sum_i^n x_i \frac{f_i}{N}$$

If $f_i/N = p(x_i)$, this expression becomes

$$\sum_i^n x_i \, p(x_i)$$

which by definition equals the population mean μ. Furthermore, if $f_i/N = p(x_i)$ and $\bar{x} = \mu$, it must be similarly true that

$$s^2 = \sum_i^n (x_i - \bar{x})^2 \frac{f_i}{N}$$

is equal to

$$\sum_i^n (x_i - \mu)^2 \, p(x_i) = \sigma^2$$

Thus, when the observations in our sample define (or exhaust) our population,

$$Z = \frac{X - \mu}{\sigma} = \frac{X - \bar{x}}{s}$$

Furthermore, if it is true for an exhaustive sample that

$$\bar{x} = \mu = E(X)$$

and
$$s^2 = \sigma^2 = V(X)$$

it must be similarly true that
$$\bar{z} = E(Z) = 0$$

and
$$s_Z{}^2 = V(Z) = 1.0$$

Recall from Chapter Three that I once passed a dreary afternoon in Wendover, Nevada, waiting for a service station to improvise a radiator hose for my eccentric Italian sports car. The most irritating aspect of the delay was not the breakdown (breakdowns were the only reliable feature of the car and therefore never unexpected), but the inordinate difficulty in replacing a relatively common part. Most American drivers have come to expect that any well-stocked service station will be able to supply replacement headlights, hoses, and fanbelts. This is because the American automotive parts industry has *standardized* the specifications for such items. In this context, as in most ordinary conversation, standardization means that the objects have been rendered identical, or interchangeable, with respect to various important properties, such as voltage rating, diameter, or length. The same is true of its meaning in statistics. Two important properties of any distribution are its location and dispersion, and we have just seen that standardization renders the "specifications" of these properties, μ and σ, identical for all distributions.

We must emphasize, however, that although standardization may change the *parameters* of a probability distribution, it has no effect on the probability *function*. Standardizing a random variable may therefore affect the shape of its distribution, but the probability relationships will remain unchanged. Thus, if

$$z_i = \frac{x_i - \mu}{\sigma}$$

then
$$p(z_i) = p(x_i)$$

C. The Standardized Binomial Random Variable: $Z = \dfrac{X - \mu}{\sigma}$ when $X : B(N, p)$

In later chapters we shall have occasion to standardize values of random variables drawn from many types of probability distributions. In Chapter Eight, especially, we shall be concerned with standardizing values of binomially distributed random variables, and it is therefore important that the student acquire a facility for this particular transformation.

The reader should recall from Chapter Five that the population mean μ is often expressed as the expectation $E(X)$. Similarly, σ^2 often appears as $V(X)$. The choice of notation is largely a matter of convenience. Thus, when dealing with standard deviations, σ is less cumbersome notation than $\sqrt{V(X)}$. Conversely, certain proofs are simplified by using the algebra of expectations, in which case the expectation notation is preferred. Mathematically, however, μ is identical to $E(X)$ and σ^2 to $V(X)$, and each may be substituted for the other at any time. It is apparent, therefore, that the definitional formula for the standardized random variable

$$ Z = \frac{X - \mu}{\sigma} $$

may be alternatively expressed as

$$ Z = \frac{X - E(X)}{\sqrt{V(X)}} $$

Recall from Chapter Five that if X is a binomial random variable with sample size N and probability of success p, then

$$ E(X) = Np $$

And from Chapter Six, if $X : B(N, p)$, then

$$ V(X) = Npq $$

For any binomial random variable X, therefore, the standardized variable Z becomes

$$ Z = \frac{X - \mu}{\sigma} $$

$$ = \frac{X - E(X)}{\sqrt{V(X)}} $$

$$ = \frac{X - Np}{\sqrt{Npq}} $$

To standardize any value x_i of a binomially distributed random variable, we must therefore compute $E(X) = Np$ and $V(X) = Npq$ and solve the equation

$$ z_i = \frac{x_i - Np}{\sqrt{Npq}} . $$

D. Problems, Review Questions, and Exercises

1. A university admissions officer has two applicants who took the Scholastic Aptitude Test in different years. The population in which the first applicant took the test had a mean $\mu = 480$ and a standard deviation $\sigma = 85$. For the year in which the second applicant took his examination, $\mu = 450$ and $\sigma = 80$. If the raw scores for the two applicants are $x_1 = 630$ and $x_2 = 580$,

    a. What Z-scores were obtained by the applicants?

(*Note.* Save your answer to **1a** for use in later problem sets.)

An individual's standard score y on the SAT is computed by multiplying his Z-score by 100 and adding 500. That is, $Y = 500 + 100Z$.

    b. What is μ_Y?

    c. What is σ_Y?

    d. What Y-scores were obtained by the two applicants?

2. If X is temperature in degrees Farenheit and Y is temperature in degrees centigrade,

$$Y = \frac{X - 32}{9/5}$$

    a. Why is 32 subtracted from degrees Farenheit?

    b. Why is $X - 32$ divided by 9/5.?

3. What standard score corresponds to an observed value of 8 for the random variable discussed in problem **4** of Chapter Two?

4. What standard score corresponds to an observed value of 8 for the random variable defined by the experiment in problem **1** of Chapter Four?

the normal distribution

One cannot proceed very far in genetics, psychology, economics, geography, or any other discipline that treats intrinsically variable data without eventually encountering references to the "normal" or "Gaussian" distribution. Such references invariably apply to probability or relative frequency distributions that assume a symmetrical, bell-shaped configuration with a heavy concentration of observations falling in a narrow range about the mode and with the frequencies of observations decreasing as their values depart more extremely from the mode. To say that a random variable is "normally distributed," however, is not a catch-all description for any distribution that has these geometric properties. Like any other mathematical curve, the normal curve is a graphic representation of a function and is therefore described precisely only by the set of mathematical operations that define the functional relationship. The function defining the normal curve is:

$$\phi(X) = \frac{1}{\sqrt{2\pi\sigma^2}} e^{-(X-\mu)^2/2\sigma^2}$$

In the absence of adequate explanation, the apparent ubiquity of random variables distributed according to such a formidable mathematical expression must seem to reflect either Divine intervention, suspicious coincidence, or naive experimental assumptions of the most tenuous variety. Unfortunately for the more skeptical among you, none of these conclusions is necessarily true. But then, neither is it true that empirically

135

observed random variables are *in fact* normally distributed. It *is* the case, however, that under certain conditions many relative frequency distributions *approximate* the normal curve. When such conditions prevail, the experimenter treats his data as if this distribution conformed to the normal curve. He therefore draws conclusions about his data that, although never precisely true, are "close enough for folk music."

These sorts of "as if" assumptions are made frequently in our daily lives. We all know, for example, that the Earth is shaped rather like a grapefruit; it is somewhat flattened at the poles and bulges slightly at the equator. Over short distances, however, the curvature of the Earth is almost imperceptible, and relatively small areas of terrain are often treated as planar surfaces. In building a house, for example, one generally assumes his foundation to be erected on a "level" surface. In other contexts, such as ballistics, one can often obtain satisfactory results by assuming the Earth to be a perfect sphere. No terrestrial surface is, in fact, either a perfect plane or a spherical section, but these geometric forms are convenient *models*, and conclusions based on them may be sufficiently accurate for many purposes. In a similar fashion, the normal curve is a model for many types of relative frequency distribution. When we say that a random variable is "normally distributed," therefore, we generally mean that the relative frequency of any interval Δx_i will be very close to the corresponding probability value computed from the normal curve even though the curve, like the plane or the sphere, is a purely idealized mathematical abstraction.

It is similarly true that a number of theoretical probability distributions *become* normal as various of their parameters approach infinity. These include Student's t-distribution, the χ^2-distribution, and, as we shall presently discover, the binomial distribution. This topic is treated more fully in Chapter Ten and is mentioned here only so that the student may begin his study of the normal curve with some appreciation for its wide range of application.

In introductory texts of this sort, new mathematical concepts are often introduced by presenting the reader with a formula and then breaking the expression down into its component parts to facilitate explanation. It is our experience, however, that the initial exposure to a complex mathematical expression may produce a state of "symbol shock" in the unprepared reader, and we have therefore tried to avoid this format. Indeed, we have reversed the procedure whenever possible and developed formulas synthetically rather than explain them analytically. In Chapter Three, for example, we generated the formula for binomial probability by first abstracting the operations required to solve a particular sort of problem and then formalizing the operations in mathematical notation. The product moment correlation coefficient is developed in much the same way in Chapter Nine. The mathematical expression for the normal curve, however, involves several mathematical concepts that are purely conceptual. Unlike the concepts of counting and measuring, they do not represent abstractions of ordinary quantitative operations that we routinely perform on concrete objects. In much of the present chapter, therefore, we abandon our preferred strategy of induction and abstraction and rely instead on a rather traditional, formally deductive presentation.

The reader may be assured, however, that the mathematical expression defining the normal curve is seldom actually encountered in statistics. Indeed, the function most generally appears as a deceptively tangible curve representing a high idealized relationship between probability and the value set of a random variable. In this context, the curve is a type of histogram and simply represents a further development of the graphic techniques discussed in Chapter Two. Indeed, our presentation introduces the normal distribution as the *limiting* case of a histogram representing the probability distribution of a binomial random variable. Like most idealized curves, however, the normal curve is a function of a *continuous* random variable, while a binomial random variable can assume only *discrete* values. Unfortunately, many of the conventions we have established thus far are appropriate only to discrete random variables, and our first task, therefore, is to adapt some of these to the histographic representation of continuous probability distributions.

In this regard, it may be noted that the preceding chapters in Part Two have in general concentrated on the distribution properties of discrete random variables. Thus, we concluded Chapters Five and Six by deriving the expectation and variance of a binomial random variable, and the expectation and variance of a standardized binomial distribution occupied the last part of Chapter Seven. Chapter Eight, therefore, is our first encounter with the distribution properties of a continuous random variable and represents something of a new orientation in our treatment of distribution properties.

A. The Probability Distribution of a Continuous Variable: Probability Density

When we introduced the notion of continuous variation in Chapter Four we began our discussion by considering the frequency distribution of maze-running errors in a sample of laboratory rats. Recall that the number of errors was distributed as shown in Table 8.1.

Table 8.1 Frequency Distribution for Maze-Running Errors

Δx_i	116–135	136–155	156–175	176–195
f_i	9	28	41	22

If, for the moment, we consider that our 100 animals constitute our entire population, then the relative frequency $f_i/100$ of each class Δx_i becomes the probability associated with that class, and we obtain the probability distribution shown in Table 8.2.

Table 8.2   Probability Distribution for Maze-Running Errors

Δx_i	116–135	136–155	156–175	176–195
$p(\Delta x_i)$	.09	.28	.41	.22

Recall now from Chapter Two that any such probability distribution may be represented in the form of a histogram, where the height of each bar is proportional to the probability of the interval whose limits define the base of the bar. Thus, the probability distribution of errors in our population of 100 rats may be represented as in Figure 8.1.

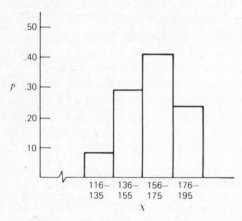

Figure 8.1   Histogram for probability distribution in Table 8.2.

In this example we have grouped all of our observations into four classes, or intervals. If we assume that the random variable is uniformly distributed in each class, the probability associated with the entire interval Δx_i must be divided equally among all of the values in that class. Thus, each value in the interval will assume the same probability. In the first interval, for example, $p(\Delta x_i) = .09$ is uniformly distributed across the 20 values

$$\{116, 117, \ldots 135\}$$

Therefore, each value in the interval assumes a probability of

$$\frac{.09}{20} = .0045$$

and in general, the probability of any value $x_j \in \Delta x_i$ is equal to

$$p(\Delta x_i)/n_i$$

where n_i is the number of values in the interval $x\Delta_i$.

In contrast to our discrete distribution of errors, we also presented in Chapter Four the distribution of a continuous variable, weight. Representing each interval by its real class limits, the frequency distribution of weights of Flint Judo Club competitors is shown in Table 8.3.

Table 8.3   Frequency Distribution for Weights of Judo Competitors

Δx_i	115.5–135.5	135.5–155.5	155.5–175.5	175.5–195.5
f_i	3	9	13	7

Once again, if we consider the Flint Judo Club to be our entire population, each frequency f_i may be divided by $N = 32$, and we thus obtain the distribution of probabilities shown in Table 8.4.

Table 8.4   Probability Distribution for Weights of Judo Competitors

Δx_i	115.5–135.5	135.5–155.5	155.5–175.5	175.5–195.5
$p(\Delta x_i)$	.09	.28	.41	.22

And as before, we may wish to represent our probability distribution graphically, as in Figure 8.2. If, however, we now attempt to compute the probability associated with any particular value x_i, we find ourselves in a curious dilemma. Consider the values in the first interval Δx_1. As in the preceding example, we assume that all values from 115.5 to 135.5 are uniformly distributed, and we

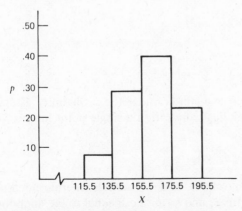

Figure 8.2   Histogram for probability distribution in Table 8.4.

again note that this interval carries a probability of .09. Therefore, each value in the interval should be assigned a probability of $.09/n_1$, where n_1 is the number of values in the interval:

$$p(x_i) = p(\Delta x_1)/n_1$$

However, we know from our definition of a continuous random variable that any continuous interval includes an uncountable number of values. Thus, n_1 is uncountable, and any division by an uncountable quantity is mathematically defined as zero. In the present example, therefore,

$$p(x_i) = 0$$

and in general, the probability of any single value x_i of a continuous random variable is equal to zero. This does not, of course, mean that the probability of finding a player weighing, say, 145 pounds is necessarily zero. A weight of 145 would be indicated on an ordinary gymnasium scale for anyone weighing between 144.5 and 145.5, and the probability of observing a player in this *interval* is finite and calculable. When we say that $p(145) = 0$, we mean that there exists a zero probability of observing an individual weighing *precisely* 145.000 . . . with an *infinite* number of zeros to the right of the decimal. Nonetheless, our graphic representation indicates a positive, nonzero probability for every value from 115.5 to 195.5.

We resolve this paradox by representing the probability associated with any continuous interval Δx_i as the *area* enclosed by the bar rather than as the *height* of the bar. Each bar is, of course, a rectangle, and it will be recalled from plane geometry that for any rectangle

$$\text{Area} = (\text{height})(\text{base})$$

In the case of the rectangular bar associated with the interval Δx_i,

$$\text{Area} = p(\Delta x_i)$$

$$\text{Base} = \Delta X$$

Therefore

$$p(\Delta x_i) = (\text{height})(\Delta X)$$

With this convention in mind, let us again consider the probability associated with a single value x_i. By mathematical convention a single point has no width. Thus, for the X-value x_i,

$$\Delta X = 0$$

If the reader wishes, he may think of the value x_i as a degenerate interval Δx for which the upper and lower limits x_a and x_b are both equal to the midpoint x_i. Recall from Chapter Four that the class width ΔX of any continuous interval

140

may be computed by subtracting its lower class limit from its upper class limit. Thus, for our degenerate "interval" Δx:

$$\Delta X = x_i - x_i$$

$$= 0$$

By the formula given above,

$$p(\Delta x) = (\text{height})(0)$$

$$= 0$$

Graphically, the "bar" representing the probability of Δx becomes a single vertical line, and we know from plane geometry that a line has zero area. In this type of representation, therefore, the probability of a single value x_i is indeed zero, thus rectifying the inadequacy we found in representing probability as the height of a histogram.

Finally, we can see that if

$$p(\Delta x_i) = (\text{height})(\Delta X)$$

then

$$\text{height} = \frac{p(\Delta x_i)}{\Delta X}$$

This ratio, which corresponds to the height of our graph, is defined as *probability density* and is symbolized by the Greek letter ϕ (*phi*), as opposed to p, which we customarily use to represent probability.

It is easy to forget that theoretical developments are often initiated by very practical considerations. The reader is therefore reminded that the concept of probability density evolved out of our efforts to extend the use of histograms to distributions of continuous data. When a probability distribution is graphically represented in terms of probability density, the probability of observing any X-value in the interval Δx_i, that is, $p(\Delta x_i)$, is represented as the area bounded by the class limits of the interval, the X-axis, and the top of the graph, which is generically called a "curve" despite its often steplike appearance. We know from Chapter Two that the total probability associated with the entire value set of a random variable must be 1.0, and it follows that the total area bounded by the curve must also equal 1.0.

In such a representation, we know from the definition given above that the height of the curve is defined as probability density. This implies that for any value of the random variable, x_i, there exists a corresponding probability density value, which is expressed as the height of the curve at that point, and our representation thus satisfies the definition of a function as given in Chapter Three. Therefore, when probability is graphically represented as the area under the curve instead of as the height of the curve, the mathematical function defining the curve is called a *probability density function* and is symbolized as $\phi(X)$, which is read "phi of X."

> The nonnegative function $\phi(X)$ is a probability density function if the entire area under the curve from x_1 to x_n is equal to 1.0 and if the probability associated with any interval x_a to x_b can be represented as the area under $\phi(X)$ between x_a and x_b.

For those who are acquainted with calculus notation, $Y = \phi(X)$ is a probability density function if for $\phi(X) \geq 0$,

$$\int_{x_1}^{x_n} \phi(X) \, dX = 1.0$$

and if

$$p(x_a \leq x \leq x_b) = \int_{x_a}^{x_b} \phi(X) \, dX$$

It should not be assumed from the foregoing discussion that probability density is appropriate only to display of continuous variables. Indeed, most statisticians have adopted the convention of representing *all* probability distributions in terms of probability density. That is, in most histographic displays, probability is represented as the area under the curve instead of as the height of the curve. And hereafter, we, too, shall conform to this convention. Unless otherwise noted, therefore, the height of any probability histogram will represent $\phi(X)$ rather than $p(X)$, and the probability of any interval Δx_i will be given by the area defined by its class limits, the X-axis, and the top of the graph, as discussed above.

In this regard, recall from Chapter Two that probabilities associated with the values of a discrete random variable may be graphically represented as a set of vertical lines. This convention is illustrated in Figure 8.3, where X is the number of heads in four tosses of a fair coin. Earlier in this chapter we suggested that a

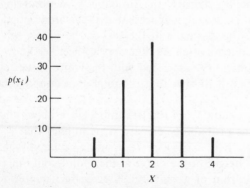

Figure 8.3   Line graph representing probability distribution for number of heads obtained in four tosses of a fair coin.

single point x may be considered a degenerate interval Δx of zero width. Similarly, a straight line may be considered a degenerate rectangle with zero area. In the present example, therefore, the total area enclosed by our "rectangles" must equal zero and cannot be used to represent total probability, which must necessarily equal 1.0.

In Chapter Four, however, we explained that a histogram can be constructed for a discrete random variable by expressing each value x_i in terms of its real limits $(x_i - 0.5)$ and $(x_i + 0.5)$. As in Figure 8.4, each rectangle, or bar, is based

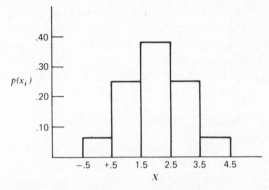

Figure 8.4   Histogram representing probability distribution for number of heads obtained in four tosses of a fair coin.

on these fictitious limits, and the width of each bar, ΔX, is therefore equal to their difference.

$$\Delta X = (x_i + 0.5) - (x_i - 0.5)$$

$$= 1.0$$

Once again from Chapter Two, we know that in a conventional histogram the height of each bar is equal to the probability of the X-value falling at the midpoint of its base:

$$\text{Height} = p(x)$$

With a base equal to 1.0, however, the *area* of each rectangle must also equal the probability. Thus, for the interval $x_i - 0.5$ to $x_i + 0.5$:

$$\text{Area} = (\text{height})\,(\Delta X)$$

$$= p(x_i)(1.0)$$

$$= p(x_i)$$

If the area of each rectangle represents probability, the histogram satisfies our definition of a probability density function, and the height of the rectangle becomes probability density.

B. The Limit of a Binomial Probability Density Function

In the last section we developed two important notions. First, we generated the concept of probability density and demonstrated how a continuous probability distribution may be histographically displayed as a probability density function. Second, we established the convention of graphically representing *discrete* probability distributions as probability density functions. In this section we shall apply this new convention to the binomial random variable $X:B(N, 0.5)$. This is not, however, presented simply as an exercise in descriptive statistics. On the contrary, we shall be concerned with a very important theoretical question: What happens to a binomial density function as the sample size N increases?

To begin our inquiry, let us consider the binomial experiment, toss a fair coin twice. If we define the random variable X as the number of heads observed in our experiment, we obtain the probability distribution shown in Table 8.5.

Table 8.5 Probability Distribution for $X:B(2, 0.5)$

x_i	0	1	2
$p(x_i)$	.25	.50	.25

If we wish to represent this distribution in the form of a histogram, however, we must perform our bit of mathematical legerdermain and treat X as though it were a continuous variable. That is, every discrete value x_i becomes an interval of unit width, Δx_i, with real class limits $x_i - 0.5$ and $x_i + 0.5$. The probability values, of course, remain unchanged (Table 8.6).

Table 8.6 Probability Distribution for $X:B(2, 0.5)$ with X Expressed in unit-width Intervals

Δx_i	−0.5 to 0.5	0.5 to 1.5	1.5 to 2.5
$p(\Delta x_i)$	.25	.50	.25

The histographic representation of this distribution is given in Figure 8.5(a). By the convention which we adopted at the end of the last section, we have represented our distribution as a probability density function. The values on the vertical axis are therefore probability density values and not probabilities. The probability of each interval, however, appears in the rectangle based on that interval and is, of course, equal to its area.

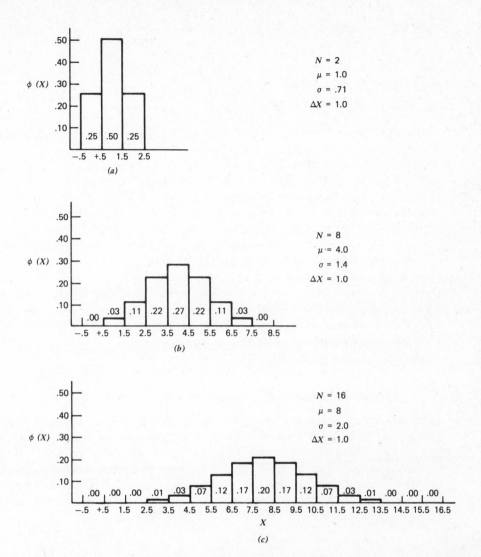

Figure 8.5 Histograms for probability distributions of three binomial random variables.

Let us proceed in the same fashion and consider the distribution of heads in eight tosses of a fair coin (Table 8.7). Space prohibits us from representing each value x_i by the limits of its fictitious interval, but it must be understood that $x = 0$ should be considered to be the unit interval -0.5 to $+0.5$; the value $x = 1$ is represented by the interval $+0.5$ to $+1.5$, and so forth. In Figure 8.3(b) we use these class limits to define intervals over which we construct rectangles and thus display our probability distribution as a histogram.

Table 8.7   Probability Distribution for $X:B(8, 0.5)$

x_i	0	1	2	3	4	5	6	7	8
$p(x_i)$	.00	.03	.11	.22	.27	.22	.11	.03	.00

Finally, Figure 8.5c is the histogram representing the probability distribution of the number of heads observed in 16 tosses of a fair coin, that is, $X:B(16, 0.5)$.

If we compare the three histograms in Figure 8.5 we see that as N increases, the distribution of X spreads out. This is only to be expected; the width of a histogram reflects the amount of dispersion in the distribution, and for a binomial distribution $\sigma^2 = Npq$. Thus, as N increases, so does the variance and so must the physical, or geometric, width of the histogram. Recall, however, that the *total* area under the curve represents the *total* probability of all values of X and must therefore always be equal to 1.0. Because a histogram is constructed of rectangles, its total enclosed area is proportional to its average height multiplied by its base. If, therefore, the baseline increases with sample size while the area remains constant, the height of the histogram must *decrease* proportionally. And, indeed, we see that the largest value of $\phi(X)$ in the first distribution ($N = 2$) is .50; the largest value of $\phi(X)$ in the second distribution ($N = 8$) is .27; and the largest value of $\phi(X)$ in the last distribution ($N = 16$) is only .20.

1. The Standardized Binomial Distribution and the Standard Normal Curve

In the last section we saw that the shape of a binomial probability density function changes markedly as N increases. Variance increases as a function of sample size, and the histogram consequently spreads out. Conversely, the height of the function decreases, thereby maintaining a constant total area under the curve of 1.0. The effects of increased sample size are quite different, however, when we standardize the values of our binomial random variable. Recall from Chapter Seven that the standardized random variable Z is defined as

$$\frac{X - \mu}{\sigma}$$

Recall, too, that if X is binomially distributed, then

$$Z = \frac{X - Np}{\sqrt{Npq}}$$

Because our discrete probability distributions are represented in terms of probability density, however, each value x_i is represented by an interval Δx_i of unit width:

$$\Delta x_i = x_a \qquad \text{to} \qquad x_b$$
$$= x_i - 0.5 \qquad \text{to} \qquad x_i + 0.5$$

146

To standardize our distribution, we must therefore standardize *the class limits* of these intervals:

$$z_a = \frac{x_a - Np}{\sqrt{Npq}} = \frac{(x_i - 0.5) - Np}{\sqrt{Npq}}$$

$$z_b = \frac{x_b - Np}{\sqrt{Npq}} = \frac{(x_i + 0.5) - Np}{\sqrt{Npq}}$$

In our first distribution, X is a binomial random variable with $N = 2$ and $p = 0.5$. Thus

$$\mu = Np = 2(0.5) = 1.0$$

$$\sigma = \sqrt{Npq} = \sqrt{2(0.5)(0.5)} = .707$$

Substituting these values into our expressions for z_a and z_b, we obtain the standardized limits listed below with their corresponding X-values:

Real Class Limits for $X:B(2, 0.5)$	Standardized Real Class Limits for $X:B(2, 0.5)$
-0.5	-2.12
$+0.5$	$-.71$
$+1.5$	$+.71$
$+2.5$	$+2.12$

If we identify each interval by its real class limits, the standardized probability distribution of our first random variable, $X:B(2, 0.5)$, becomes that shown in Table 8.8.

Table 8.8 Probability Distribution for Standardized Binomial Random Variable ($N = 2$, $p = 0.5$)

Δz_i	-2.12 to -0.70	$-.70$ to $+.70$	$+.70$ to $+2.12$
$p(\Delta z_i)$	.25	.50	.25

The histogram constructed from these class limits is given in Figure 8.6a, and by applying the same standardization procedures to the distributions for $X:B(8, 0.5)$ and $X:B(16, 0.5)$ we obtain the distributions represented in Figures 8.6b and 8.6c.

If we compare our three standardized distributions in Figure 8.6, we see, in contrast to the distributions in Figure 8.5, that the heights of the histograms remain relatively unchanged. The maximum value of $\phi(Z)$ is approximately equal to .38 for all three, and were we to compute the *average* height of each histogram across all of its intervals, we would find our three averages to be

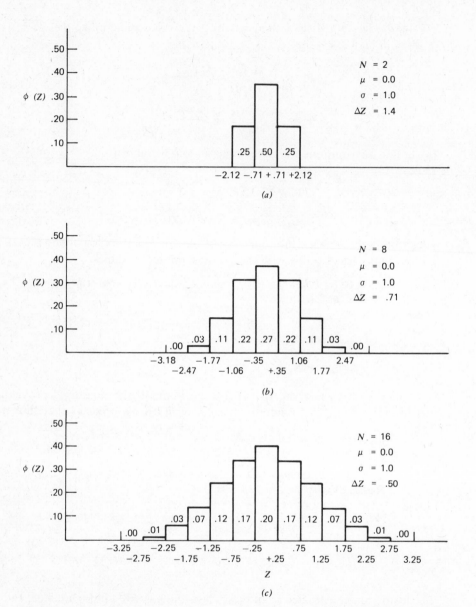

Figure 8.6 Histograms for probability distributions of three standardized binomial random variables.

precisely equal. For standardized binomial distributions, therefore, the height of the density function *does not* decrease as sample size *N* increases. The reason for this may not be immediately apparent, but it is not difficult to understand. When we compared the histograms for our three unstandardized distributions, we found that height decreases with increased sample size to compensate for greater dispersion, thus maintaining a constant area under the curve. From Chapter Seven, however, we know that the variance of any standardized variable is always equal to 1.0. The variance σ^2 of the standardized binomial distribution must therefore remain fixed at 1.0, and its total dispersion, or width, therefore remains constant despite changes in sample size. With no increase in dispersion, therefore, there is no compensatory decrease in height.

These considerations notwithstanding, we know that the number of elements in the value set of a binomial random variable is intimately related to the number of binomial trials N. Thus, for $N = 2$, we have three possible X-values; for $N = 8$, we have nine, and so forth. Furthermore, each of these values must be represented on the baseline of the histogram. In a standardized histogram, where the baseline remains constant, these values must therefore become arrayed more compactly as sample size and, concomitantly, the number of X-values increases. As N increases, therefore, each unit-width interval Δx_i expressed in standard units must become narrower. This is easily seen if we consider

$$\Delta x_i = x_a \qquad \text{to} \qquad x_b$$

$$= x_i - 0.5 \qquad \text{to} \qquad x_i + 0.5$$

If we standardize this interval we have already established that

$$z_a = \frac{x_i - 0.5 - Np}{\sqrt{Npq}}$$

$$z_b = \frac{x_i + 0.5 - Np}{\sqrt{Npq}}$$

Thus, the class width ΔZ of our standardized interval must equal

$$z_b - z_a = \frac{x_i + 0.5 - Np}{\sqrt{Npq}} - \frac{x_i - 0.5 - Np}{\sqrt{Npq}}$$

$$= \frac{1}{\sqrt{Npq}}$$

For a fixed value of p, therefore, the standardized width of any binomial unit interval Δx_i is inversely proportionate to the square root of sample size. This is graphically illustrated in Figure 8.7, in which portions of three standardized binomial distributions are constructed on the same Z-axis. Each bar in Figure 8.7 is defined by the standardized limits of a unit-width X-interval. As sample size increases, therefore, the individual bars become narrower, and each vertical "step" becomes smaller. Eventually, the steplike discontinuities in the graph will disappear altogether, and the histogram will become a smooth function. The curve that our standardized binomial density function approaches is illustrated in Figure 8.8 superimposed on the probability density function of Z for $X{:}B(16, 0.5)$.

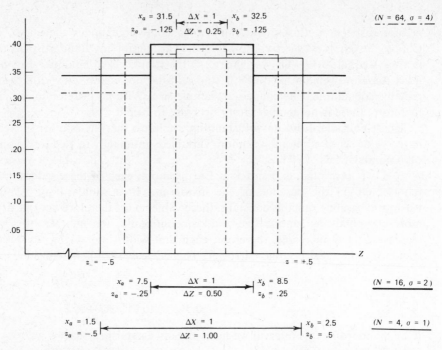

Figure 8.7   Unit-width X-intervals and corresponding Z-intervals for three standardized binomial distributions.

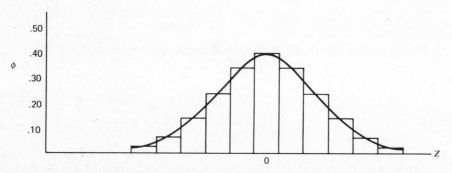

Figure 8.8   Standard normal probability density function and distribution of standardized binomial random variable.

This curve is known as the *standard normal probability density function* and is given by the formula

$$\phi(Z) = \frac{1}{\sqrt{2\pi \, V(Z)}} \, e^{-Z^2/2}$$

Formally, then, if $X:B(N,p)$,

$$\underset{N \to \infty}{\text{limit}}\ \phi(Z) = \frac{1}{\sqrt{2\pi\, V(Z)}}\, e^{-Z^2/2}$$

where

$$Z = \frac{X - \mu}{\sigma}$$

$$= \frac{X - Np}{\sqrt{Npq}}$$

2. Properties of the Standard Normal Probability Density Function

In the last section we demonstrated that as sample size increases, the probability density function of a standardized binomial random variable loses its characteristic steplike appearance and becomes a smooth curve. Moreover, we asserted without proof that the mathematical function defining this curve is

$$\phi(Z) = \frac{1}{\sqrt{2\pi\, V(Z)}}\, e^{-Z^2/2}$$

It is beyond the scope of this text to prove that this function is, in fact, the limit of the standardized binomial density function. We can, however, consider the expression in some detail and thereby derive a number of the curve's more important properties.

To begin, we note that the first term

$$\frac{1}{\sqrt{2\pi\, V(Z)}}$$

is composed entirely of constants. The Greek letter π (*pi*) is defined in mathematics as the ratio of the circumference of a circle to its diameter. It is an irrational constant approximately equal to 3.14159. $V(Z)$, of course, is the variance of a standardized random variable and is always equal to 1.0. Whatever the value of our random variable Z, therefore, the value of the first term

$$\frac{1}{\sqrt{2\pi\, V(Z)}}$$

remains fixed, and for the sake of convenience we therefore represent it by the letter K. Our entire function thus reduces to

$$\phi(Z) = K\, e^{-Z^2/2}$$

The letter e, however, also symbolizes a constant. It is the base of the natural logarithm system and, like π, is an irrational number. Expressed to five decimal places, e is equal to 2.71828.

We see, then, that everything in our expression reduces to constants except the exponent of e:

$$-\frac{Z^2}{2}$$

As the random variable Z assumes different values, therefore, the corresponding changes in $\phi(Z)$ are entirely determined by this exponent. Hence, this fraction is the "working" part of the formula for the standard normal curve, and everything else is excess baggage. Indeed, the exponent has two essential characteristics from which we develop all of the distribution properties discussed in this section.

First the reader will observe that the random variable Z enters the exponent as a squared quantity. Thus, our density function is actually a function of Z^2, $\phi(Z) = f(Z^2)$. In this regard, we know that for any value z

$$z^2 = (-z)^2$$

Thus

$$f(z^2) = f[(-z)^2]$$

and

$$\phi(z) = \phi(-z)$$

This means that the probability density associated with any value z is precisely equal to the probability density associated with the corresponding negative value, $-z$. This property was also apparent in the standardized binomial distributions we considered earlier. In the first Z-distribution ($N = 2$), for example, the height of the graph $\phi(Z)$ across the interval -2.12 to $-.71$ is precisely equal to $\phi(Z)$ for the interval $+.71$ to $+2.12$. In the standardized probability distribution based on an N of 8, the probability density associated with Z-values in the interval -1.77 to -1.06 is equal to the probability density for the interval $+1.06$ to $+1.77$, and so forth. In our standardized binomial distributions, therefore, the positive half of any histogram is a mirror image of the negative half. The same is true for our standard normal distribution, and we therefore say that the function, or curve, is *symmetrical about zero*.

The second important feature of our exponent is that it is negative. To appreciate the implications of this, let us consider the general case of a constant raised to a negative power. Recall from high school algebra that by definition

$$a^{-X} = \frac{1}{a^X}$$

As the value of X increases, therefore, a^{-X} must decrease. Conversely, a^{-X} must be greatest when X is equal to zero. Although our expression

$$e^{-Z^2/2}$$

may seem more complex than a^{-X}, they are of the same form, and $e^{-Z^2/2}$ must therefore equal

$$\frac{1}{e^{Z^2/2}}$$

and our function $\phi(Z)$ becomes equal to

$$K \frac{1}{e^{Z^2/2}}$$

It follows, therefore, that the standard normal curve reaches its maximum value (height) when $z = 0$ and decreases as the absolute value of Z increases.

Knowing that $\phi(Z)$ is greatest at $z = 0$, we may proceed to determine its value at this point:

$$\phi(Z) = K e^{-Z^2/2}$$

$$\phi(0) = K e^{-0}$$

$$= K \frac{1}{e^0}$$

However, any number raised to the power of 0 is defined as 1:

$$a^0 = 1$$

Therefore

$$\phi(0) = K \frac{1}{1}$$

$$= K$$

As defined above,

$$K = \frac{1}{\sqrt{2\pi V(Z)}}$$

$$= \frac{1}{\sqrt{(2)(3.14159)(1.0)}} \qquad \text{(approx.)}$$

$$= \frac{1}{\sqrt{6.28138}}$$

$$= \frac{1}{2.506} \qquad \text{(approx.)}$$

$$= .40 \qquad \text{(approx.)}$$

We see, then, that $\phi(Z)$ is unimodal at $z = 0$, that it is symmetrical about its mode, and that its maximum value $\phi(0)$ is approximately .40. We therefore know that $\phi(Z)$ must decrease as Z departs from zero, but we have established nothing about the manner of this decrease. That is, besides its unimodality and symmetry, we still know little about the actual shape of the curve. Thus, all of the curves illustrated in Figure 8.9 satisfy the properties we have derived so far.

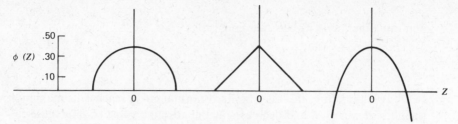

Figure 8.9   Three distributions exhibiting unimodality at z = 0, *symmetry about mode, and maxima of approximately* .40.

To gain some further insight into this matter, consider once again the general expression

$$a^{-X} = \frac{1}{a^X}$$

We have already determined that a^{-X} decreases as X increases. We also observe, however, that a^{-X} is always some fraction of 1. Thus, as X approaches infinity, a^{-X} may indeed become infinitesimally small, but it will never quite equal zero. We therefore say that zero is the *limit* of a^{-X} as X approaches infinity:

$$\lim_{X \to \infty} a^{-X} = 0$$

We also note that $1/a^X$ is positive. For any finite value of X, therefore, a^{-X} must be a positive, nonzero quantity. And, it must be similarly true that for finite values of Z

$$e^{-Z^2/2}$$

assumes only positive, nonzero values and approaches zero only in the limit as the exponent $Z^2/2$ becomes infinitely large. It should not escape the reader's attention that $Z^2/2$ approaches infinity as Z approaches *either* $-\infty$ *or* $+\infty$. Hence

$$\lim_{Z \to \pm\infty} \phi(Z) = 0$$

Graphically, therefore, the tails of the standard normal curve flatten out and become parallel to the horizontal axis as Z approaches positive or negative infinity. Indeed, as Z becomes infinitely large ($\pm$), the curve becomes congruent with the horizontal axis, $\phi(Z) = 0$. You may recall from high school algebra that such a curve is said to have a *horizontal asymptote* at $y = 0$, or in this case at $\phi(Z) = 0$.

It is apparent, therefore, that our function must yield a meaningful quantity $\phi(z)$ for any finite Z-value, positive or negative. Formally, then the *domain of definition* for $\phi(Z)$ is the entire real line. This implies that the standard normal random variable Z can assume any real value from $-\infty$ to $+\infty$. The value set of Z is therefore

$$\{-\infty \ldots + \infty\}$$

To summarize, therefore, we have established that

$$\phi(Z) = \frac{1}{\sqrt{2\pi V(Z)}} e^{-Z^2/2}$$

is a positively valued function that is defined across the entire real line. It is unimodal and symmetrical about its mode, which occurs at $z = 0$. Furthermore, we know that the maximum value of the function is approximately .40 and that $\phi(Z)$ approaches zero asymptotically as Z approaches $\pm \infty$.

Thus far, our consideration of the standard normal probability density function has emphasized properties of a purely mathematical nature. This focus was intended to familiarize the reader with the geometric properties of the curve and to provide him with at least some understanding of its mathematical expression, which he might otherwise dismiss at totally beyond comprehension. Consequently, the probability characteristics of the density function have received relatively little attention. The reader is therefore reminded that our primary concern with the standard normal curve is in its application as a probability distribution, and with this in mind we turn our attention to properties that are more properly the concern of the statistician.

To begin, let us briefly discuss the location and dispersion of the standard normal probability distribution. As indicated above, the standard normal curve is the theoretical limit of the standardized binomial distribution. From Chapter Seven, we know that our standardized binomial random variable must have an expectation of 0 and a variance of 1; moreover, we know that these values are not influenced by sample size and must therefore remain constant even as N approaches infinity. It therefore follows that for any standard normal probability distribution

$$\mu = 0$$

$$\sigma^2 = 1$$

This means that the mode and expectation of the standard normal distribution are both equal to 0. This equivalence is implicit in the concept of symmetry, and we may therefore say that the standard normal curve is symmetrical about its expectation, $\mu = 0$.

As indicated above, the geometric properties developed earlier in this section were discussed, albeit somewhat informally, primarily in the context of pure mathematics. Nonetheless, these properties have important implications in terms of probability relationships, as we may discover by examining the probabilities associated with some selected intervals Δz_i. We know from earlier discussion that the standard normal curve is unimodal at $z = 0$ and that both tails of the curve approach zero asymptotically. Thus, in Figure 8.10, We see that our curve drops off sharply as Z departs from zero and then levels out very quickly, approaching the horizonal axis, $\phi(Z) = 0$, at an ever decreasing rate. We see, therefore, that most of the area is concentrated in a narrow interval centered about $z = 0$. Thus, while the total area under the curve must be equal

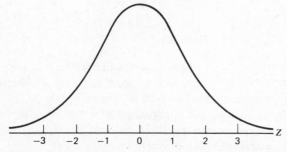

Figure 8.10   The standard normal distribution.

to 1.0, over 68 percent of this total is found in the interval -1 to $+1$. More particularly,

$$p(-1 \leq z \leq +1) = .6816$$

Similarly, more than 95 percent of the curve lies between -2 and $+2$:

$$p(-2 \leq z \leq +2) = .9545$$

And finally, less than 1 percent of the total area under the curve lies in the tails beyond $z = \pm\, 3$. Stated more positively,

$$p(-3 \leq z \leq +3) = .9980$$

Thus, although the domain of the standard normal curve extends, in fact, from $-\infty$ to $+\infty$, its practical limits are generally considered to be -3 and $+3$.

C. The Cumulative Probability Distribution of the Standard Normal Curve

At the beginning of Chapter Eight we demonstrated that it is conceptually meaningless to speak of the probability associated with a specific value x_i of a continuous random variable. For this reason it is not possible to construct probability tables for continuous random variables. We know, however, that it *is* possible to compute the probability of a continuous *interval* Δx_i. In this connection, recall from Chapter Three that we defined the cumulative probability of x_j as $\Sigma\, p(x_i)$ for all values $x_i \leq x_j$. Thus, $F(x_j)$ is the probability associated with the interval

$$\Delta x = x_1 \qquad \text{to} \qquad x_j$$

We may, therefore, represent the probability distribution of a continuous random variable as a *cumulative* probability table. On the basis of these considerations, the standard normal probability distribution is always tabled in cumulative form. The entries in Table **C** of Appendix V are the cumulative

156

probabilities associated with all two-place standard normal values from $z = -3.00$ to $z = +3.00$.

To use Table **C** find the row that corresponds to your Z-value expressed to the first decimal place. Thus, to find, say, the cumulative probability of $z = 1.25$, we first find 1.2 in the far left-hand column, headed z_j. This number identifies the row that includes the cumulative probabilities for the Z-values 1.20, 1.21 . . . 1.29. Our second step is to find the column that gives the number in the second decimal place of our Z-value. In the present example, we want the column headed .05. The entry at the intersection of this row and column is the cumulative probability of 1.25 (i.e., $1.2 + .05$) and is equal to .8944. That is, according to Table **C**,

$$F(1.25) = .8944$$

Graphically, this means that 89.44 percent of the area under the standard normal curve is included between $z = -\infty$ and $z = 1.25$. In Figure 8.11, the hatched portion of the curve corresponds to the cumulative probability of $z = 1.25$.

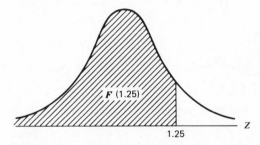

Figure 8.11   Cumulative probability of $z = 1.25$ *under standard normal distribution.*

1. Manipulating Cumulative Probabilities of the Standard Normal Curve

At the end of Chapter Two we illustrated how a cumulative probability distribution may be used to compute the probability of a specific value x_i. With a continuous random variable, of course, the probability of any particular value is zero, but the same principle may be used to determine the probability of any continuous interval Δx_i. We may therefore use our table of standard normal cumulative probabilities to determine the probability of any interval z_a to z_b. Suppose, for example, that z_a to z_b is an interval in the positive half of the standard normal curve and that $+z_a < +z_b$. The area representing the probability of this interval.

$$p(z_a \leq z \leq z_b)$$

is identified by hatching in Figure 8.12.

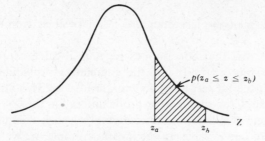

Figure 8.12   Probability of interval in positive half of standard normal distribution.

To find this probability using Table **C**, we first find the cumulative probability of the larger value, z_b. The cumulative probability $F(z_b)$ corresponds to the hatched portion of the curve in Figure 8.13. We then look up the cumulative probability of the lesser value, z_a. As before, $F(z_a)$ is represented by the hatched area in Figure 8.14.

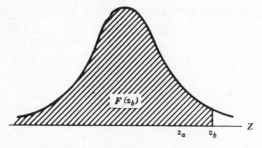

Figure 8.13   Cululative probability of upper class limit for interval in positive half of standard normal distribution.

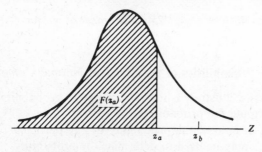

Figure 8.14   Cumulative probability of lower class limit for interval in positive half of standard normal distribution.

Examining the areas in Figures 8.12 to 8.14, it may be seen that if we subtract the cumulative probability of z_a from the cumulative probability of z_b, the difference is the strip of area representing the probability of our interval z_a to z_b:

$$p(z_a \leq z \leq z_b) = F(z_b) - F(z_a)$$

And, in general, the probability of any interval is found by subtracting the cumulative probability of the *lower* limit of the interval from the cumulative probability of the *upper* limit of the interval.

By way of example, suppose we wish to find the probability of observing a Z-value between $z = 1.25$ and $z = 2.00$:

$$p(1.25 \leq z \leq 2.00)$$

To obtain this probability, we subtract the cumulative probability $F(1.25)$ from the cumulative probability $F(2.00)$:

$$p(1.25 \leq z \leq 2.00) = F(2.00) - F(1.25)$$

From Table **C** we find that

$$F(2.00) = .98$$

$$F(1.25) = .89$$

Therefore

$$p(1.25 \leq z \leq 2.00) = .98 - .89$$

$$= .09$$

If our interval lies in the negative half of the curve, the procedure is identical. Consider, for example, the probability of observing z between $-z_b$ and $-z_a$ (Figure 8.15).

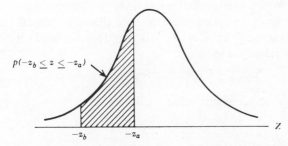

Figure 8.15   Probability of interval in negative half of standard normal distribution.

In performing our calculation, however, it must be emphasized that the terms "large" and "small" refer to algebraic value and not absolute or numerical value. Thus, a large negative number is *smaller* than a small negative number. For example, -2.00 is smaller than -1.25 and would therefore be the lower limit of the interval -2.00 to -1.25. We have stipulated that z_b is numerically larger than z_a, and to find the area in question, therefore, we take

$$p(-z_b \leq z \leq -z_a) = F(-z_a) - F(-z_b)$$

To determine probabilities represented by areas under the tails of the curve, we employ the same principles, but the actual manipulations are somewhat different. Suppose, for instance, we wish to find the probability of observing a Z-value in the negative tail of the curve below, say, $-z_a$. This probability is again represented as the hatched portion of the curve in Figure 8.16.

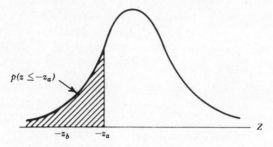

Figure 8.16   Probability in negative tail of standard normal distribution.

We see from Figure 8.16 that our interval includes all Z-values less than or equal to $-z_a$. Thus, we are seeking

$$p(z \leq -z_a)$$

However, we know by the definition of cumulative probability that

$$p(z \leq -z_a) = F(-z_a)$$

To determine this probability, therefore, we simply find $F(-z_a)$ in our table of cumulative probabilities. While this procedure may seem different from that used in the preceding examples, the reader should note that the interval under consideration is $-\infty$ to $-z_a$, and that $z = -\infty$ is therefore the lower limit of our interval. Thus, we are really computing

$$F(-z_a) - F(-\infty)$$

Because $-\infty$ is the smallest value in our value set, however,

$$F(-\infty) = 0$$

and

$$F(-z_a) - F(-\infty) = F(-z_a) - 0$$
$$= F(-z_a)$$

A similar procedure is involved in determining the area under the *positive* tail of the standard normal curve. To illustrate, let us consider the probability of observing a Z-value beyond z_a. That is, $p(z \geq z_a)$. If we enter Table **C** at z_a, our cumulative probability value corresponds to the *unshaded* portion of the curve in Figure 8.17. To find the hatched area, therefore, we must subtract this value, $F(z_a)$, from the *total* area under the curve. By our definition of a probability density function, we know that this total area is 1.0. Therefore

$$p(z \geq z_a) = 1 - F(z_a)$$

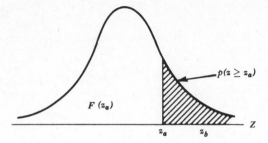

Figure 8.17   Probability in positive tail of standard normal distribution.

In this case our interval extends from z_a to $+\infty$, and our solution is actually

$$F(+\infty) - F(z_a)$$

However, $+\infty$ is the largest element in our value set, and

$$F(+\infty) = 1.0$$

Therefore

$$F(+\infty) - F(z_a) = 1.0 - F(z_a)$$

All of the preceding examples have required that we determine the cumulative probability associated with a given Z-value, z_j. It is also possible, however, to reverse the procedure and, beginning with a specific cumulative probability, to find its corresponding Z-value. That is, we may use the cumulative probability table to find z_j given only that $F(z_j)$ is equal to some specific value p. To use Table **C** in this manner, we simply find p among the entries and read z_j from the corresponding row and column headings. Suppose, for example, we wish to know that Z-value has a cumulative probability of .10. Entering Table **C**, we find that the cumulative probabilities increase from left to right and from top to bottom. Scanning them in this order we discover that no entry is precisely equal to .1000, but we do find .1003. If we are willing to accept a difference of .0003 as rounding error, we may accept the corresponding Z-value as z_j. Proceeding with our solution, we see that .1003 lies at the intersection of the row marked -1.2 and the column headed .09. Therefore

$$z_j = -1.29$$

The example given above was cast in terms of finding a Z-value corresponding to a particular cumulative probability, $F(z_j) = p$. We know, however, that

$$F(z_j) = p(z \leq z_j)$$

and the same problem might therefore have been presented in terms of finding z_j such that the probability of observing a value z *less than* or equal to z_j is p. Conversely, the problem might have required that we find z_j such that the probability of observing a Z-value *greater than* z_j is p. However, when p is defined as the probability of observing values greater than (or equal to) z_j, its area lies in the positive tail of the curve, and determination of z_j therefore involves some additional considerations.

In Figure 8.18, the hatched portion of the curve corresponds to

$$p(z \leq z_j) = p$$

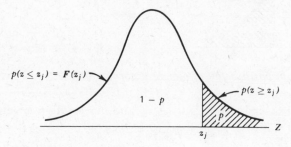

Figure 8.18 Probability values involved in finding z_j such that $p(z \geq z_j) = p$. *See text for explanation.*

The remainder of the curve must therefore equal

$$1 - p$$

We also see, however, that this unshaded area, which is equal to $1 - p$, must represent

$$p(z \leq z_j)$$

which by definition is the cumulative probability

$$F(z_j)$$

When p is in the positive tail of the curve, therefore,

$$F(z_j) = 1 - p$$

and the problem reduces to one of finding the Z-value with a cumulative probability equal to $1 - p$. To illustrate, let us find the value z_j such that

$$p(z \geq z_j) = .10$$

If $p(z \geq z_j) = .10$, it must be true that

$$p(z \leq z_j) = .90$$

or

$$F(z_j) = .90$$

To determine the value we seek, therefore, we enter Table **C** until we find the nearest acceptable value to .9000. This turns out to be .8977, and the corresponding Z-value is $+1.29$.

162

D. The Nonstandard Normal Curve

As the reader is now well aware, the formula for the standard normal curve is

$$\phi(Z) = \frac{1}{\sqrt{2\pi\,V(Z)}}\,e^{-Z^2/2}$$

We also know, however, that

$$Z = \frac{X - \mu}{\sigma}$$

and our exponent may thus be expressed as

$$-\left(\frac{X - \mu}{\sigma}\right)^2 \!/2$$

or, if we remove σ from the parentheses and put it in the denominator,

$$-(X - \mu)^2/2\sigma^2$$

In either case, we see that by substituting $(X - \mu)/\sigma$ for Z in the exponent, our density function becomes a function of X rather than Z, and $\phi(Z)$ becomes $\phi(X)$ and $V(Z)$ becomes $V(X)$, or σ^2. The density function thus becomes

$$\phi(X) = \frac{1}{\sqrt{2\pi\,\sigma^2}}\,e^{-(X-\mu)^2/2\sigma^2}$$

As the reader may recall from the beginning of the chapter, this is the expression for the normal, or Gaussian, probability distribution. We have just seen that its derivation from the standard normal density function requires only a simple substitution of equivalent terms. Such substitution has no effect on the functional relationship, and the unstandardized normal curve must therefore possess all of the mathematical properties that characterize the standard normal curve. It is symmetrical and unimodal, and its tails, like those of the standard normal curve, approach the horizontal axis, $\phi(X) = 0$, asymptotically. Moreover, the function reaches its maximum value when the exponent

$$-\left(\frac{X - \mu}{\sigma}\right)^2 \!/2 = \frac{-Z^2}{2}$$

is equal to zero. The reader is cautioned, however, that this does *not* occur when $x = 0$, but rather when $x = \mu$. That is, when X assumes a value equal to its own expectation, the exponent becomes

$$-\left(\frac{\mu - \mu}{\sigma}\right)^2 \!/2 = 0$$

Like the standard normal distribution, therefore, the nonstandard distribution is symmetrical about its own expectation μ

The preceding paragraph may have called the reader's attention to the fact that the value of the exponent is not only determined by the value of X, but by the values of μ and σ. Thus, computation of any probability density $\phi(x_i)$ requires that we assign numerical values to μ and σ. Thus, the expectation and variance are *parameters* of the normal distribution, and we therefore indicate that a variable X is normally distributed by the following notation:

$$X:N(\mu, \sigma^2)$$

It should be apparent, therefore, that there are really an *infinite* number of normal distributions, one for each possible value of μ and σ. Thus,

$$X:N(100, 225)$$

defines a very different distribution from

$$X:N(40, 16)$$

The probability, for example, of finding an X-value between 85 and 115 is very large under the first distribution,

$$p(85 \leq x \leq 115) = .68$$

but is less than .001 under the second distribution. It would be quite impossible, therefore, to table cumulative probability distributions for every normal curve that a statistician is likely to encounter. On the other hand, we would need only one such table if all normal distributions assumed the same values for μ and σ. Fortunately, this may be accomplished by standardization, which renders the expectation of any random variable equal to 0 and its variance equal to 1. Moreover, this transformation has no effect on the probability function that describes a distribution, and we therefore know that for any normally distributed random variable

$$X:N(\mu, \sigma^2)$$

it must be true that

$$Z:N(0, 1)$$

When the random variable is standardized, the independent variable becomes Z, and our normal density function becomes

$$\phi(Z) = \frac{1}{\sqrt{2\pi\,\sigma^2}}\, e^{-(Z-\mu)^2/2\sigma^2}$$

For a standardized variable, however, $\mu = 0$ and $\sigma^2 = 1$, and our expression thus reduces to

$$\phi(Z) = \frac{1}{\sqrt{2\pi}}\, e^{-Z^2/2}$$

which the reader should recognize as a slightly simplified expression of the standard normal probability density function, which we have been discussing throughout this chapter. We see, then, that the standard normal curve represents the probability distribution of any normally distributed random variable that has undergone standardization. Or, more generally, it represents the distribution of any normally distributed random variable with expectation equal to 0 and variance equal to 1. If, therefore, we wish to determine the probability of any interval x_a to x_b where $X{:}N(\mu, \sigma^2)$, we need only standardize our X-values and proceed as explained earlier in the chapter.

Suppose, for example, that X is a normally distributed random variable with expectation equal to 900 and variance equal to 2500:

$$X{:}N(900, 2500)$$

Furthermore, let us suppose that we wish to determine the probability of observing an X-value less than or equal to 950,

$$p(x \leq 950)$$

First, we know by definition that

$$p(x \leq 950) = F(950)$$

We therefore standardize $x_a = 950$ and submit our standardized value z_a to our table of standard normal cumulative probabilities.

$$z_a = \frac{x_a - \mu}{\sigma}$$

$$= \frac{950 - 900}{\sqrt{2500}}$$

$$= \frac{50}{50}$$

$$= 1.00$$

From Table **C** we find that

$$F(1.00) = .84$$

Therefore

$$p(z \leq 1.00) = .84$$

$$p(x \leq 950) = .84$$

As an additional exercise, let us determine the probability of observing an X-value in the interval 950 to 1000. Recall from our earlier discussion of cumulative probabilities that

$$p(950 \leq x \leq 1000) = F(1000) - F(950)$$

To determine these cumulative probabilities, however, we must first convert them to standard values. If we let

$$x_b = 1000$$

then

$$z_b = \frac{1000 - \mu}{\sigma}$$

$$= \frac{1000 - 900}{50}$$

$$= \frac{100}{50}$$

$$= 2.00$$

Similarly, we let

$$x_a = 950$$

and from the last example we know that

$$z_a = 1.00$$

Therefore

$$p(950 \leq x \leq 1000) = p(1.00 \leq z \leq 2.00)$$

$$= F(2.00) - F(1.00)$$

We know from our preceding example that $F(1.00) = .84$, and from Table **C** we find that $F(2.00)$ equals .98. Therefore

$$F(2.00) - F(1.00) = .98 - .84$$

$$= .14$$

and

$$p(950 \leq x \leq 1000) = .14$$

1. The Normal Approximation to the Binomial

Although the binomial probability function involves only simple algebraic operations, the actual computation of probabilities can become extremely tedious as sample size increases. In contrast, we have seen that computation of probabilities using the cumulative probability table for the standard normal distribution is a relatively simple routine. In this regard, we know that the standardized binomial becomes normal as N increases, and it is therefore often expedient to use the standard normal distribution to approximate binomial probabilities. To compute a normal approximation, one simply determines the real limits of his binomial interval Δx, standardizes them, and submits them to Table **C** as if his variable were normally distributed. Consider, for example, the binomial random variable $X:B(16, 0.5)$, and suppose we wish to determine the probability of obtaining either 6, 7, 8, 9, or 10 successes. That is,

$$p(6 \leq x \leq 10)$$

We know from Chapter Three that we would compute this probability by calculating

$$\frac{16!}{r! \, (16 - r)!} \, (0.5)^r (0.5)^{16-r}$$

for each of our five values of r and sum the results:

$$p(x = 6) = .1222$$
$$p(x = 7) = .1746$$
$$p(x = 8) = .1964$$
$$p(x = 9) = .1746$$
$$p(x = 10) = .1222$$

$$p(6 \leq x \leq 10) = .7900$$

Alternatively, we might simply define our interval 6 to 10 and find the normal approximation to the binomial. In effect, this procedure involves the pretext that our random variable is *in fact* normally distributed but that our measurements have permitted only assignment of integer values to our sample points. Thus, we proceed as if the integer x_i has been assigned to every value in the continuous interval $x_i - 0.5$ to $x_i + 0.5$. This implies, of course, that any binomial interval x_i to x_j is defined by the *apparent* limits of the *real* interval $x_i - 0.5$ to $x_j + 0.5$. Before we standardize the limits of our interval 6 to 10 and enter Table **C**, we must therefore *correct for continuity* by extending the interval 0.5 units beyond these apparent limits. Thus, we set x_a equal to 5.5 and x_b equal to 10.5.

$$z_a = \frac{5.5 - Np}{\sqrt{Npq}}$$

$$= \frac{5.5 - 8}{2}$$

$$= -\frac{2.5}{2}$$

$$= -1.25$$

Similarly,

$$z_b = \frac{10.5 - 8}{2}$$

$$= +1.25$$

Our standardized interval Δz therefore becomes -1.25 to $+1.25$, and the probability we seek is

$$p(-1.25 \leq z \leq +1.25)$$

which by our discussion of the cumulative probability distribution must equal

$$F(1.25) - F(-1.25)$$

From Table **C** we find that

$$F(1.25) = .8944$$

$$F(-1.25) = .1056$$

Therefore

$$p(-1.25 \leq z \leq +1.25) = .7888$$

Thus, even an N of 16 is "close enough to infinity" so that the probability generated by our normal model differs by only .0012 from the true probability value. It must be noted, however, that we exercised considerable care in selecting our example. First, we specified that p equal 0.5. As p becomes either larger or smaller than 0.5, the resulting binomial distribution loses its symmetry (becomes *skewed*) and the normal approximation becomes increasingly poor. This skewness may be reduced by increasing the size of the sample, but for values of p near either 0 or 1, a binomial distribution based on even a very large sample may not be satisfactorily approximated by the normal distribution. Second, we chose an interval that was very close to the expectation, $\mu = 8$. However skewed the distribution, the probabilities of intervals falling near the mean are better approximated by the normal model than probabilities of intervals in the extremes. The normal curve, therefore, best approximates binomial probabilities of intervals near the middle of distributions based on moderate values of p. The more one departs from these constraints, the larger must be his sample. As N becomes infinitely large, of course, the location of the interval and the size of p become unimportant.

E. Problems, Review Questions, And Exercises

1. Assuming SAT scores to be normally distributed, what is the probability of obtaining a score as high or higher than the two applicants in problem **1** of Chapter Seven?
2. Two college applicants have obtained SAT scores in the 93rd percentile and 91st percentile respectively. What are their Y-scores?
3. Consider again the experiment defined in problem **1** of Chapter Four.
   a. Use the binomial probability function to compute the probability of obtaining more than 7 heads.

b. Compute the same probability using the normal approximation to the binomial.

c. Does **2b** yield a close approximation of the exact probability computed in **2a**?

4. Consider again problem **5** of Chapter Three.

a. Use the normal approximation to the binomial to compute the probability of obtaining 6 or fewer hits.

b. Does **3a** yield a close approximation of the exact probability computed using the binomial probability function? If not, why?

the concept of covariation

A. Joint Distributions of Data: The Scatter Diagram

In Chapter Eight our treatment of distribution properties shifted emphasis from discrete to continuous random variables. Our discussion has nonetheless been confined to parameters and statistics that describe properties appropriate a distribution of *one* random variable. In Chapter Nine, the discussion extends to the properties that characterize the distribution of *two* random variables observed simultaneously. That is, we consider some of the parameters and statistics that describe a *joint distribution* of two random variables.

To get a feel for the meaning of a joint distribution, let us suppose that each person in a large sample wrote his height and weight on a piece of paper. We could tabulate the data as shown in Table 9.1.

Table 9.1 Heights and Weights for *N* Subjects

Subject	Height	Weight
001	5 feet 7 inches	160 lb
002	6 feet 4 inches	219 lb
003	6 feet 0 inches	180 lb
. . .	. . .	. . .
N	5 feet 1 inch	135 lb

Now let our two variables define a Cartesian plane, with weight represented on the *X*-axis and height represented on the *Y*-axis (Figure 9.1).

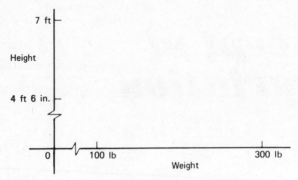

Figure 9.1   Height and weight as axes of Cartesian plane.

Each point in the plane represents a combination, or joint occurrence, of some height and some weight. Using height and weight values as coordinates, therefore, we can plot a point for each subject. Our plot, which is formally known as a *scatter plot* (or scatter diagram or scattergram), would look something like Figure 9.2.

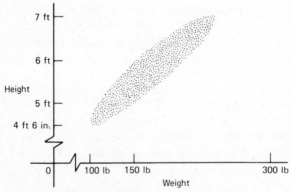

Figure 9.2   Scatter plot of height and weight.

1. The Concept of a Statistical Relation

If we examine the scattergram of heights and weights we notice various characteristics. First, the plot is roughly eliptical, with the major axis of the elipse considerably longer than the minor axis. That is, the elipse is markedly longer than it is wide.

172

This property has a good deal of importance in terms of our ability to estimate the value of one variable from knowledge about the value of the other variable. Let us suppose that we drew a subject at random from our sample. Let us further suppose that we knew neither his height nor his weight, but that for some reason it was necessary to make an estimate of his height. With no information other than the fact that he is a member of our original sample, the best we could say is that his height is somewhere between 4 feet 6 inches and 7 feet. If, on the other hand, we knew that his weight was, say, 150 pounds, the scatter plot suggests a more sensible estimate. Although our entire sample ranges in height from 4 feet 6 inches to 7 feet we see that individuals weighing 150 pounds range only from 5 to 6 feet (Figure 9.3).

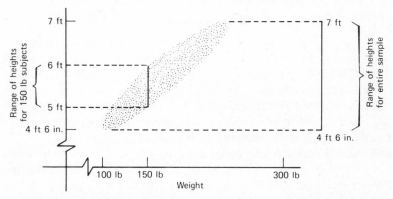

Figure 9.3   Range of heights for entire sample compared with range of heights for Ss weighing 150 pounds.

Thus, our best estimate is now the interval 5 to 6 feet. Therefore, knowing our subject's weight has allowed us to reduce the range of our estimate from $2\frac{1}{2}$ feet (4 feet 6 inches to 7 feet) to just 1 foot (5 to 6 feet). This means that our estimate of his height has become more precise on the basis of *knowing his weight*. This is the essence of a statistical relation: that knowing the value of one variable reduces uncertainty about (or increases the precision of estimate of) the value of a second variable. More formally,

> If there exists a set of pairs of values (*x*,*y*) such that knowing the value of one member of any pair, x_i, increases the precision of estimate of the value of the second member, y_i, the two variables are statistically related.

Exactly what we mean by "increasing the precision of estimate" or "reducing uncertainty about" will be discussed in detail when we introduce the notion of correlation. Even without the formal mathematics, however, it should be apparent that the degree to which knowing x_i allows us to reduce the range of our estimate for y_i is related to the *width* of the scatter-plot. That is, the narrower the plot, the smaller the range of *Y*-values associated with any given value of *X*.

Therefore, as the magnitude, or strength, of the statistical relation increases, the width of the scattergram decreases in comparison to its length. Thus, when the plot degenerates to a single line (or other curve), prediction is perfect, and each value of X is associated with one and only one value of Y. The reader will recall that these are the conditions that define a function. Therefore, if X is a perfect predictor of Y, then

$$Y = f(X)$$

The second property that one might notice is that the points seem to cluster symmetrically about an imaginary straight line, which we have been loosely calling the major axis. This tells us that if X were a perfect predictor of Y, the function that defined their relation would be a linear function. That is,

$$f(X) = a + bX$$

Not only are X and Y related, therefore, but the relationship is *linear*. It is possible, of course, to think of a great many relationships that are not linear. If Y is the height of a fly ball and X is the distance from the hitter, the relationship is curvilinear (specifically, parabolic); first the ball goes up then it comes down. The methods and techniques that we shall be considering, however, are appropriate only to linear relationships. And, even when linearity is not specified, it must be understood to be implied unless otherwise noted.

Finally, we see that the major axis lies on a diagonal from the lower left of the plane to the upper right of the plane. Thus, as one moves along the axis from left to right, *both* X and Y are seen to increase. This implies, of course, that small values of X are usually associated with small values of Y, and that large values of X are generally associated with large values of Y. This should really surprise no one. We rather expect someone who is 4 feet 6 inches tall to weigh less than someone who is 6 feet 5 inches tall. Surprising or not, however, it tells us that our linear relationship is *positive*.

Now let us consider the joint distribution of two other variables, present age and remaining life expectancy (Figure 9.4).

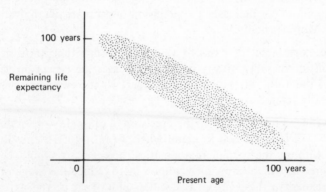

Figure 9.4   Scatter plot of present age and remaining life expectancy.

Once again we see that the scatter plot is rather elongated and that we can thus make a more precise estimate of a person's remaining life expectancy if we know his age than if we do not know his age. Also, the plot suggests that the relationship is again linear.

However, this time the major axis runs from the upper left to the lower right, indicating that large values of one variable are associated with small values of the other variable. Thus, the younger a person is, the more years he has ahead of him, or as Joan Baez cheerfully reminds us: "The older you get, the sooner you're going to die." In more general (and less depressing) terms, as one moves along the axis one variable increases while the other decreases. This defines our linear relationship as *negative*.

Lest the reader be concerned with my unconscious motivations for selecting this example, let him be assured that the considerations were purely economic. Each year it becomes more expensive to purchase additional life insurance.

Finally, let us consider two variables that should have absolutely *no* relationship. In this regard we shall adapt an example from Guenther's *Concepts of Statistical Inference* (1965) and examine a scatterplot of I.Q. and shoe width (Figure 9.5).

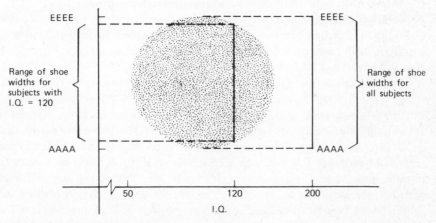

Figure 9.5   Scatter plot of show widths for entire sample compared with range of show widths for Ss with I.Q. equal to 120.

In this example the plot looks vaguely like a circle that has been flattened slightly at the top and bottom. Thus, the range of observed shoe widths for subjects with, say, an I.Q. of 120 is pretty much the same as the range for the entire sample. And, knowing a person's I.Q. permits us to make an estimate of his shoe width that is *no more precise* than we would make simply on the basis of knowing that he is a member of our sample. This is the sort of scatter diagram obtained when one plots the joint distribution of two *independent* variables.

B. Numerical Description of a Linear Relationship: The Covariance

Now that the student has some intuitive feel for the notion of a statistical relation, we may quantify our considerations and develop some numerical index that describes the strength or magnitude of a linear relation.

In this connection we shall find it useful to examine some of the geometric properties of the three scatter diagrams that we have discussed thus far. To facilitate our comparison two conventions will be introduced. First, the scatterplots will be represented in outline only. For the moment this is simply a typographical and heuristic convenience with no mathematical implications that need concern us. Second, we shall change the position of our X and Y axes so that their origin (the point defined by the coordinates 0,0) falls at the geometric center of each scatter diagram. This sort of transformation is known as translation of axes. The skeptical reader may be suspicious of what appears to be some mathematical sleight-of-hand, but it will be recalled from Chapter Seven that changing the location of a distribution on the real line affects neither the form nor the probability relationships of the distribution. Translation of axes simply involves changing the location of a joint distribution with respect to both of its dimensions simultaneously, and as before none of the relational properties of the distribution are affected (Figure 9.6).

By translating our axes we have divided each plane into four quadrants, which we designate in anticlockwise order from 1 to 4. Moreover, if we examine our three scattergram outlines in terms of their geometric areas, we can detect systematic differences in the distribution of their respective areas across the four quadrants. For example, the joint distribution of heights and weights is concentrated in quadrants 1 and 3, with a relatively small proportion of its area falling in quadrants 2 and 4. Furthermore, it may be seen that as the magnitude of this relation increases, that is, as the plot becomes *narrower*, the proportion of area in quadrants 1 and 3 increases, and the proportion of area in quadrants 2 and 4 decreases.

If we look at the representation of our negative relation, present age versus life expectancy, we see that the reverse is true. The bulk of the area falls in quadrants 2 and 4, while only a small proportion falls in quadrants 1 and 3. And, once again, as the scatter plot narrows and precision of estimate increases, the disparity of the distribution becomes more pronounced.

Finally, if we look at the plot of I.Q. and shoe width we see what happens when there is no relation. When the plot is as wide as it is long, the area seems evenly distributed among all four quadrants, and we see no apparent difference between the proportion of area falling in quadrants 1 and 3 and the proportion falling in quadrants 2 and 4.

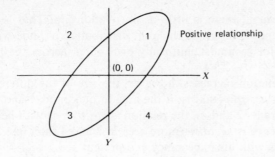

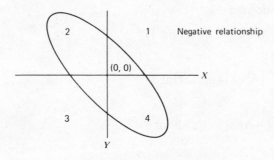

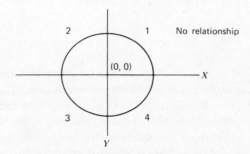

Figure 9.6   Scatter plots illustrating positive relationship, negative relationship, and no relationship.

Therefore, we should be able to describe both the magnitude and the direction of a statistical relation with some numercial index that is based on the algebraic difference between the area of the scatterplot falling in quadrants 1 and 3 and the area falling in quadrants 2 and 4. That is, some statistic that reflects

(Area in quadrant 1 + Area in quadrant 3)

minus

(Area in quadrant 2 + Area in quadrant 4)

When this difference is small, there is little or no relationship. That is, we have a scatter plot like the one representing the joint distribution of shoe width and I.Q.; the area of the plot is almost uniformly distributed across the four quadrants.

As the strength of the relation increases, however, the size of the difference increases. If the difference is *positive* (that is, if the area in quadrants 1 and 3 is larger than the area in quadrants 2 and 4), the relation is *positive*, as with height and weight. But if the difference is *negative* (more area in 2 and 4 than in 1 and 3), the relation is *negative*, as with life expectancy and present age.

Alternatively, we might simply assign positive values to the areas in quadrants 1 and 3 and negative values to the areas in quadrants 2 and 4. This would permit us to take the simple algebraic sum

(Area in quadrant 1)

+

(Area in quadrant 3)

+

($-$Area in quadrant 2)

+

($-$Area in quadrant 4)

which is arithmetically identical to the difference given above.

1. The Sample Covariance: C_{XY}

In moving from a geometric model to a completely abstract numerical statistic, we must translate some geometric concepts into arithmetic operations. This is especially true of the concept of area. Recall from plane geometry that area is most generally computed by multiplying some horizontal distance by some vertical distance. Thus, the area of a rectangle equals its width multiplied by its height; the area of a parallelogram equals its base multiplied by its height; and, the product of base and altitude gives the area of a trapezoid.

Therefore, we shall consider "area" to be reflected in the product of each subject's score on the "horizontal" axis and his score on the "vertical" axis. That is, the numerical value that we shall consider analogous to the geometric concept of area will be the subject's X-value multiplied by his Y-value. To compute the "area" bounded by the scattergram in any given quadrant, therefore, we simply take the sum of these products for all subjects falling in that quadrant.

With these stipulations in mind, we may utilize the strategy presented above and formally develop the notion of the covariance without introducing any new considerations.

First, we must translate our axes to the geometric center of the distribution. At this point it must be noted that the geometric center of the distribution is the intersection of the means of the two variables: mean height and mean weight, mean I.Q. and mean shoe size, and so on. The geometric center is the point $(\bar{x}, \bar{y})$. Translating the axes so that the origin (0,0) falls at the point corresponding to $(\bar{x}, \bar{y})$ is therefore simply a matter of transforming our variables so that $\bar{x} = 0$ and $\bar{y} = 0$. It will be recalled that this may be accomplished by converting all scores to deviation scores. That is, each subject's X-score is converted from x to $(x - \bar{x})$ and each subject's Y-score becomes $(y - \bar{y})$.

It is this transformation that allows us to assign positive values to the products of scores (i.e., "areas") in quadrants 1 and 3 negative values to the products in quadrants 2 and 4, which is the first step in the computation of the algebraic sum given above.

With the zero point now located in the center of the distribution, both the X-coordinate and the Y-coordinate for *every* subject falling in quadrant 1 must be positive. That is, for every subject in quadrant 1 *both* $(x - \bar{x})$ and $(y - \bar{y})$ must be greater than 0. This means that the product of the deviation scores for any subject in quadrant [1]

$$(x_i - \bar{x})(y_i - \bar{y})$$

must also be positive.

In quadrant 2, however, the X-deviation scores $(x - \bar{x})$ are negative, while the $(y - \bar{y})$ scores are positive. Therefore, the product of deviation scores for any subject falling in quadrant 2 will be negative.

In quadrant 3, both coordinates are negative, and therefore their product will be positive.

In quadrant 4, the values of $(x - \bar{x})$ are positive while the values of $(y - \bar{y})$ are negative. Products of deviation scores for subjects falling in quadrant 4 will therefore be negative (Figure 9.7).

If we now sum all of these products we have a value that reflects the difference between the area of the scattergram in quadrants 1 and 3 and the area in quadrants 2 and 4. When the sum is close to zero, it means that there are almost as many negative products as positive products, which implies that there is almost as much area in quadrants 2 and 4 as there is in 1 and 3. Once again, this is what we would expect for two independent variables.

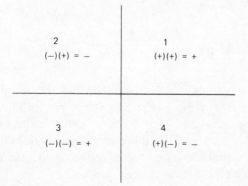

2	1
$(-)(+) = -$	$(+)(+) = +$
3	4
$(-)(-) = +$	$(+)(-) = -$

Figure 9.7 Signs of products in four quadrants of Cartesian plane for standardized variables.

When the sum is positive, however, it means that we have more positive products than negative products. Thus, the relation is positive, and there is more area in quadrants 1 and 3 than there is in 2 and 4.

Finally, when the sum is negative we have a negative relation with more area in quadrants 2 and 4 than in quadrants 1 and 3.

Clearly, then, the sum of the products $(x_i - \bar{x})(y_i - \bar{y})$ reflects the algebraic difference between the area of the scatter plot falling in quadrants 1 and 3 and the area falling in 2 and 4. As such, it must also reflect the magnitude and direction of the statistical relation. It does, however, have an undesirable property that the reader may have anticipated. Unless the sum is *exactly* zero, the absolute magnitude of the sum will increase as a function of the sample size *irrespective* of the strength of the statistical relation. Quite simply, as N increases, the number of products increases, and thus the number of terms in the sum increases.

This can be corrected by the simple expedient of dividing our sum by N and thus considering the *average* product of deviation scores instead of the *sum* of the products. This mean cross product is defined as the sample covariance and is expressed as

$$C_{XY} = \frac{1}{N} \sum_{i}^{N} (x_i - \bar{x})(y_i - \bar{y})$$

From this expression we see that the sample covariance satisfies at least the first two of our criteria for a good statistic. It is single-valued, and its computation requires only algebraic operations. To demonstrate its conformity to our other criteria, however, requires that we express the covariance in frequency notation. As defined above, the sample covariance involves N products $(x_i - \bar{x})$ $(y_i - \bar{y})$, each of which is based on one observation of two jointly distributed random variables X and Y. If our sample is large, however, it will generally be true that many of the products $(x_i - \bar{x})(y_i - \bar{y})$ will assume the same value. That is, although there are N observations and N products in the sum, these products will represent only n different values. Therefore

$$C_{XY} = \frac{1}{N} \sum_{i}^{N} (x_i - \bar{x})(y_i - \bar{y})$$

$$= \frac{1}{N} \sum_{i}^{n} [(x - \bar{x})(y - \bar{y})]_i f_i$$

In this notation, the random variable under consideration is the entire product $[(X - \bar{x})(Y - \bar{y})]$, which is why we require only one subscript i and it appears outside the brackets. The frequency with which we observe the value $[(x - \bar{x})(y - \bar{y})]_i$ is indicated by f_i. Expressed in this convention, therefore, we see that the sample covariance includes every observed value of the random variable and the frequency with which it appears in our sample.

2. The Population Covariance: cov (XY)

If we further modify the last expression developed above by bringing the constant $1/N$ into the summation, C_{YX} becomes

$$\sum_{i}^{n} [(x - \bar{x})(y - \bar{y})]_i \frac{f_i}{N}$$

When expressed in this form, it is apparent that the sample covariance is very similar to the sample variance

$$s^2 = \sum_{i}^{n} (x_i - \bar{x})^2 \frac{f_i}{N}$$

In this regard, recall that the definitional formula for the population variance was derived by conceptualizing the population variance as the expected *squared deviation* of X from its population mean:

$$V(X) = E[X - E(X)]^2$$

Similarly, the population variance is conceptualized as the expected *product of deviations* of X and Y from their respective population means and is therefore defined by expression

$$\text{cov}(XY) = E[X - E(X)] [Y - E(Y)]$$

C. The Covariance of Two Standardized Variables: The Pearson Product—Moment Correlation Coefficient

1. The Sample Correlation Coefficient: r_{XY}

The covariance is thus an index of linear relationship with many useful properties. We know that if two variables are independent the covariance will be zero. We also know that any departure from zero indicates a relationship, and that the strength of the relation is directly proportional to the size of the covariance. Finally, we know that the sign of the covariance tells us the direction of the relation, positive or negative.

One property we have not considered, however, is comparability. If the relationship of X and Y is the same magnitude as the relationship of two other variables X' and Y', will the covariance give the same value in both cases?

We can explore this question by examining the sample covariances associated with two such pairs of variables. Recall our first example of height and weight. Let height in *feet* equal X and weight in pounds equal Y. The sample covariance of height and weight is thus expressed as

$$C_{XY} = \frac{1}{N} \sum_{i}^{N} (x_i - \bar{x}) (y_i - \bar{y})$$

Now let X' represent height measured in *inches*. Naturally, the magnitude of the relation of height and weight is not affected by measuring height in different units, and therefore we should wish our covariance to remain constant. However,

$$C_{X'Y} = \frac{1}{N} \sum_i^N (x_i' - \bar{x}')(y_i - \bar{y})$$

If X = height in feet and X' = height in inches, then every value $x_i' = 12x_i$. Moreover,

$$\bar{x}' = \frac{\sum_i^N x_i'}{N}$$

$$= \frac{\sum_i^N 12x_i}{N}$$

By the rules of summation given in Appendix I, however, the constant 12 may be factored out of the sum. Therefore

$$\bar{x}' = \frac{12 \sum_i^N x_i}{N}$$

$$= 12\bar{x}$$

and

$$C_{X'Y} = \frac{1}{N} \sum_i^N (x_i' - \bar{x}')(y_i - \bar{y})$$

$$= \frac{1}{N} \sum_i^N (12x_i - 12\bar{x})(y_i - \bar{y})$$

$$= \frac{1}{N} \sum_i^N 12(x_i - \bar{x})(y_i - \bar{y})$$

Once again, the constant 12 may be factored out of the summation. Thus

$$C_{X'Y} = (12) \frac{1}{N} \sum_i^N (x_i - \bar{x})(y_i - \bar{y})$$

$$= 12\, C_{XY}$$

Thus, the covariance of height and weight is 12 times greater if height is measured in inches than if height is measured in feet. We see, therefore, that the magnitude of a covariance is dependent on the units in which the variables are expressed. This property has an obvious and very unfortunate implication. A nonzero covariance cannot be interpreted. If the covariance is equal to zero, we know that no linear relation exists. But, if the covariance takes on any other value, we cannot make *any* statement about the *magnitude* of the relation. It is always the case that a covariance can be increased by using smaller units (such as inches) or decreased by using large units (such as miles), and thus we have no way to determine whether our covariance indicates a weak relation, a moderate relation, or a very strong relation.

This property would cease to be a problem, however, if the covariance were always computed on variables that were expressed in the same metric, that is, if any pair of variables could always be converted to some common, or standard, scale of measurement. And, of course, the reader will recall from Chapter Seven that by standardizing a variable we quite explicitly convert its scale to standard units. If we therefore take the sample covariance of *standardized* scores we provide ourselves with an index that is independent of the units of measurement and that may thus be meaningfully used to interpret the magnitude of a linear relation.

$$C_{z_X z_Y} = \frac{1}{N} \sum_{i}^{N} [(z_X)_i - \bar{z}_X] [(z_Y)_i - \bar{z}_Y]$$

Recall from Chapter Seven, however, that the sample mean of a standardized variable is always equal to zero. Thus

$$\bar{z}_X = 0$$

and

$$\bar{z}_Y = 0$$

Therefore

$$C_{z_X z_Y} = \frac{1}{N} \sum_{i}^{N} (z_X)_i (z_Y)_i$$

For sample data, however,

$$(z_X)_i = \frac{x_i - \bar{x}}{s_X}$$

and

$$(z_Y)_i = \frac{y_i - \bar{y}}{s_Y}$$

We may therefore substitute these expressions back in our formula for $C_{z_X z_Y}$ and

$$C_{z_X z_Y} = \frac{1}{N} \sum_{i}^{N} \left[\frac{x_i - \bar{x}}{s_X} \right] \left[\frac{y_i - \bar{y}}{s_Y} \right]$$

$$= \frac{1}{N} \sum_{i}^{N} \frac{(x_i - \bar{x})(y_i - \bar{y})}{s_X s_Y}$$

$$= \frac{\sum_{i}^{N} (x_i - \bar{x})(y_i - \bar{y})}{N s_X s_Y}$$

This is the definitional formula for the Pearson product-moment correlation coefficient. When computed for sample data, the Pearson correlation is designated by the letter r, which is subscripted with the two random variables whose relation it describes. Thus

$$r_{XY} = \frac{\sum_i^N (x_i - \bar{x})(y_i - \bar{y})}{N \, s_X s_Y}$$

Like the sample covariance, the sample correlation coefficient may be converted to frequency notation:

$$r_{XY} = \frac{\sum_i^n [(x - \bar{x})(y - \bar{y})]_i f_i}{N \, s_X s_Y}$$

In this form, it is again apparent that our statistic satisfies all four of our criteria for a good statistic.

a. Limits of the Correlation Coefficient. If we examine the formula for the sample correlation coefficient with the factor $1/N$ taken outside the summation sign, we notice that it differs from the sample covariance only by the denominator, $s_X s_Y$:

$$r_{XY} = \frac{1}{N} \sum_i^N \frac{(x_i - \bar{x})(y_i - \bar{y})}{s_X s_Y}$$

$$C_{XY} = \frac{1}{N} \sum_i^N (x_i - \bar{x})(y_i - \bar{y})$$

Therefore

$$r_{XY} = \frac{C_{XY}}{s_X s_Y}$$

We know from our presentation of covariance that when two variables are entirely unrelated.

$$C_{XY} = 0$$

It follows, therefore, that two unrelated variables should also produce a correlation coefficient equal to 0.

$$r_{XY} = \frac{C_{XY}}{s_X s_Y}$$

$$= \frac{0}{s_X s_Y}$$

$$= 0$$

At the other extreme we have the situation where two variables are perfectly related. That is, where $Y = f(X)$ and f is a linear function. It is not difficult to demonstrate that if the relation is positive, $r_{XY} = 1.0$.

$$r_{XY} = \frac{1}{N} \sum_i^N \frac{(x_i - \bar{x})(y_i - \bar{y})}{s_X s_Y}$$

To assure that X and Y are, in fact, perfectly related, let us suppose that they represent precisely the same measure. That is, $x_i = y_i$ for every value of i. This relation is, of course, positive. Under this condition we may substitute X for Y every time Y appears in the preceding expression:

$$r_{XY} = \frac{1}{N} \sum_i^N \frac{(x_i - \bar{x})(x_i - \bar{x})}{s_X s_Y}$$

$$= \sum_i^N \frac{1}{N} \left[\frac{x_i - \bar{x}}{s_X} \right]^2$$

$$= \sum_i^N \frac{z_i^2}{N}$$

Recall that $\bar{z} = 0$. Zero may, of course, be subtracted from any expression without changing its value. Therefore

$$z_i = z_i - \bar{z}$$

and

$$r_{XY} = \sum_i^N \frac{(z_i - \bar{z})^2}{N}$$

$$= s_Z^2$$

$$= 1.0$$

Therefore, if X and Y are perfectly and positively related,

$$r_{XY} = 1.0$$

Although the condition $x_i = y_i$ assures a correlation of 1.0 (provided, of course, that $s^2 > 0$), this does not imply that x_i *must* equal y_i to yield a perfect correlation. In our proof we demonstrated that

$$\sum_i^N \frac{z_i^2}{N} = 1.0$$

and it should therefore be apparent that r will equal 1.0 for any set of pairs (x,y) for which $(z_X)_i = (z_Y)_i$. We would, for example, obtain a perfect correlation for N pairs of simultaneous temperature recordings measured in centigrade (X) and in Farenheit (Y). Even though X and Y would assume equal values only at $-40°$, every pair of *standardized* recordings would be identical.

By similar arguments, it can be demonstrated that if X and Y are perfectly and negatively related,

$$r_{XY} = -s_Z^2$$
$$= -1.0$$

2. The Population Correlation Coefficient: ρ_{XY}

From the last section we may define the sample correlation coefficient of two variables X and Y as the sample covariance of the two variables in standard form:

$$r_{XY} = C_{Z_X Z_Y}$$

Similarly, the population correlation coefficient of X and Y is equal to the population covariance of Z_X and Z_Y:

$$\rho_{XY} = \text{cov}(Z_X Z_Y)$$

By definition,

$$\text{cov}(Z_X Z_Y) = E[Z_X - E(Z_X)][Z_Y - E(Z_Y)]$$

However, we know from Chapter Seven that the expectation of a standardized random variable is always equal to zero. Thus

$$E(Z_X) = E(Z_Y) = 0$$

and

$$\text{cov}(Z_X Z_Y) = E(Z_X Z_Y)$$

This expectation of the product of standardized scores is defined as the population correlation coefficient:

$$\rho_{XY} = E(Z_X Z_Y)$$

Like the sample correlation coefficient, ρ is defined from -1.0 to $+1.0$, with -1.0 representing a perfect negative relationship, 0 representing no relationship, and $+1.0$ indicating a perfect positive relationship.

D. Interpreting Correlation: The Coefficient of Determination

When we introduced the notation of a statistical relation we defined it in terms of increasing the precision of estimate of a variable or reducing the uncertainty about a value of the variable. It should be expected, therefore, that the correlation coefficient is somehow descriptive of this increase in precision or reduction of uncertainty. This expectation is well founded, but before we can formally express the relationship of correlation to precision, we must examine the properties of a joint distribution more fully.

186

1. The Marginal Probability Distributions of X and Y

Let us once again consider the joint distribution of heights and weights. However, let us now consider that our sample is comprised of an infinitely large population. In our earlier example we focused entirely on the joint distribution of height and weight and bypassed any treatment of the two variables considered separately. Nonetheless, it should be apparent that both height and weight are variables in their own right. And now that we are thinking in population terms we know that there is a probability distribution associated with each. We can, therefore, represent these distributions in the margins of the X, Y plane. In accord with this convention we call these representations the *marginal distributions* of X and Y (Figure 9.8). These distributions can, of course, be described in terms of location and variability, and if we indicate the expectation of each distribution with a dotted line, we note that their intersection (μ_X, μ_Y) falls, as we should anticipate, at the geometric center of the joint distribution.

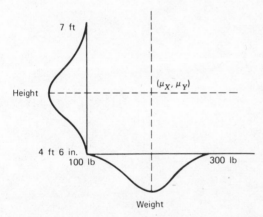

Figure 9.8   Marginal distributions of height and weight.

The marginal variances, however, are of even more concern to us. Recall that in our discussion of the scatterplot of heights and weights we used the range of the Y-values to indicate the "best" estimate we could make about the height of any randomly selected subject. It should be recalled that the range is a crude measure of variability, and thus we were, in fact, making a statement about the dispersion of heights in our sample. In Chapter Six it was demonstrated that the variance is usually a more suitable measure of dispersion than the range. Thus, in point of fact, we use the marginal variance rather than the range as an index of the degree of dispersion. In this context we can see that expected variation is what is meant by "uncertainty." That is, if we had to guess the height of a randomly selected subject, our uncertainty about the accuracy of our guess would be directly proportional to the variability of heights in our population.

2. The Joint Probability Distribution of (X, Y)

Considering the joint distribution of height and weight, we see that the scatter-plot now contains an infinite number of entries and thus becomes a continuous surface rather than a collection of discrete points. We therefore represent it as an enclosed area of roughly the same shape as our plot of data points as given above. This surface is a two-dimensional representation of the joint probability distribution of X and Y, that is the distribution of values of (X, Y) (Figure 9.9).

We observe in our joint probability distribution the same property that gave rise to our initial definition of a statistical relation: For any given value of X there are many values of Y. As before, for example, we see that the heights associated with $x = 150$ pounds vary approximately from 5 to 6 feet. Recall, however, that our surface is a probability distribution. Therefore, the narrow strip associated with any value x_i is not simply a range of Y values. Instead it represents *both* the set of Y-values associated with x_i *and* their probabilities. Therefore, it is a probability distribution in its own right and is known as the *conditional probability distribution* of Y on x_i.

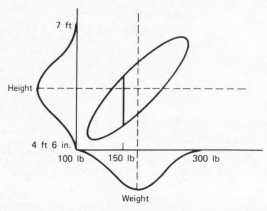

Figure 9.9   Marginal and joint distributions of height and weight.

3. Conditional Probability Distributions

To obtain a better intuitive understanding of conditional distributions, let us consider in some detail the geometric nature of the joint distribution. Although the joint distribution of (X, Y) is generally depicted as a two-dimensional ellipse, it actually represents relationships among *three* variables, X, Y, and the probabilities (or probability densities) associated with all pairs of values (x_i, y_i). Thus, our two-dimensional correlation surface is really a projection of a three-dimensional solid. This is illustrated in Figure 9.10, where X and Y define a plane perpendicular to the page, and the vertical axis represents probability density $\phi(X, Y)$. It is apparent from this illustration that probability density is greatest for pairs of values that fall near the center of the joint distribution

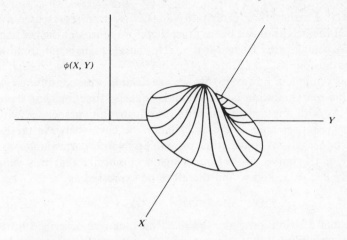

Figure 9.10 Joint distribution of (X,Y) *in three dimensions. (Adapted with permission from Dixon and Massey,* Introduction to Statistical Analysis, *New York: McGraw-Hill, 1969.)*

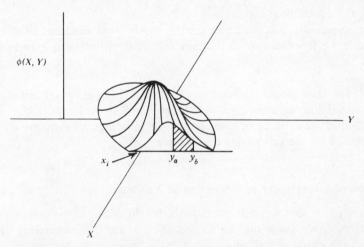

Figure 9.11 Conditional distribution of Y | x$_i$ *as cross section of joint distribution. (Adapted with permission from Dixon and Massey,* Introduction to Statistical Analysis, *New York: McGraw-Hill, 1969.)*

(μ_X, μ_Y). As one moves away from the center, the probability density decreases quite sharply at first and then decreases more gradually as one nears the periphery of the distribution.

Suppose now that we "slice" into our solid in the plane of the page along a line corresponding to some value x_i. The result is illustrated in Figure 9.11, and we see that the cross-sectional profile of our slice is a smooth, unimodal curve. As indicated above, our "slice" intersects the XY plane at $x = x_i$. This intersection is the baseline of our curve and must represent the set of all Y-values associated with x_i. Furthermore, probability density is again represented on the vertical axis, and the height of our curve must therefore represent the probability density values associated with Y when $x = x_i$. This curve, then, is our conditional probability distribution of Y on x_i.

The reader is reminded, however, that we are dealing with continuous variables and that this representation is a probability density function, not a probability function. As with any density function, the probability associated with an interval y_a to y_b is represented as the area under the curve between these two values. In the figure given above, this probability value corresponds to the shaded portion of the curve, but we must again emphasize that this value is appropriate only if $x = x_i$ and would therefore be expressed as

$$p[(y_a \leq y \leq y_b)|x = x_i]$$

The conditional distributions, like the marginals, can be described in terms of location and variability. We can, for example, compute the expectation of a conditional probability distribution. The expectation of Y given x_i is denoted

$$\mu_{Y|x_i}$$

and is read, the conditional expectation of Y on x_i.

Of more immediate concern to us are the conditional variances of Y on X. For any value x_i, the variance of the conditional distribution is expressed by

$$\sigma^2_{Y|x_i}$$

and is read, the conditional variance of Y on x_i. This parameter tells us the amount of variability we can expect to observe in Y if we know that $x = x_i$, and thus it reflects the degree of uncertainty surrounding any estimate we may make of Y based on the knowledge that x equals x_i.

4. Reduction of Uncertainty as Reduction of Variance

Let us again suppose that we must estimate the height of one of the individuals in our population. If we know that he weighs 150 pounds the uncertainty of our estimate is now proportional to $\sigma^2_{Y|150}$. This is in contrast to the situation described above in which his weight was unknown, and our uncertainty was thus proportional to the marginal variance of Y, σ_Y^2.

While it is relatively unimportant to a descriptive treatment of correlation, inferential use of correlation requires the assumption that the conditional variances of Y on x are all equal. That is,

$$\sigma^2_{Y|x_i} = \sigma^2_{Y|x_j}$$

for every value of i and j. This assumed property of the joint probability distribution is called *homoscedasticity* and permits us to dispense with subscripts and represent the common conditional variance of Y as

$$\sigma^2_{Y|x}$$

190

Thus far, then, we have developed the notion that the marginal variance of Y reflects the degree to which we would be uncertain about an estimate of Y in the absence of any information about X. We have also suggested that the conditional variance of Y should reflect our degree of uncertainty associated with an estimate of Y when the value of X is known.

Therefore, the difference between the marginal variance and the conditional variance should indicate the amount by which our uncertainty of estimate is reduced by specifying a value for X. If, for example, the marginal variance of heights in our population is equal to 100 inches and the conditional variance of height given weight is equal to 10 inches,

$$\sigma_Y^2 - \sigma_{Y|x}^2 = 100 - 10 = 90$$

Thus, the uncertainty of our estimate is reduced by 90 units. However, imagine that our marginal variance were 10,000 units and the conditional variance equal to 9910 units. Once again,

$$\sigma_Y^2 - \sigma_{Y|x}^2 = 10,000 - 9910 = 90$$

but reducing a variance of 10,000 by 90 units constitutes less than a 1 percent reduction, where in the previous example paring away 90 units reduced the variability by 90 percent. Thus, it should be apparent that the difference between the marginal variance and the conditional variance becomes meaningful only if considered in relation to the magnitude of the marginal variance. Thus, we take the ratio

$$\frac{\sigma_Y^2 - \sigma_{Y|x}^2}{\sigma_Y^2} = \frac{100 - 10}{100} = .90$$

which tells us that the uncertainty of our estimate has been reduced by 90 percent. In more positive terms, this means that our precision of estimate has been *increased* by 90 percent. Thus, when there is no increase of precision and the conditional variance remains as large as the marginal variance, the numerator is equal to zero, and the ratio must also be zero. In contrast, when the joint distribution is a perfect line, it has no width and the conditional variance is equal to zero. The ratio then becomes

$$\frac{\sigma_Y^2}{\sigma_Y^2}$$

which must always equal exactly 1.0.

This ratio, therefore, has a lower limit of 0 and an upper limit of 1.0. The correlation coefficient, however, is defined from -1.0 to $+1.0$, and it is thus apparent that our ratio of variances is not precisely equal to the correlation. In point of fact, the ratio presented above is the *square* of the population correlation coefficient:

$$\rho^2 = \frac{\sigma_Y^2 - \sigma_{Y|x}^2}{\sigma_Y^2}$$

This ratio may be expressed as the difference between two fractions and thus simplified:

$$\rho^2 = \frac{\sigma_Y^2 - \sigma_{Y|x}^2}{\sigma_Y^2}$$

$$= \frac{\sigma_Y^2}{\sigma_Y^2} - \frac{\sigma_{Y|x}^2}{\sigma_Y^2}$$

$$= 1 - \frac{\sigma_{Y|x}^2}{\sigma_Y^2}$$

The last expression is called the *coefficient of determination* and is not usually presented in terms of precision of estimate. Most generally the coefficient of determination is interpreted as the proportion of variance σ_Y^2 attributable to linear correlation.

The essence of covariation is that change in X is accompanied by some concomitant change in Y, which implies that X and Y have common sources of variability. That is, some proportion of the marginal variance of Y must be attributable to factors that also influence the variability of X. If X is held constant, therefore, any residual variation in Y must be due to factors that are entirely unrelated to X. Therefore the ratio

$$\frac{\sigma_{Y|x}^2}{\sigma_Y^2}$$

indicates the proportion of Y variability that must be accounted to sources unique to Y. Conversely the complement of this ratio

$$1 - \frac{\sigma_{Y|x}^2}{\sigma_Y^2}$$

gives the proportion of the marginal variance of Y that we may attribute to those factors common to both variables. This expression, the coefficient of determination, may therefore be interpreted as the proportion of Y variance attributable to covariation with X. More succinctly it is generally called the proportion of variance attributable to correlation.

E. Problems, Review Questions, And Exercises

The following are the heights and weights for a sample of six men in the Flint Judo Club:

Height (inches)	Weight (pounds)
62	140
65	155
67	165
69	154
72	175
74	220

1. Compute (a) the sample covariance C_{XY} for height in *inches* and weight and (b) the sample covariance for height in *feet* and weight.
   c. Does the statistical relationship of height and weight in our sample change when the scale of measurement changes?
   d. If not, why are 1a and 1b different?
2. Standardize the values of height in inches, height in feet, and weight.
   a. Compute the covariance $C_{z_X z_Y}$ of height in inches and weight.
   b. Compute the covariance $C_{z_X z_Y}$ of height in feet and weight.
   c. If you compare your answers for 2a and 2b what can you infer about the covariance of standardized variables?
3. Compute the correlation coefficient r_{XY} for height in feet and weight.
4. What definition of r is suggested by comparing your answers to 2a, 2b and 3?

part three
statistical
inference

At the beginning of Part Two we stated that observation and description are two of the most fundamental activities of science. Nonetheless, they represent only one aspect of scientific inquiry. In addition to answering the question "What do we know?" the scientist also assumes the responsibility for the question "What does it mean?" Thus, no matter how meticulous his observations and no matter how carefully quantified his experimental results, the scientist's data are always subject to the question, so what? What do your observations tell us about your random variable *in general*? What, indeed, do your observations tell us about nature *in general*?

Suppose, for example, that a behavioral scientist is interested in the the types of people who regularly engage in strenuous athletics. He therefore collects data on 50 volunteers selected at random from participants in the local YMCA-YWCA recreation program and finds that the mean I.Q. for his group is 112. By itself, this result is simply an empirical curiosity. To assume scientific importance it must somehow be generalized beyond the individuals whom the experimenter actually observed. That is, besides *describing* the distribution properties of his *sample*, the scientist must make some *inference* about the distribution of his random variable in the *population* from which the simple was drawn.

Very frequently, therefore, a statistical inference is a statement about the parameter values of a random variable, and in this regard, statisticians have developed two general methods that permit the scientist to *estimate* these values. The first of these is *point estimation*, which involves only a single value, or point. Using techniques of this sort, for example, the investigator in our YMCA-YWCA problem might infer from his data that the mean of the population from which he drew his volunteers is equal to some specific value $\hat{\mu}$. If, however, his estimate is computed from continuous data, we will see in Chapter Ten that the probability must be zero that $\hat{\mu}$ is precisely equal to the parameter value μ. Alternatively, therefore, he might estimate that the value of his parameter is included in a specific interval Δx. This would permit him to assign a nonzero probability value to his estimate and is an example of *interval estimation.*

Although estimation techniques enable the scientist to make statements of specified precision about the properties of population distributions, our understanding of the universe would not proceed very quickly if this were the entire scope of scientific endeavor. While such ad hoc findings may be interesting, they are of relatively little importance unless the scientist can make some additional inferences about the *general processes* that underlie his experimental results and the parameter estimates generated by these results. It is generally understood, therefore, that scientific observations are prompted by some prior *conceptualization* of the state of nature. This conceptualization may be an elegant and general theory, such as Darwin's theory of evolution and natural selection, or it may be a more modest formulation that describes relationships among a few variables operating under a highly specific set of circumstances. Indeed, it may even be a proposition that is based entirely on prior empirical observation instead of theoretical considerations of any sort. In any case, this conceptualization will generate implications about one or more random variables, and if the formulation is sound, these implications should be confirmed by observations on the random variable(s) in question.

In addition to estimation techniques, therefore, statisticians have developed methods of *hypothesis testing* that permit the investigator to decide whether or not his implications are supported by his empirical observations. To illustrate the considerations involved in testing hypotheses, let us suppose that the scientist in our example had collected various snippits of data that seemed to suggest that people who are physically fit are generally more intelligent than the population at large, which has a mean I.Q. of 100. On the face of it, his observed sample mean $\bar{x} = 112$ would seem to confirm this conclusion, but it must be remembered that this figure was based on only 50 observations. Thus, even if the mean I.Q. of the physically fit *population* were precisely equal to the general population mean of 100, he could still have obtained a *sample* mean as large as 112 on the basis of *random chance*. In order to draw any conclusion, therefore, he must determine the probability that the 12-point difference between his sample and the general population was simply due to random error. If this probability is very large, he will hesitate to conclude the physically fit people are more intelligent than people in general. If the probability is very small, however, he will have considerably more confidence in this conclusion.

As we indicated in the introduction to the text, all problems involving estimation and hypothesis testing have four characteristics in common. First, the investigator hopes to make some general statement about an attribute manifested by all members of a specific *population*. In our two examples, the attribute was intelligence, and the population was all physically fit persons.

Second, the members of the population exhibit *variability* with respect to the attribute in question. In the general population, the variance of I.Q. scores determined by the Wechsler Adult Intelligence Scale is 225, and we should expect similar variability in the I.Q. scores of physically fit persons. Much of this variability, of course, reflects the hereditary and experiential influences that contribute to relatively stable individual differences in intelligence. In addition to these sources of *systematic* variability, I.Q. scores are also affected by differences in mood, level of motivation, fatigue, and so forth. In general, such transient influences are not subject to direct experimental control, and the experimenter must therefore assume that they contribute only *random* variability to his I.Q. scores.

Third, the entire population is not available for study, and the experimeter must therefore base his inference on a relatively small *sample* of observations. If his inference is to be appropriate, therefore, he must be certain that his observations are not subject to influences that will introduce systematic differences between his sample and the population they represent.

If we consider the selection of a sample to constitute a random experiment, it follows that the set of scores observed for one sample may be quite different from the collection of scores obtained in a second sample. This implies, of course, that distribution properties are likely to differ from sample to sample, and we should therefore expect that the distribution properties of any particular sample may differ from those, of the parent population. No matter how carefully chosen his sample therefore, any generalization the investigator might make about the distribution of I.Q. scores in the physically fit population would be subject to some degree of *residual uncertainty*. This is the fourth characteristic common to all inference problems.

Although statistical estimation and hypothesis testing share these four properties, they nonetheless have one essential difference. In our estimation problem, the scientist had no preconceptions about the parameter value of his parent population. In the second problem, however, the investigator had a specific hypothetical value, 100, against which he wished to test his data. Thus, estimation serves primarily to extend numerical *description* from the sample to the population, while hypothesis testing permits the experimenter to test specific *deductions* about the properties of population distributions. In this regard, we indicated in the introduction to the text that contemporary scientific philosophy places more importance on explanation than on description. Thus, Part Three is weighted heavily in favor of hypothesis testing, with estimation covered entirely in Chapter Ten.

A. Statistics as Random Variables
B. The Sampling Distribution of the Mean
   1. The Expectation of $\overline{X}$: $E(\overline{X}) = \mu_{\overline{x}}$
   2. The Variance of $\overline{X}$: $V(\overline{X}) = \sigma_{\overline{x}}^2$
C. Point Estimation
   1. Desirable Properties of Point Estimators
   2. Estimating μ
   3. Estimating σ^2
D. The Central Limit Theorem
   1. Computation of $p(\Delta \overline{x}_i)$ from the Normal Curve
E. Interval Estimation
   1. Confidence Intervals for μ
   2. Confidence Intervals for μ Based on the CLT
F. Problems, Review Questions, and Exercises

sampling distributions and estimation

Just as our discussion of probability theory in Part One provided a foundation for the development of distribution properties in Part Two, so are these distribution properties the basis of statistical inference. Indeed, all of the remaining chapters except Chapter Fourteen discuss inferences about either population means or population variances. Throughout Part Two, however, we were concerned primarily with the distribution properties of random variables that we generated either empirically or conceptually by assigning numerical values to *individual observations*. In contrast, most of the techniques developed in Part Three apply to random variables generated by assigning numerical values to *entire distributions* of data. That is, our random variables are the statistics we developed in Part Two.

A. Statistics as Random Variables

Recall from Chapter Four that the *parameters* of any *population* distribution are constant. The same is not true, however, for the *statistics* that describe distributions of *sample* data. That is, if we draw several random samples from the same population, we expect the values of our statistics to vary somewhat from sample to sample. Consider, for instance, a population in which some random variable X has an expectation

$$E(X) = \mu$$

Suppose now that we randomly select a sample of 25 objects from the population and compute the sample mean:

$$\bar{x}_1 = \frac{\sum\limits_{i}^{25} x_i}{25}$$

We replace these objects and draw yet another random sample of 25 objects, computing their mean as before:

$$\bar{x}_2 = \frac{\sum\limits_{i}^{25} x_i}{25}$$

Under ordinary circumstances, we would not expect our two samples to include precisely the same list of observations, and selection of a random sample therefore constitutes a *random experiment* as defined in Chapter Two. Furthermore, it is most likely that $\bar{x}_1$ and $\bar{x}_2$ would take different values. Generalizing our example, selection of N such samples would generate an array of N sample means consisting of n different values $\bar{x}_i$. It should thus be apparent that the mean of any *one* sample drawn from our population could potentially assume any of these n different values. By the definition given in Chapter Two, therefore, the sample mean must be a *random variable* $\overline{X}$ defined by the value set

$$\{\bar{x}_2, \bar{x}_2, \ldots \bar{x}_n\}$$

Like any other random variable, moreover, each value $\bar{x}_i$ is associated with a particular probability $p(\bar{x}_i)$. The random variable $\overline{X}$ is therefore described by some probability distribution, which we call the *sampling distribution of the mean*. And in general,

> The probability distribution associated with any statistic is called the sampling distribution of the statistic.

B. The Sampling Distribution of the Mean

As we indicated at the beginning of the chapter, techniques of statistical inference are seldom applied to distributions of individual values x_i. Instead, we are usually concerned with the sampling distribution of some statistic that describes our sample and that *could* be computed for all samples we *might* conceivably draw from the same population. The most important of these is the sampling distribution of the mean, and we shall therefore consider the distribution properties of $\overline{X}$ in some detail.

1. The expectation of $\overline{X}$: $E(\overline{X}) = \mu_{\overline{X}}$

Let us once again consider a random variable X. To determine the expectation of the sample mean $E(\overline{X})$ it is necessary only that we define the expectation of X as μ. That is.

$$E(X) = \sum_i^n x_i\, p(x_i)$$

$$= \mu$$

By definition we know that

$$\bar{x} = \frac{\sum\limits_i^N x_i}{N}$$

Therefore

$$E(\overline{X}) = E\left(\frac{\sum\limits_i^N x_i}{N}\right)$$

However, by the rules of summation given in Appendix I, the constant $1/N$ may be factored out of the summation, and the expression therefore becomes

$$E(\overline{X}) = E\left(\frac{1}{N}\sum_i^N x_i\right)$$

And, by the rules of expectation in Appendix II, the constant $1/N$ may be further factored out of the expectation. Thus

$$E(\overline{X}) = \frac{1}{N} E\left(\sum_i^N x_i\right)$$

$$= \frac{1}{N} E(x_1 + x_2 + \ldots + x_n)$$

Recall, however, that the concept of expectation is appropriately applied only to *random variables*. The last equation must therefore be meaningless. By our definition of the sample mean, the first term inside the parentheses, x_1, represents the numerical value assigned to the first observation. The second value to be observed is x_2, and so forth. Thus, each term x_i is a *fixed* numerical value, and the sum

$$(x_1 + x_2 + \ldots + x_N)$$

is therefore a *constant*. It must be noted, however, that each observation assumes a fixed value only *after* the observation has been taken. As we indicated above, random sampling is a type of random experiment, and by definition, therefore, each observation may presumably yield any of m possible results. At the outset of the experiment, these m results include every member of the sample space, and the first observation could therefore assume any of the n values associated with these results. *Before* the experiment is actually conducted, then, the first observation is a random variable X_1 that includes every X-value in the population. That is,

$$X_1 = \{x_1, x_2, \ldots x_n\}$$

Furthermore, if the second observation is likewise defined as a potential, rather than empirical, outcome, then it too becomes a random variable X_2. If we sample with replacement, moreover, the sample space is unchanged after the first observation, and X_2 must be defined by precisely the same value set as X_1:

$$X_2 = \{x_1, x_2, \ldots x_n\}$$

And if we similarly define every term in our expectation as the *a priori* outcome of the observation it represents, our expression becomes

$$E(\overline{X}) = \frac{1}{N} E(X_1 + X_2 + \ldots + X_N)$$

By the rules of expectation, this must be equal to

$$\frac{1}{N} [E(X_1) + E(X_2) + \ldots + E(X_N)]$$

Our only remaining task, therefore, is to determine the appropriate values for $E(X_1)$, $E(X_2)$, and so forth. In this regard, recall that the value set of X_1 was defined by the value set of the entire population. That is,

$$X_1 = \{x_1, x_2, \ldots x_n\}$$

By definition, therefore,

$$E(X_1) = \sum_1^n x_i \, p(x_i)$$

which is precisely equal to the population mean, $E(X)$:

$$E(X) = \sum_1^n x_i \, p(x_i)$$

$$= \mu$$

Thus

$$E(X_1) = E(X) = \mu$$

As indicated above, sampling with replacement guarantees that the second observtion X_2 may potentially assume precisely the same set of values. Moreover, this procedure also insures that the probabilities associated with these n values remain unchanged. The distributions of X_1 and X_2 must therefore be identical, and it follows that

$$E(X_2) = E(X_1) = \mu$$

And, in general,

$$E(X_1) = E(X_2) = \ldots = E(X_N) = E(X) = \mu$$

Therefore

$$E(\overline{X}) = \frac{1}{N} \underbrace{(\mu + \mu + \ldots + \mu)}_{N \text{ times}}$$

$$= \frac{1}{N} (N\mu)$$

$$= \mu$$

Thus, the expectation of the sample mean $\mu_{\overline{x}}$ is equal to μ, the mean of the population from which the sample observations are drawn.

2. The Variance of $\overline{X}$: $V(X) = \sigma_{\overline{X}}^2$

Once again, X is defined as a random variable with expectation equal to μ. To determine the variance of $\overline{X}$, however, we must further define the variance of X, $V(X)$, as equal to σ^2. As before,

$$\bar{x} = \frac{\sum_{i}^{N} x_i}{N}$$

Therefore

$$V(\overline{X}) = V\left(\frac{\sum_{i}^{N} x_i}{N}\right)$$

$$= V\left(\frac{1}{N} \sum_{i}^{N} x_i\right)$$

By the rules of expectation, a constant that is factored out of a variance is squared. Thus

$$V(\overline{X}) = \frac{1}{N^2} V\left(\sum_{i}^{N} x_i\right)$$

$$= \frac{1}{N^2} V(x_1 + x_2 + \ldots + x_n)$$

Again, however, the expression in parentheses is a constant, and it will be recalled from the rules of expectation that the variance of a constant is always equal to 0. To avoid this impasse, we assume as before that our empirical observations are yet to be made, and each term in the sum becomes a random variable:

$$V(X) = \frac{1}{N^2} V(X_1 + X_2 + \ldots + X_N)$$

In our derivation of $\mu_{\bar{x}}$ we stipulated that observations be replaced after every draw. Each value x_i therefore retained the same probability across all N observations, thus assuring that $X_1, X_2, \ldots X_N$ were identically distributed. Moreover, if $p(x_i)$ remains constant from each observation to the next, it must necessarily be true that no observation affects the probabilities associated with subsequent observations. By definition, therefore, the random variables $X_1, X_2, \ldots X_N$ are independent of one another, and by the rules of expectation

$$V(\overline{X}) = \frac{1}{N^2} [V(X_1) + V(X_2) + \ldots + V(X_N)]$$

Because our variables are identically distributed, it must be true that

$$V(X_1) = V(X_2) = \ldots = V(X_N) = V(X) = \sigma^2$$

Thus

$$V(\overline{X}) = \frac{1}{N^2} \underbrace{(\sigma^2 + \sigma^2 + \ldots + \sigma^2)}_{N \text{ times}}$$

$$= \frac{1}{N^2} (N\sigma^2)$$

$$= \frac{\sigma^2}{N}$$

The variance of the sample mean $\sigma_{\bar{x}}{}^2$ is therefore equal to σ^2, the variance of the population from which sample observations are drawn, *divided by* the number of observations in each sample.

C. Point Estimation

In Chapter Four we introduced the distinction between sample statistics and population parameters. We know, of course, that sample statistics are computed directly from the observations at hand, while determination of a population parameter is based on probability considerations. In general, therefore, computing the value of a parameter requires considerable information, such as the probability function or the probability $p(x_i)$ associated with every element of the value set. In practical situations such information is frequently unavailable, and we must often, therefore, rely on our sample data to approximate the parameter values of our parent population. Such approximations are called *point estimates*, because they involve only a single value, or point.

It will also be recalled from Chapter Four that Roman letters, such as s and r, are customarily used to indicate statistics and that Greek letters, such as σ and ρ, indicate parameters. Because it is based on sample data, an estimator is a type of statistic. However, estimators often replace unknown parameter values in transformations and probability functions, so we shall indicate estimators with Greek letters. To distinguish estimators from parameters, however, they will be identified with a caret (^), or "hat". Thus, an estimator of the population mean would be indicated by

$$\hat{\mu}$$

which is read, "estimate of *mu*," or more simply, "*mu* hat."

1. Desirable Properties of Point Estimators

When we introduced the notion of a statistic in Chapter Four, we indicated the possibility of formulating many different expressions to describe any particular distribution property. Similarly, we could probably conceive of many ways to estimate the value of any population parameter. As with descriptive statistics, however, there are various criteria that make some estimators more acceptable than others.

i. *An estimator should be unbiased.* $\hat{\theta}$ is an unbiased estimator of the parameter θ if the expected value of $\hat{\theta}$ is equal to θ. That is, $\hat{\theta}$ is unbiased if $E(\hat{\theta}) = \theta$.

ii. *An estimator should be consistent.* $\hat{\theta}$ is a consistent estimator of the parameter θ if as sample size approaches infinity, probability approaches 1.0 that $|\hat{\theta} - \theta|$ is less than any arbitrarily small value ϵ. That is, $\hat{\theta}$ is consistent if

$$\lim_{N \to \infty} p(|\hat{\theta} - \theta| < \epsilon) = 1.0$$

where N is sample size and ϵ is any arbitrarily small number. In more concrete terms, this means that a consistent estimator provides an ever better approximation of the parameter as sample size increases. From this informal definition, it would intuitively seem that unbiased estimators should also be consistent, and students therefore have occasional difficulty distinguishing between these two properties. The distinction becomes apparent, however, if we realize that while all unbiased estimators are indeed consistent, it is *not true* that all consistent estimators are unbiased. Suppose, for example, that $\hat{\theta}$ is an estimator of θ such that

$$E(\hat{\theta}) = \theta + \frac{1}{N}$$

By definition, therefore, $\hat{\theta}$ is a biased estimator of θ. As sample size N increases, however, we see that the bias term $1/N$ becomes negligible, and the probability thus approaches 1.0 that the difference $(\hat{\theta} - \theta)$ is less than any positive value we may care to specify. Thus, $\hat{\theta}$ is a consistent estimator of θ even though it is biased.

iii. *An estimator should be efficient.* If $\{\hat{\theta}_1, \hat{\theta}_2, \ldots \hat{\theta}_n\}$ are estimators of θ, $\hat{\theta}_i$ is the most efficient if the variance of its sampling distribution $V(\hat{\theta}_i)$ is less than the variance of any other estimator, $V(\hat{\theta}_j)$. Like any statistic, an estimator is a random variable and may therefore be described in terms of the location and dispersion of its sampling distribution. The most efficient of several estimators is simply the one that displays the least variation from sample to sample and that therefore has the smallest variance.

To compare these various properties, consider the sampling distributions of two estimators $\hat{\theta}_1$ and $\hat{\theta}_2$ (Figure 10.1). We see that the distribution of $\hat{\theta}_1$ is con-

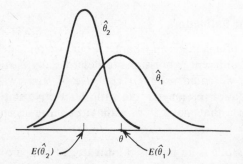

Figure 10.1 Sampling distributions of two estimators

siderably more variable than the distribution of $\hat{\theta}_2$ and that $\hat{\theta}_2$ must be the more efficient estimator. However, the expectation of $\hat{\theta}_1$ is equal to the parameter, θ, and $\hat{\theta}_1$ is therefore unbiased. Customarily, unbiasedness is considered the most important property of an estimator, but this is not always true. We shall see below that bias can often be corrected if our estimator is consistent. Thus, faced with a choice between an unbiased estimator and a consistent estimator with a correctable bias, the experimenter should select the more efficient of the two.

Sampling Distributions and Estimation

2. Estimating μ

Earlier in this chapter we proved that $E(\overline{X}) = \mu$. By definition, therefore, $\overline{X}$ must be an unbiased estimator of μ. Because it is unbiased the sample mean must also be consistent. Furthermore, it is easily proven that for any given sample size N, $\overline{X}$ is more efficient than any other unbiased estimator of the population mean.

Let us define $\hat{\mu}$ as an unbiased estimator of the population mean that differs from $\overline{X}$ by some constant value c. That is,

$$\hat{\mu} = \overline{X} + c$$

We know from Chapter Six that the variance of any variable X may be expressed as

$$V(X) = E[X - E(X)]^2$$

The variance of our estimator $\hat{\mu}$ is therefore given by

$$V(\hat{\mu}) = E[\hat{\mu} - E(\hat{\mu})]^2$$

Because $\hat{\mu}$ is an unbiased estimator of μ, however,

$$E(\hat{\mu}) = \mu$$

Thus

$$V(\hat{\mu}) = E(\hat{\mu} - \mu)^2$$
$$= E(\hat{\mu}^2 - 2\hat{\mu}\mu + \mu^2)$$

Substituting $(\overline{X} + c)$ for $\hat{\mu}$, this becomes

$$V(\hat{\mu}) = E[(\overline{X} + c)^2 - 2(\overline{X} + c)\mu + \mu^2]$$
$$= E(\overline{X}^2 + 2\overline{X}c + c^2 - 2\overline{X}\mu - 2c\mu + \mu^2)$$
$$= E(\overline{X}^2) + E(2\overline{X}c) + E(c^2) - E(2X\mu) - E(2c\mu) + E(\mu^2)$$

The only variables in this expression are $\overline{X}^2$ and $\overline{X}$, and it thus becomes

$$V(\hat{\mu}) = E(\overline{X}^2) + 2cE(\overline{X}) + c^2 - 2\mu E(\overline{X}) - 2c\mu + \mu^2$$

By definition we know that

$$E(X) = \mu$$

Furthermore, we proved earlier in this chapter that

$$E(\overline{X}) = \mu$$

Thus

$$V(\hat{\mu}) = E(\overline{X}^2) + 2c\mu + c^2 - 2\mu^2 - 2c\mu + \mu^2$$

Collecting and rearranging terms, this becomes

$$V(\hat{\mu}) = E(\overline{X}^2) + \mu^2 + c^2$$

or

$$V(\hat{\mu}) = E(\overline{X}^2) - [E(\overline{X})]^2 + c^2$$

From Chapter Six we know that the computational formula for the variance of X is $E(X^2) - [E(X)]^2$. Hence, the first two terms in our right-hand expression must equal $V(\overline{X})$:

$$E(\overline{X}^2) - [E(\overline{X})]^2 = V(\overline{X})$$

and

$$V(\hat{\mu}) = V(\overline{X}) + c^2$$

Thus

$$V(\overline{X}) = V(\hat{\mu}) - c^2$$

and $V(\overline{X})$ must therefore be smaller than the variance of any other unbiased estimator $V(\hat{\mu})$.

3. Estimating σ^2

We know from Chapter Six that for a finite population, the sample variance s^2 of an exhaustive sample is equal to the population variance σ^2. Intuitively, it might therefore seem that s^2 is an unbiased estimator of σ^2, but in point of fact $E(s^2) \neq \sigma^2$. To prove this, we first derive a simple computational form of the sample variance:

$$
\begin{aligned}
s^2 &= \frac{\sum\limits_i^N (x_i - \bar{x})^2}{N} \\[2mm]
&= \frac{\sum (x^2 - 2x\bar{x} + \bar{x}^2)}{N} \\[2mm]
&= \frac{\sum x^2}{N} - \frac{2\bar{x}\sum x}{N} + \frac{N\bar{x}^2}{N} \\[2mm]
&= \frac{\sum x^2}{N} - 2\bar{x}\frac{\sum x}{N} + \bar{x}^2 \\[2mm]
&= \frac{\sum x^2}{N} - 2\bar{x}^2 + \bar{x}^2 \\[2mm]
&= \frac{\sum\limits_i^N x_i{}^2}{N} - \bar{x}^2
\end{aligned}
$$

In this form, the expectation of the sample variance becomes

$$E(S^2) = E\left(\frac{\sum x^2}{N} - \bar{x}^2\right)$$

$$= E\left(\frac{\sum x^2}{N}\right) - E(\bar{x}^2)$$

Both terms in parentheses, however, are constants, and we have precisely the same dilemma we encountered in our derivations of $E(\overline{X})$ and $V(\overline{X})$. Once again, therefore, we assume that the observations from which we compute s^2 are yet to be made, and our expression becomes

$$E(S^2) = E\left(\frac{\sum X^2}{N}\right) - E(\overline{X}^2)$$

Consider the first term in this expression:

$$E\left(\frac{\sum X^2}{N}\right) = \frac{1}{N} E(\sum X^2)$$

From the algebra of expectations, we know that the expectation of a sum is equal to the sum of the expectations. Thus

$$E\left(\frac{\sum X^2}{N}\right) = \frac{1}{N} \sum [E(X^2)]$$

To solve for $E(X^2)$ we recall the computational form of the population variance given in Chapter Six:

$$\sigma^2 = E(X^2) - [E(X)]^2$$

$$= E(X^2) - \mu^2$$

Therefore

$$E(X^2) = \sigma^2 + \mu^2$$

and

$$\frac{1}{N} \sum [E(X^2)] = \frac{1}{N} \sum (\sigma^2 + \mu^2)$$

$$= \frac{1}{N} N (\sigma^2 + \mu^2)$$

$$= \sigma^2 + \mu^2$$

Now let us consider the second term, $E(\overline{X}^2)$. If we express the computational form of the variance in expectation notation, it becomes

$$V(X) = E(X^2) - [E(X)]^2$$

Thus, if our random variable is the sample mean $\overline{X}$,

$$V(\overline{X}) = E(\overline{X}^2) - [E(\overline{X})]^2$$

or

$$\sigma_{\overline{X}}^2 = E(\overline{X}^2) - \mu^2$$

Thus

$$E(\overline{X}^2) = \sigma_{\overline{X}}^2 + \mu^2$$

It follows, therefore that

$$E(S^2) = E\left(\frac{\sum X^2}{N}\right) - E(\overline{X}^2)$$

$$= (\sigma^2 + \mu^2) - (\sigma_{\overline{X}}^2 + \mu^2)$$

$$= \sigma^2 - \sigma_{\overline{X}}^2$$

Earlier in the chapter we proved that

$$\sigma_{\overline{X}}^2 = \frac{\sigma^2}{N}$$

Thus

$$E(S^2) = \sigma^2 - \frac{\sigma^2}{N}$$

$$= \sigma^2\left(1 - \frac{1}{N}\right)$$

$$= \sigma^2\left(\frac{N-1}{N}\right)$$

We see, therefore, that the expectation of the sample variance is smaller than the population variance by a factor of $(N - 1)/N$. It should be apparent, however, that this bias factor approaches 1.0 as N becomes infinitely large. By increasing sample size, therefore, the difference between σ^2 and s^2 may be reduced to any arbitrary value and S^2 is therefore a *consistent* estimator of σ^2 even though it is biased. More importantly, however, knowing the bias factor permits us to derive an unbiased estimate of σ^2:

$$E(S^2) = \sigma^2\left(\frac{N-1}{N}\right)$$

Therefore

$$\frac{N}{N-1}E(S^2) = \sigma^2$$

The term $N/(N-1)$ is a constant and may therefore be moved into the expectation.

$$E\left(S^2 \frac{N}{N-1}\right) = \sigma^2$$

By definition, therefore,

$$S^2 \frac{N}{N-1}$$

must be an unbiased estimator of σ^2.

$$s^2 \frac{N}{N-1} = \frac{\sum_i^N (x_i - \bar{x})^2}{N} \cdot \frac{N}{N-1}$$

$$= \frac{\sum_i^N (x_i - \bar{x})^2}{N-1}$$

This expression is the definitional formula for the unbiased estimator of the population variance and shall be identified by the symbol $\hat{\sigma}^2$ throughout the remainder of this text.

D. The Central Limit Theorem

Earlier in the chapter we established that the expectation of $\overline{X}$ is equal to μ and that the variance of $\overline{X}$ is equal to σ^2/N, where μ and σ^2 are the expectation and variance of X. Recall from Chapter Four, however, that a complete description of a random variable requires not only the values of its parameters, but specification of its probability function. To determine the probability function associated with the sampling distribution of the mean, we must introduce the Central Limit Theorem (CLT). The CLT is actually a collection of theorems that apply the same fundamental principle under a variety of mathematical conditions, but one of its more general expressions states in part that

> If a random variable Y is expressible as the sum of N independent random variables $(X_1 + X_2 + \ldots + X_N)$, then Y becomes normally distributed as N approaches infinity.

The proof of the Central Limit Theorem is beyond the scope of this text, but we have already encountered one of its more important implications. Recall that in Chapters Five and Six we demonstrated that a binomial random variable may be represented as the sum of N independent indicator random variables. That is, if $X:B(N, p)$, then

$$X = I_1 + I_2 + \ldots + I_N$$

where

$$I = \{0, 1\}$$

As N, the number of indicator variables, increases, therefore, the binomial random variable satisfies the conditions of the Central Limit Theorem, and the distribution of X approaches the normal. Our earlier development of the normal distribution as the limiting case of the binomial distribution is, therefore, simply one application of the Central Limit Theorem. Similarly, we shall see in Chapter Thirteen that the χ^2 (*chi*-square) distribution is also defined by a sum of N independent random variables and, like the binomial distribution, becomes normal as N increases without limit.

Indeed, the CLT may be applied to a wide variety of economic, biological, sociological, and geographic problems in which some variable Y is conceptualized as the sum of many independent variables. Biologists, for example, assume that most genetic characters that exhibit *continuous* variation, such as weight, body measurements, and intelligence, are primarily determined by the additive effects of many genes. Many of these *polygenes* exist in two forms, one of which contributes a measurable increment to the character and one of which does not. The potential contribution of any particular genetic locus may therefore be represented by either a 1 or a 0, and each gene is therefore quantified as an indicator variable. Sexual reproduction assures that any individual will receive an independent assortment of his parents' chromosomes, each of which may include both contributing ($= 1$) and noncontributing ($= 0$) genes. The total amount of the character he possesses will be determined by the sum of these genetic "values," and like the sum of a large number of independent indicator variables, therefore, most polygenic characters are approximately normally distributed.

Of more immediate concern to us are the implications of the Central Limit Theorem for the sampling distribution of the mean. To derive the expectation of the mean, we found it necessary to express $E(\overline{X})$ as the expectation of a sum of N random variables.

$$E(\overline{X}) = \frac{1}{N} E(X_1 + X_2 + \ldots + X_N)$$

We may, of course, move the constant $1/N$ into the expectation, thus giving us

$$E(\overline{X}) = E[(1/N)(X_1 + X_2 + \ldots + X_N)]$$

and it must therefore be true that

$$\overline{X} = 1/N (X_1 + X_2 + \ldots + X_N)$$

In deriving $V(\overline{X})$ we further established that the variables $X_1, X_2, \ldots X_N$ are independent of one another, and we therefore see that the mean $\overline{X}$ is expressible as the sum of N independent random variables. By the Central Limit Theorem, therefore, the distribution of $\overline{X}$ must approach the normal as the number of observations in our sample increases.

If we combine this conclusion with our derivations of $\mu_{\overline{X}}$ and $\sigma_{\overline{X}}^2$, we obtain a somewhat more common, if less general, formulation of the Central Limit Theorem:

> If X is a random variable with expectation μ and variance σ^2, the distribution of means $\bar{x}_i$ from samples comprised of N independent observations becomes normal with expectation $\mu_{\bar{x}} = \mu$ and variance $\sigma_{\bar{x}}^2 = \sigma^2/N$ as N approaches infinity.

As we move into inferential statistics, we shall see that the broad scientific applicability of statistics is based largely on the Central Limit Theorem. We have emphasized several times that, discounting manual solutions, the computation of any probability $p(\Delta x_i)$ requires the probability function associated with X. In many scientific problems, of course, it is unrealistic to expect that the probability distribution of our random variable will conform precisely to some convenient mathematical model. In such cases, application of statistical techniques can be highly questionable. Irrespective of the population distribution of X, however, we can often invoke the Central Limit Theorem and assume that means of large samples drawn from the population will be normally distributed. We may thus rely on the normal distribution to determine probabilities associated with intervals $\Delta \bar{x}_i$. As with the binomial approximation to the normal, moreover surprisingly small samples are often large enough to permit confident use of the normal distribution to compute such probabilities. To illustrate this, Bailey (1971) used an electronic computer to generate sampling distributions of $\bar{X}$ based on samples of various sizes for random variables with uniform, J-shaped, and U-shaped probability distributions. These computer simulations are illustrated in Figure 10.2, and it is apparent that even when the distribution of X is far from normal, one requires only modest samples for the distribution of $\bar{X}$ to assume the symmetrical, bell-shaped form that characterizes the normal distribution.

If the distribution of X is highly irregular (discontinuous, multimodal, etc.) one needs a minimum of 25 to 30 observations in each sample in order to assume a normal distribution of $\bar{X}$. For random variables that are more nearly normally distributed, however, one requires considerably smaller samples. If, for example, the distribution of X is unimodal and symmetrical, the table of cumulative normal probabilities will yield very accurate estimates of $p(\Delta \bar{x}_i)$ for samples of 10 or more observations. Indeed, if the distribution of X is *exactly* normal, than $\bar{X}$ is normally distributed for *any* size sample. This is readily apparent if we consider that for the smallest possible sample ($N = 1$), $\bar{x}$ must always equal x_i and the distributions of X and $\bar{X}$ must therefore be identical.

1. Computation of $p(\Delta \bar{x}_i)$ from the Normal Curve

To illustrate the application of the Central Limit Theorem to the sampling distribution of $\bar{X}$, let us consider a random variable X whose distribution is un-

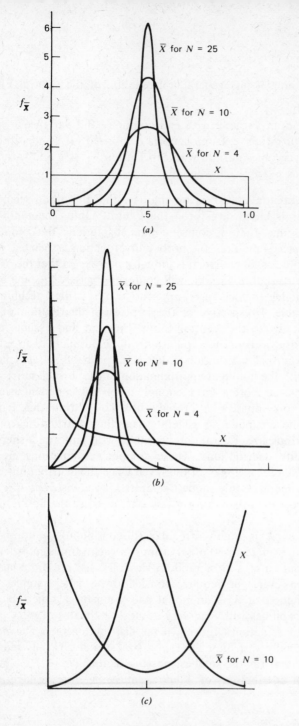

Figure 10.2   Sampling distributions of $\overline{X}$ for various sample sizes based on uniform, J-shaped, and U-shaped distributions of X. (Adapted with permission from Bailey, Probability and Statistics: Models for Research, New York: Wiley, 1971.)

known, but whose expectation $\mu = 900$ and whose variance $\sigma^2 = 2500$. If we draw samples of 25 observations, we may assume on the basis of the CLT that the distribution of

$$\overline{X} = \{\bar{x}_1, \bar{x}_2, \dots \bar{x}_n\}$$

is approximately normal. Suppose now that we wish to determine the probability of observing a mean that is equal to or greater than 910. That is, what is

$$p(\bar{x} \geq 910)$$

This probability is represented as the hatched portion of Figure 10.3. We see, then, that the area in question falls in the positive tail of the curve, and from Chapter Eight we therefore know that

$$p(\bar{x} \geq 910) = 1 - F(910)$$

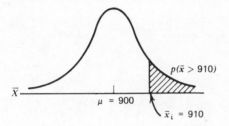

Figure 10.3 Probability of values in upper tail of sampling distribution of $\overline{X}$.

It will also be recalled that if a random variable X is normally distributed, then the cumulative probability of any value x_i may be determined by standardizing the value and submitting it to a table of cumulative probabilities of the standard normal curve. That is,

$$F(x_i) = F(z_i)$$

where

$$z_i = \frac{x_i - \mu}{\sigma}$$

Our first step, therefore, is to standardize 910, the value in question. It must be remembered, however, that 910 is not an X-value, but a value of $\overline{X}$, and the expectation and variance appropriate to the Z-transformation are $\mu_{\bar{x}}$ and $\sigma_{\bar{x}}$. Thus

$$z_{\bar{x}} = \frac{910 - \mu_{\bar{x}}}{\sigma_{\bar{x}}}$$

We know that $\mu_{\bar{x}} = \mu$, and μ was given above as 900. Thus

$$z_{\bar{x}} = \frac{910 - 900}{\sigma_{\bar{x}}}$$

We also know that

$$\sigma_{\bar{x}}{}^2 = \frac{\sigma^2}{N}$$

It follows, therefore, that $\sigma_{\bar{x}}$, the *standard error of the mean*, must equal

$$\sqrt{\sigma^2/N} = \frac{\sigma}{\sqrt{N}}$$

We specified that each sample contain 25 observations and that σ^2, the population variance of X, is equal to 2500. The standard error of the mean, $\sigma_{\bar{x}}$, thus becomes

$$\sqrt{2500/25} = \frac{50}{5} = 10$$

Hence

$$z_{\bar{x}} = \frac{910 - 900}{10}$$

$$= \frac{10}{10}$$

$$= 1.0$$

From Table **C** we see that

$$F(1.0) = .84$$

or

$$p(z \leq 1.0) = .84$$

For our unstandardized random variable $\overline{X}$, then,

$$F(910) = p(\bar{x} \leq 910)$$

$$= .84$$

Conversely,

$$p(\bar{x} \geq 910) = 1 - F(910)$$

$$= 1 - .84$$

$$= .16$$

By permitting us to assume a normal distribution of $\overline{X}$, therefore, the Central Limit Theorem extends rigorous application of probability theory to situations where the probability function of the random variable X is unknown. For the moment, this additional flexibility is simply an aid to statistical description, but some indication of its full theoretical importance becomes apparent in the next section on interval estimation.

E. Interval Estimation

In the introduction to this textbook, we defined inferential statistics as a body of mathematical conventions and techniques that permit us to make generalizations about populations from sample data and to determine the residual uncertainty intrinsic to such generalizations. By now, this definition should be considerably more meaningful than it was in the introduction. The reader knows, for instance, that our generalizations involve statements about the values of population parameters. He also knows that the sample data that generate these statements are usually summarized in statistics of one sort or another. And in the 10 chapters since his first encounter with this definition, the reader has also had some exposure to the concept of uncertainty. In Chapter Nine, for example, reduction of uncertainty was discussed in terms of reduction of variance. It should not be surprising, therefore, that the residual uncertainty surrounding any statistical generalization is related to the variance of the statistic on which the generalization is based. Finally, the reader knows that statisticians generally express the certainty or uncertainty of any quantitative event in terms of probability. He might anticipate, therefore, that the precision, or uncertainty, of our estimate might be expressed as a probability, such as $p(\bar{x} = \mu)$.

Recall from Chapter Eight, however, that for any particular value x_i of a continuous random variable X,

$$p(x = x_i) = 0$$

Furthermore, if X is a continuous random variable, then $\overline{X}$ is also continuous, and the probability must therefore be zero that any particular sample mean $\bar{x}$ is precisely equal to μ, the value of the population mean:

$$p(\bar{x} = \mu) = 0$$

Indeed, most statistics are either continuous or (like the sample mean of a discrete random variable) become continuous as sample size increases. It is generally the case, therefore, that

$$p(\hat{\theta} = \theta) = 0$$

where $\hat{\theta}$ is an estimator of the parameter θ. Even for an unbiased and highly efficient point estimator, therefore, we can never generate a meaninfgul probability statement to reflect the uncertainty of our estimate.

Even though a nonzero probability cannot be assigned to a *single value* of a continuous variable, it will also be recalled from Chapter Eight that a nonzero probability value may be meaningfully assigned to any continuous *interval* Δx. Statisticians have therefore developed techniques for computing interval estimates of parameters and procedures for determining the likelihood that such a *confidence interval* includes the value of the parameter in question.

1. Confidence Intervals for μ

Confidence intervals may be computed for a variety of parameters, including the population standard deviation σ and the population correlation coefficient ρ. In this section, however, we shall consider only interval estimation of the population mean μ. For any standardized normal random variable Z, we know from Table **C** that

$$p(-1.96 \leq z \leq 1.96) = .95$$

Thus

$$p\left(-1.96 \leq \frac{x - \mu}{\sigma} \leq 1.96\right) = .95$$

This expression involves two inequalities:

$$-1.96 \leq \frac{x - \mu}{\sigma}$$

and

$$\frac{x - \mu}{\sigma} \leq 1.96$$

Both sides of any inequality may be multiplied by a positive constant without altering the direction of the inequality. That is, if $a \leq b$, then $ca \leq cb$. We may therefore multiply both sides of our two inequalities by the constant σ:

$$-1.96\,\sigma \leq x - \mu$$

$$x - \mu \leq 1.96\,\sigma$$

Furthermore, both sides of any inequality may be multiplied by a *negative* constant, but the *direction* of the inequality changes. That is, if $a \leq b$, then $-ca \geq -cb$. If, therefore, we multiply both sides of our inequalities by -1, we obtain

$$1.96\,\sigma \geq \mu - x$$

and

$$\mu - x \geq -1.96\sigma$$

These inequalities may, of course, be transposed, yielding

$$\mu - x \leq 1.96\sigma$$

and

$$-1.96\sigma \leq \mu - x$$

220

Finally, any value may be added to or subtracted from both sides of an inequality without changing the direction of the relationship. Thus, if $a \leq b$, then $(a + c) \leq (b + c)$. Thus, adding x to both sides of our two inequalities,

$$\mu \leq x + 1.96\sigma$$

and

$$x - 1.96\sigma \leq \mu$$

The probability associated with our two inequalities is, of course, unchanged by our algebraic manipulations. If we once again combine them into a single expression, it must therefore be true that

$$p(x - 1.96\sigma \leq \mu \leq x + 1.96\sigma) = .95$$

In most introductory presentations, the student is permitted to assume that this statement reads as follows: The probability is .95 that μ lies between $x - 1.96\sigma$ and $x + 1.96\sigma$. It must be noted, however, that this statement is true only so long as the limits of the interval may assume *any* pair of values in the set $X \pm 1.96\sigma$. Thus, it is true only *until* the value of x is specified. As soon as x assumes a numerical value, the limits $x \pm 1.96\sigma$ define a *fixed* interval, which either includes the constant μ (in which case the correct probability is 1.0) or does not include μ (in which case the probability is 0). And, indeed, it is generally true that *prior* events may be assigned only probability values of 1.0 or 0. This restriction does not, however, extend to subjective certainty. Thus, an experimenter who computes limits of $x \pm 1.96\sigma$ will be 95 percent certain that his interval spans the population mean. This probabilistic certainty is defined as *confidence*, and for any given value of x, we may therefore say that the confidence is .95 that μ lies in the interval $x - 1.96\sigma$ to $x + 1.96\sigma$.

2. Confidence Intervals for μ Based on the CLT

It should be apparent that by selecting appropriate coefficients for σ, one may generalize the procedure outlined in the last section and compute interval estimates of any specified confidence. From Table **C**, for example we know that

$$p(-2.33 \leq z \leq 2.33) = .99$$

Thus, confidence $= .99$ that

$$(x - 2.33\sigma \leq \mu \leq x + 2.33\sigma)$$

Similarly, confidence $= .90$ that

$$(x - 1.64\sigma \leq \mu \leq x + 1.64\sigma)$$

confidence $= .68$ that

$$(x - 1.0\sigma \leq \mu \leq x + 1.0\sigma)$$

and so forth.

Even though we may compute an interval estimate to any specified degree of confidence, however, it should be apparent that this procedure requires that X be normally distributed. Moreover, the actual width of our interval depends entirely on the standard deviation σ. Thus, if σ is a very large, any interval we compute might be so wide as to provide relatively little information. It would not be terribly useful, for example, to state with 95 percent confidence that the mean I.Q. of physically fit persons was somewhere between 83 and 141. We may overcome these deficiencies, however, by relying on the Central Limit Theorem. By the CLT, we know that $\overline{X}$ is normally distributed and that its mean $\mu_{\overline{x}}$ is equal to μ, the population mean of X. It must therefore be true that

$$p(-1.96 \leq z_{\overline{x}} \leq 1.96) = .95$$

and that confidence $= .95$ that

$$(\overline{x} - 1.90\sigma_{\overline{x}} \leq \mu \leq \overline{x} + 1.96\sigma_{\overline{x}})$$

Furthermore, we have also established that

$$\sigma_{\overline{x}} = \frac{\sigma}{\sqrt{N}}$$

Substituting $\sigma/\sqrt{N}$ for $\sigma_{\overline{x}}$ in our 95 percent confidence interval, we obtain

$$\left(\overline{x} - 1.96\,\frac{\sigma}{\sqrt{N}} \leq \mu \leq \overline{x} + 1.96\,\frac{\sigma}{\sqrt{N}}\right)$$

We see from this result that

$$1.96\,\frac{\sigma}{\sqrt{N}}$$

must decrease as sample size N increases. By increasing our sample size, therefore we may reduce our interval

$$\overline{x} - 1.90\sigma/\sqrt{N} \qquad \text{to} \qquad x + 1.96\sigma/\sqrt{N}$$

to any desired width without affecting its confidence. Despite the advantages thus derived from the CLT, it is seldom that an experimenter who must *estimate* his population mean μ will *know* his population standard deviation σ. If, however, $N \geq 30$, he may substitute the sample standard deviation s with relatively little error.

To illustrate the inferential leverage conferred by these developments, let us consider a random variable X about which nothing is known. Suppose further that an experimenter has made 36 observations and has computed

$$\overline{x} = 910$$

$$s = 50$$

Finally, let us suppose that he wishes to determine a 90 percent confidence interval for the population mean μ of his parent population. In Table **C** he finds that

$$p(z \geq 1.64) = .05$$

$$p(z \leq -1.64) = .05$$

Thus

$$p(-1.64 \leq z \leq 1.64) = .90$$

With a sample size of 36 he may assume that $\overline{X}$ is normally distributed, and he therefore knows that confidence = .90 that

$$(\bar{x} - 1.64\,\sigma_{\bar{x}} \leq \mu \leq \bar{x} + 1.64\,\sigma_{\bar{x}})$$

or

$$(\bar{x} - 1.64\,\sigma/\sqrt{N} \leq \mu \leq \bar{x} + 1.64\,\sigma/\sqrt{N})$$

He does not know the value of σ, however, so he substitutes s, and the limits of his 90 percent confidence interval thus become

$$\bar{x} - 1.64\,\frac{s}{\sqrt{N}} \quad \text{and} \quad \bar{x} + 1.64\,\frac{s}{\sqrt{N}}$$

Substituting his computed values into these two expressions, his confidence interval is

$$910 - 1.64\,(60/\sqrt{36}) \quad \text{to} \quad 910 + 1.64(60/\sqrt{36})$$

or

$$910 - 16.4 \quad \text{to} \quad 910 + 16.4$$

Thus, the experimenter is 90 percent certain that the population mean μ falls in the interval 893.6 to 926.4.

Suppose, however, that this interval is too wide to be really helpful, that the experimenter wants to "trap" the population mean in a narrower interval. To reduce the interval he draws a sample of 100 observations. For the sake of comparison, let us suppose that this sample mean remains 910 and his sample standard deviation remains equal to 60. Based on this larger sample, his confidence interval would now become

$$910 - 1.64(60/\sqrt{100}) \quad \text{to} \quad 910 + 1.64(60/\sqrt{100})$$

or

$$910 - 9.84 \quad \text{to} \quad 910 + 9.84$$

Thus, he is now 90 percent sure that this parameter falls in the interval 900.16 to 919.84.

The confidence limits in this example were, of course, chosen because 90 percent of the area under the normal curve lies between $\mu - 1.64\sigma$ and $\mu + 1.64\sigma$. It should be understood, however, that there are an *infinite* number of such intervals. For example,

$$p[(\mu - 1.48\sigma) \leq x \leq (\mu + 1.87\sigma)] = .90$$

$$p[(\mu - 2.06\sigma) \leq x \leq (\mu + 1.41\sigma)] = .90$$

Thus, we might as easily have established confidence limits of $\bar{x} - 1.48\sigma_{\bar{x}}$ and $\bar{x} + 1.87\sigma_{\bar{x}}$ or $\bar{x} - 2.06\sigma_{\bar{x}}$ and $\bar{x} + 1.41\sigma_{\bar{x}}$. Because the normal distribution is symmetrical about μ, however, it can be shown that the *shortest* interval estimate for μ is symmetrical about the point estimate $\bar{x}$. For an asymmetrically distributed random variable, the shortest interval estimate for any specified level of confidence would be correspondingly asymmetrical about the point estimate.

F. Problems, Review Questions and Exercises

1. Let X be a uniformly distributed random variable defined by the value set $X = \{1, 2, 3\}$.
   a. Compute to four decimal places the expected value μ and the variance σ^2 of this random variable.
   b. Assume X to be normally distributed and use Table **C** to compute $p(x > 2)$.
   c. What is the difference between this probability value and the exact probability given in the problem?
2. Assume that two independent observations are made on X.
   a. Tabulate the probability distribution for $\bar{X}$.
   b. Using the definitional formulas for expected value and variance, compute $\mu_{\bar{x}}$ and $\sigma_{\bar{x}}^2$.
   c. How are μ and $\mu_{\bar{x}}$ related? How are σ^2 and $\sigma_{\bar{x}}^2$ related?
   d. Assume $\bar{X}$ to be normally distributed and use Table **C** to compute $p(\bar{x} > 2.5)$.
   e. What is the difference between this probability value and the exact probability tabulated in **2a**?
3. Assume that three independent observations are made on X.
   a. Tabulate the probability distribution for $\bar{X}$.
   b. Using the definitional formulas for expected value and variance, compute $\mu_{\bar{x}}$ and $\sigma_{\bar{x}}^2$.
   c. How are μ and $\mu_{\bar{x}}$ related? How are σ^2 and $\sigma_{\bar{x}}^2$ related?
   d. Assume $\bar{X}$ to be normally distributed and use Table **C** to compute $p(\bar{x} > 2.67)$.

e. What is the difference between this probability value and the exact probability tabulated in **3a**?
4. What can you conclude from a comparison of **1e**, **2e**, and **3e**?
5. Given the value for $\sigma_{\bar{x}}^2$ from **3b**, compute a 99 percent confidence interval for μ based on the following sample of three observations: 2, 3, 3.
6. What is the probability relationship between μ and the interval you computed in problem **5**?

chapter eleven

hypothesis testing

Like estimation, hypothesis testing permits the scientist to make generalizations about populations from sample data. As the reader might expect, these generalizations are again concerned with distribution properties of random variables, and the sample data on which they are based are summarized in statistics. These *test statistics*, however, are somewhat different from the statistics we have encountered thus far. In Part Two we developed statistics that describe distribution properties of samples, and in Chapter Ten we introduced statistics to estimate the parameters that describe properties of population distributions. In general the mathematical expression for either a descriptive statistic such as s^2 or an estimator such as $\hat{\sigma}^2$ has some discernable relationship to the distribution property it describes. A test statistic, however, is often the product of complex mathematical transformations and may therefore seem only indirectly related to the distribution property about which the experimenter wishes to generalize.

Besides yielding statistical generalizations, hypothesis tests also permit the investigator to express in terms of probability the uncertainty that characterizes such generalizations. In our discussion of confidence intervals, we demonstrated that the residual uncertainty of any estimate is related to the variability of the statistic on which the statistical generalization is based. The same is true in hypothesis testing, but in this context the uncertainty is expressed in terms of *significance* rather than confidence.

A. Outline and Example of Hypothesis Testing

To get some feeling for the logic and basic mechanics of hypothesis testing, we shall briefly outline the steps involved in a statistical test and illustrate them with a highly idealized example.

1. Formulation

The first step in conducting an hypothesis test is to formulate some testable proposition about the state of nature. This proposition may concern any observable phenomenon, but to be subject to statistical test, it must be expressed in terms of the distribution properties of some random variable. Actually, the investigator presents two propositions. The first is the *test hypothesis* (H_0), which states that some parameter θ will assume a specific value θ_0 if his formulation is correct. The second is the *alternative hypothesis* (H_1), which indicates the parameter value θ_1 or values $\{\theta_1\}$ to be expected if his formulation is not correct.

Suppose, for example, that an anthropologist decides to test a belief that is currently popular among his colleagues that Neanderthal man disappeared from the Scandinavian Penninsula at the end of the last ice age and that the area remained uninhabited until later migrations of Cro-magnon man. If this belief is correct, then all ice-age human remains should be Neanderthal. Existing anthropological data indicate that Neanderthal skull capacities are normally distributed with a mean of 1300 cubic centimeters and a standard deviation of 120 cubic centimeters. That is,

$$\text{Neanderthal skull capacities: } N(\mu = 1300, \sigma^2 = 14{,}400)$$

If his colleagues are correct, therefore, ice-age skulls found in the Scandinavian Penninsula (S.P.) should have an expected capacity of 1300 cubic centimeters. Thus, his test hypothesis is

$$H_0: \mu_{\text{S.P.}} = \mu_0 = 1300$$

where μ_0 is the general notation for the expected value of a random variable under the test hypothesis.

The anthropologist himself, however, favors an alternative possibility that Cro-magnon man emigrated to the region *before* the last ice age and that the Neanderthal inhabitants were driven out, killed off, genetically absorbed, or otherwise displaced by the newcomers. If this supposition is true, then all ice-age human remains should be Cro-magnon, and according to existing records Cro-magnon skulls are normally distributed with a mean capacity of 1500 cubic centimeters and a standard deviation of 120 cubic centimeters. He therefore proposes an alternative hypothesis that the expected capacity of ice-age skulls is 1500 cubic centimeters

$$H_1: \mu_{S.P.} = \mu_1 = 1500$$

where μ_1 represents the expected value of a random variable under H_1.

2. Decisions

After formulating his hypotheses, an experimenter must decide what sort of evidence would *disconfirm* one hypothesis and provide *support* for the other hypothesis. First, of course, he must select his units of observation. In our example, the hypotheses are statements about ice-age skulls in the Scandinavian Penninsula, and the anthropologist therefore decides to go to the Scandinavian Penninsula and find a skull from the last ice age.

It must be remembered, however, that all statistical procedures require numerical data, so the experimenter must also decide how his observation(s) will be expressed or summarized as a single value of some random variable T, which we call the *test statistic*. Any skull has many quantitative properties, including weight, orbital angle, and number of teeth. The hypotheses, however, mention only skull capacity, and the anthropologist therefore decides that the capacity of his specimen will be his test statistic.

Even if the test hypothesis is correct, however, the test statistic is nonetheless a random variable, and its observed value may therefore depart considerably from the value expected under the test hypothesis. The distribution of Neanderthal skulls capacities, for example, has a standard deviation of 120 cubic centimeters and our anthropologist could very conceivably find a Neanderthal skull that is much larger (or smaller) than 1300 cubic centimeters and that might consequently be mistaken for a Cro-magnon skull. In any statistical test, therefore, the experimenter must decide *how much* his data must depart from his expectations before he can safely regard his test hypothesis as unsupported. Formally, he defines an interval that is *minimally* likely to include his observed test statistic t_o if his *test* hypothesis is correct and *maximally* likely to include the statistic if the *alternative* hypothesis is correct. He then stipulates that if the observed value of his test statistic falls in this *critical region*, the test hypothesis will be rejected in favor of the alternative hypothesis. In our example, therefore, the anthropologist decides that he will not reject his test hypothesis in favor of H_1 unless the capacity of his specimen skull is at least 1385 cubic centimeters. Thus, the set

$$\{x \mid x \geq 1385\}$$

is his critical region, and the lower limit of this interval, 1385, is called the *critical value* of the test statistic.

3. Data Collection

The data for any hypothesis test may be the result of a formal experiment, a set of measurements, a collection of naturalistic observations, or even a computer simulation. But however the data are generated, they are always assumed to reflect the *true* state of nature. If a scientist's observations contradict his formulation, therefore, the experimenter must abandon his formulation, not his data.

In the present example, of course, the relevant datum is produced by measurement. The anthropologist mounts an expedition to the Scandinavian Penninsula and discovers a skull that radio carbon dating indicates to be 50,000 years old, well into the last ice age. The skull is thoroughly cleaned and found to have a capacity of 1380 cubic centimeters. That is,

$$t_o = 1380$$

4. Conclusions

The significance test per se is conducted by simple inspection. If the observed value of the test statistic falls in the critical region, the test hypothesis is abandoned in favor of the alternative hypothesis. If the statistic does not fall in the critical region, the alternative hypothesis is rejected.

The appropriate conclusion in our example is quite apparent. The skull capacity of 1380 cubic centimeters is less than the critical value, 1385 cubic centimeters, and does not, therefore, fall in the critical region. Hence, the anthropologist must reject his alternative hypothesis. Note that we do *not* say that the investigator *accepts* the *test* hypothesis. The test hypothesis states

$$\mu_{\text{S.P.}} = 1300$$

On the basis of his observation, however, the *most likely* value of the population mean of Scandinavian Penninsula Skulls has become 1380 cubic centimeters. Thus, all he can say is that the evidence does not support H_1.

B. Foundations of Hypothesis Testing

In the preceding section, we outlined a statistical test, but we did not discuss any of the philosophical or statistical considerations on which these procedures are based. In this section, therefore, we shall examine the four principal aspects of the statistical test in the perspectives of both scientific epistemology and probability theory.

1. Formulation: Theory, Model, and Hypotheses

As we have indicated above, a statistical test begins with the formulation of a testable proposition about the state of nature. Typically such a formulation starts with some conceptualization that expresses relationships among variables. The variables may derive from physics, biology, sociology, economics, or any other discipline that makes observations of the real world and attempts to systematize these observations within an explanatory conceptual framework. For traditional philosophers of science, such a conceptualization constitutes a *theory*. And, frequently, our formulation is indeed based on some very general scientific theory, such as the Wynne-Edwards theory of animal social organization or the Taylor-Wegener theory of continental drift. More often, however, tests are generated by conceptualizations of more limited scope, such as Watson and Crick's theoretical model of DNA structure or B. F. Skinner's conception of operant learning. But the overwhelming majority of statistical hypotheses derive from pure speculation, scientific hunches, or prior empirical observation of the variables under consideration. Nonetheless, we shall remain consistent with current usage and consider all such hypothetical relationships to be "theories."

In popular thought, it is generally believed that when theories are "proven correct," they cease to be theories and suddenly become facts. Implict in this thinking lurks the suspicion that "theories" are somehow less substantial or less trustworthy than the "cold, hard facts." As popular as this conception is, with the journal editor no less than the parson's wife, it reflects a gross misunderstanding of the nature and role of theory. When we speak of *facts* we refer to our empirical observations of the phenomenal world. When we speak of *theories* we mean the concepts and principles that we abstract from these observations in order to organize and explain them. Thus, facts may be empirically verified, but theories exist only in the mind of the theorist and are therefore never susceptible to proof or disproof. It is therefore as meaningless to speak of the "truth" or "correctness" of a theory as it is to ask about the color of a musical chord or the flavor of a rainbow. This is not to say, however, that any theory is as good as any other. Indeed, philosophers of science have established a number of criteria by which theories may be evaluated. A good theory, for example, should be *internally consistent*; it should never generate implications that contradict one another. Another important criterion is *parsimony*. That is, theoretical explanations should employ the fewest possible concepts and assumptions. Any theory can, of course, be made parsimonious at the expense of versatility or explanatory power. Thus, if two theories are equally parsimonious, the one that explains the wider range of phenomena is considered the more powerful of the two and is therefore the more acceptable.

Unlike the philosopher of science, however, the statistician requires only two things of a theory. First, it must be *predictive*. That is, it must permit inferences about natural phenomena that the investigator has not yet observed, but that are observable either in principle or in fact. Second, it must include enough *quantitative* facts to generate a formal statement about the distribution properties of some random variable.

In our anthropology example, for instance, theoretical relationships among various historical, geographical, and paleontological variables suggested the prediction that only Neanderthal skills would be found in the Scandinavian Penninsula. This conceptualization, moreover, included factual information that Neanderthal skull capacities are normally distributed with a mean of 1300 cubic centimeters and a standard deviation of 120 cubic centimeters.

From the theory, we generate a *statistical model.* A model is an implication of the theory that is formalized in terms of a statement about the distribution properties of some random variable. The random variable described in the model must, of course, represent some property or attribute about which the theory makes a prediction. In our example, the major premise of the theoretical prediction is that ice-age skulls found in the Scandinavian Penninsula are Neanderthal. The minor premise is that Neaderthal skull capacities are distributed $N(1300, 14,400)$. The formal implication is that the capacities of Scandinavian Penninsula skulls remaining from the ice age are distributed $N(1300, 14,400)$.

The model may, of course, be as easily generated from a set of prior observations as from a theory. Thus, the investigator may have already observed a large number of skulls from this region and found them to be normally distributed with a mean of 1300 cubic centimeters and a standard deviation of 120 cubic centimeters.

Once the model has been formalized, the investigator must indicate which of its distribution properties he will submit to statistical test. He therefore formulates a *test hypothesis* (H_0), which specifies the value of some parameter(s) or function of some parameter(s) to be expected *IF* his model is indeed an adequate representation of nature. In our example, the test hypothesis concerns expected value:

$$H_0: \mu_{\text{S.P.}} = \mu_0 = 1300 \text{ cubic centimeters}$$

The experimenter is, of course, never positive that his model *is* an adequate representation of nature. That is, the distribution properties of the random variable may be quite different from those specified in the model. Indeed, if this were not a possibility, there would be no need to conduct the test. He must therefore formulate an *alternative hypothesis* (H_1), which specifies the parameter value(s) to be expected if his model (and ultimately the theory on which the model rests) does *not* concur with the true state of nature. Occasionally, as in our example, the experimenter will have a competing model that provides this alternative. Indeed, the competing model may even generate a *simple* alternative hypothesis. That is, the alternative hypothesis, like the test hypothesis, may specify some *particular* value of the parameter in question. This was the case in our example, where

$$H_1: \mu_{\text{S.P.}} = \mu_1 = 1500 \text{ cubic centimeters}$$

More often, however, the competing model will indicate only the *direction* in which the data may be expected to depart from the value given in H_0 if the test hypothesis is to be rejected. Our example would have been more realistic for instance, if the anthropologist had not known that Cro-magnon skulls had an expected capacity of 1500 cubic centimeters, but only that they were larger than Neanderthal skulls. In such cases, H_1 is cast as a *composite* hypothesis that specifies some *range* of values that the parameter is likely to assume if H_0 is rejected. Under these circumstances, the alternative hypothesis is our example would have been

$$H_1: \mu_{\text{S.P.}} > \mu_0 = 1300 \text{ cubic centimeters}$$

Unfortunately, it is generally the case that the alternative is *not* grounded in any well-defined model, but simply expresses the possibility that H_0 is untrue. It is most likely, therefore, that our anthropologist could have justified only the alternative supposition that some group *other than* Neanderthal man inhabited the Scandinavian Penninsula during the last ice age. Thus, his alternative hypothesis H_1 would have indicated only that the expected capacity of Scandinavian Penninsula skulls was *not* equal to 1300 cubic centimeters:

$$H_1: \mu_{\text{S.P.}} \neq \mu_0 = 1300 \text{ cubic centimeters}$$

Because this sort of hypothesis does not specify the direction of anticipated departure from the value given in H_0, it is called a *nondirectional* hypothesis.

2. Decisions: Units of Observations, Test Statistics, and Rejection Rules

After formulating his hypotheses, the experimenter has three decisions to make. First, he must select his *units of observation*. Second, he must decide how his observation(s) will be quantitatively expressed. That is, he must decide on a *test statistic*. Third, he must decide prior to conducting his observations what sort of data would lead him to reject one of his hypotheses. This decision is usually cast in terms of a *rejection rule*.

a. Units of Observation. In general, the appropriate units of observation for any statistical test are members of the population specified in the hypotheses, and in our example the investigator therefore decided to examine an ice-age skull from the Scandinavian Penninsula. It must be emphasized, however, that a statistical test is always conducted to establish a general conclusion about an *entire population*, such as the population of all ice-age Scandinavian Penninsula skulls, and we saw in Chapter Ten that the certainty of such an inference is related to sample size. In real life, therefore, our anthropologist would not have set out to excavate just a single skull; he would have undoubtedly decided to recover as many specimens as time, expense, and good fortune permitted.

b. Test Statistic. In selecting a test statistic, it must always be remembered that a statistical test simply determines whether or not one's observations are consistent with the model generated by the theory. Implicitly or explicitly, therefore, the theory must specify the distribution of the test statistic. In the present example, the theory predicts the distribution of capacities for ice-age skulls found in the Scandinavian Penninsula. This permitted the anthropologist to define the capacity of one skull as his test statistic. Like a descriptive statistic, however, a test statistic should consider all observed values of the random variable. If, as suggested above, therefore, the experimenter were to draw a sample of $N > 1$ observations, a single value x would no longer be an appropriate test statistic. We know from Chapter Ten, however, that a single value x may be considered the mean of a sample of $N = 1$ observations. Thus, the test statistic in our example was really $\overline{X}$.

In this regard, recall from the introduction of our example that the theory under investigation predicts that the expectation of X is 1300 and the variance of X is 14,400. Moreover, we know from the Central Limit Theorem that $\overline{X}$ is normally distributed and that if $E(X) = \mu$ and $V(X) = \sigma^2$, then $E(\overline{X}) = \mu$ and $V(\overline{X}) = \sigma^2/N$. For a sample of, say, $N = 9$ skulls, therefore, the theory must predict that

$$E(\overline{X}) = \mu_0 = 1300$$

and

$$V(\overline{X}) = \frac{\sigma^2}{N} = \frac{14,400}{9}$$

$$= 1600$$

Thus

$$\overline{X} : N(1300, 1600)$$

and we see that the distribution of $\overline{X}$ is indeed determined completely by the theory.

This is not, however, the only property that makes $\overline{X}$ a suitable statistic for testing hypotheses about μ. Just as the sample mean possesses various characteristics that make it a good descriptive statistic and a good estimator, so does it possess various properties considered desirable of a test statistic. Unfortunately, even an elementary treatment of these characteristics is beyond the scope and level of this book and it must suffice to say that a test statistic should provide us with some information about the parameter considered in the hypotheses. As in the present example, therefore, the test statistic is sometimes an estimator of the parameter under study, but this is not always the case.

Mathematical considerations aside, it is highly desirable from a practical standpoint that the test statistic be expressed in some form that permits comparisons with entries in an available probability table. In general, we do not have tabulations for normally distributed random variables with $\mu = 1300$ and $\sigma^2 = 1600$, but from Chapter Eight we know that any normally distributed random variable may be standardized and submitted to the cumulative probability table for the standard normal random variable

$$Z = \frac{X - \mu}{\sigma}$$

Our test statistic is therefore $\overline{X}$ expressed in standard form:

$$Z_{\overline{x}} = \frac{\overline{X} - \mu_{\overline{x}}}{\sigma_{\overline{x}}}$$

$$= \frac{\overline{X} - \mu}{\sigma/\sqrt{N}}$$

Under the distribution specified by the theory in our anthropology example $\mu = \mu_0$, and for nine observations, $Z_{\overline{x}}$ therefore becomes

$$\frac{\overline{X} - \mu_0}{\sigma/\sqrt{9}} = \frac{\overline{X} - \mu_0}{120/\sqrt{9}} = \frac{\overline{X} - 1300}{40}$$

c. Rejection rules. In the last section, we established that for samples of nine observations, the appropriate test statistic in our example is

$$\frac{\overline{X} - 1300}{40}$$

The expected value of the test statistic must therefore be

$$E\left(\frac{\overline{X} - 1300}{40}\right) = \frac{E(\overline{X}) - 1300}{40}$$

Furthermore, we established that under the distribution specified in the model, $E(\overline{X}) = 1300$. If H_0 is correct, therefore, the expected value of the test statistic must be

$$\frac{1300 - 1300}{40} = 0$$

Under the alternative hypothesis, however, the population mean $\mu_{\text{S.P.}}$ is 1500. Because $\overline{X}$ is an unbiased estimator of the population mean, it follows that under the alternative hypothesis, $E(\overline{X}) = \mu_1 = 1500$, and if H_1 is correct, therefore, the expected value of his test statistic is

$$\frac{E(\overline{X}) - 1300}{40} = \frac{1500 - 1300}{40}$$

$$= 5$$

Realistically, of course, the experimenter knows that the observed value of his test statistic will never precisely equal either of these two values. The sample mean $\overline{X}$ is a normally distributed random variable, and whichever of the hypotheses is correct, his test statistic may theoretically assume any value from $-\infty$ to $+\infty$. Nonetheless, it should be apparent that if the observed test statistic

$$\frac{\overline{x}_o - 1300}{40}$$

were small, the data would tend to support the test hypothesis. If, in contrast, the value of the test statistic were very large, the data would be more consistent with H_1 and would thus suggest that H_0 should be abandoned. These considerations provide the experimenter with a method for deciding which of his hypotheses should be rejected. Quite simply, the anthropologist would establish a *rejection rule* that stipulates that if the observed value of his test statistic $z_{\bar{x}}$ exceeds some *critical value* z_α, he will reject H_0 in favor of H_1. But, if the observed test statistic does *not* exceed his critical value, he will reject H_1. In the present example, therefore, the set

$$\{z \mid z \geq z_\alpha\}$$

is defined as the *critical region* of the test statistic. That is, H_0 will be rejected if the observed test statistic

$$z_{\bar{x}} \in \{z \mid z \geq z_\alpha\}$$

The critical region may, of course, be any interval in the value set of the test statistic, but if the rejection rule is to yield a scientifically meaningful conclusion, the experimenter must choose his critical region on the basis of several considerations. To begin with, we know that

$$Z = \frac{X - \mu}{\sigma}$$

It follows, therefore, that if $\mu_{\bar{x}} = \mu_0 = 1300$ (and if $\sigma_{\bar{x}} = 40$), the test statistic

$$\frac{\bar{X} - 1300}{40}$$

is a standard normal random variable. That is, if H_0 is correct, then

$$\frac{\bar{X} - 1300}{40} = \frac{\bar{X} - \mu_{\bar{x}}}{\sigma_{\bar{x}}}$$

$$= Z_{\bar{x}}$$

If the test hypothesis is correct, therefore, we may use Table **C** to compute the probability that

$$z_{\bar{x}} \geq z_\alpha$$

for any critical value that the anthropologist might select. Suppose, for example, that he were to set $z_\alpha = 2.33$. We see from Table **C** that

$$F(2.33) = .9901$$

Thus

$$p(z_{\bar{x}} \geq 2.33) = .01 \qquad \text{(approx.)}$$

It must be emphasized, however, that this probability is appropriate *if and only if* H_0 is correct. It is therefore a conditional probability and is properly expressed as

$$p(z_{\bar{x}} \geq 2.33 \,|\, H_0 \text{ correct}) = .01$$

where

$$z_{\bar{x}} = \frac{\bar{x}_o - 1300}{40}$$

At the beginning of this section we established that under H_0 the expected value of the test statistic is zero. As z_α increases, therefore, rejection of H_0 must require an ever greater departure of the observed test statistic $z_{\bar{x}}$ from its expected value under the test hypothesis. Thus, if H_0 is correct, it follows from Chapter Eight that the probability of values in the critical region $p(z_{\bar{x}} \geq z_\alpha)$ must decrease as z_α increases. Conversely, the smaller the probability

$$p(z_{\bar{x}} \geq z_\alpha \,|\, H_0 \text{ correct})$$

the larger must be z_α and, consequently, the greater (or more *significant*) must be the departure of $z_{\bar{x}}$ from its expected value in order to reject H_0. This conditional probability is therefore defined as the *significance level* of the hypothesis test and is customarily symbolized by the Greek letter α (*alpha*).

It should be apparent from the preceding paragraph that the experimenter's choice of a critical region depends on the level of significance he desires. That is, it depends on how improbable (given H_0 correct) he feels his observed test statistic must be to justify rejection of H_0. It should be understood, however, that establishment of the significance level does not unequivocally determine the critical region. In the present example, for instance, there are an *infinite* number of intervals Δz_α for which

$$p(z_{\bar{x}} \, \epsilon \, \Delta z_\alpha \,|\, H_0 \text{ correct}) = .01$$

For example,

$$p(-1.56 \leq z_{\bar{x}} \leq -1.48) = .01$$

$$p(0.69 \leq z_{\bar{x}} \leq 0.72) = .01$$

$$p(1.85 \leq z_{\bar{x}} \leq 1.88) = .01$$

and, of course,

$$p(z_{\bar{x}} \geq 2.33) = .01$$

In selecting the most sensible interval it must be remembered that one does not simply reject the test hypothesis. He rejects H_0 *in favor of* H_1. Of all possible intervals Δz_α, therefore, the experimenter should select the one that is *most probable* if H_1 is correct. In the present example, the value of μ specified in the alternative hypothesis is *greater than* the value of μ specified in the test hypothesis. The critical region should therefore be that interval Δz_α that includes the *largest* values that the test statistic might assume. For $\alpha = .01$, this is the set

$$\{z \mid z \geq 2.33\}$$

As indicated above, the establishment of a critical region can proceed only after α has been determined. In this regard, our discussion of rejection rules has thus far implicitly assumed an arbitrary significance level $\alpha = .01$. And, indeed, it is common practice among many scientists to select, in a similarly arbitrary fashion, some "customary" significance level, such as .01, .05, or .10. In principle, however, the investigator *should* base his selection of α on consideration of the types of errors that any statistical conclusion may produce. In this regard, we have defined the test statistic in our anthropology example as

$$z_{\bar{x}} = \frac{\bar{X} - 1300}{40}$$

and have established that

$$p(z_{\bar{x}} \geq 2.33 \mid H_0 \text{ correct}) = .01$$

This tells us that if the anthropologist were to examine every possible sample of nine skulls drawn from the *population specified in the model*, his test statistic would equal or exceed 2.33 for 1 out of every 100 such samples. When H_0 is correct, therefore, the probability is .01 that the experimenter will observe a value equal to or exceeding 2.33 for *any one* particular sample drawn at random from this sample space. Consequently, if he sets his critical value z_α equal to 2.33, the probability that he will reject H_0 *even though* H_0 *is correct* must also equal .01. That is,

$$p(\text{rejecting } H_0 \mid H_0 \text{ correct}) = p(z_{\bar{x}} \geq 2.33 \mid H_0 \text{ correct})$$

If 2.33 is his critical value, however, we know by the definition of significance level that

$$p(z_{\bar{x}} \geq 2.33 \mid H_0 \text{ correct}) = \alpha$$

It follows, then, that

$$\alpha = p(\text{rejecting } H_0 \mid H_0 \text{ correct})$$

Rejecting the test hypothesis when it is in fact correct is called a *Type I error*, and in general, therefore, the level of significance

$$\alpha = p(\text{Type I error})$$

It should thus be obvious that the experimenter's choice of a significance level α must depend on least partially on how important it is to avoid a Type I error. Indeed, if incorrect rejection of H_0 were the only possible error he might make, the choice of a significance level would depend entirely on this consideration. As the reader may have anticipated, however, it is also possible to incorrectly reject the *alternative* hypothesis. Rejecting the alternative hypothesis when it is in fact correct is called a *Type II error*, and p (Type II error) must therefore equal the conditional probability

$$p(\text{rejecting } H_1 \mid H_1 \text{ correct})$$

For any particular sample size, moreover, the probability of a Type II error, like the probability of a Type I error, is determined entirely by the level of significance chosen by the experimenter. To explore this relationship, however, we must shift gears a bit and consider the anthropologist's original rejection rule in terms of the unstandardized sample mean $\overline{X}$. In this regard, we know that under the test hypothesis

$$z_{\bar{x}} = \frac{\bar{x} - 1300}{40}$$

By simple algebraic manipulation, therefore,

$$\bar{x} - 1300 = 40(z_{\bar{x}})$$

and

$$\bar{x} = 40\,(z_{\bar{x}}) + 1300$$

For the critical value $z_\alpha = 2.33$, therefore, the corresponding critical value of the sample mean, which we designate $\bar{x}_c$, must equal

$$40(2.33) + 1300 = 93.20 + 1300$$
$$= 1393.20$$

In effect, then the rejection rule states that the anthropologist must reject his alternative hypothesis if his sample mean is less than 1393.20 cubic centimeters. While this value is certainly closer to 1300 than to 1500, we know that the sample mean is normally distributed under *either* hypothesis and may therefore assume *any* real value. Even if the alternative hypothesis were correct, then, the experimenter might conceivably draw a sample with a mean as small or smaller than 1393.20 and, by his rejection rule, conclude *incorrectly* that H_1 is to be rejected. It should therefore be apparent that the probability of a Type II error is equal to the conditional probability

$$p(\bar{x} < 1393.20 \mid H_1 \text{ correct})$$

Customarily, the probability of a Type II error is symbolized by the Greek letter β (*beta*), and in the present example, therefore,

$$\beta = p(\bar{x} < 1393.20 \mid H_1 \text{ correct})$$

By the definition of cumulative probability

$$p(\bar{x} < 1393.20) = F(1393.20)$$

and the computation of β therefore requires only that $\bar{x}_c$ be standardized and its cumulative probability determined from Table **C**. However, it must be emphasized that this probability is *conditional* on the assumption that H_1 is correct, and $\bar{x}_c$ must therefore be standardized using the parameters specified in the *alternative* hypothesis. Under the alternative, $\mu = 1500$ and $\sigma = 120$, and for a sample of nine observations it follows from the Central Limit Theorem that

$$\mu_{\bar{x}} = \mu_1 = 1500$$

and

$$\sigma_{\bar{x}} = \frac{120}{\sqrt{9}} = \frac{120}{3}$$

$$= 40$$

Under the distribution specified in the alternative, therefore, the standardized value of 1393.20, designated z_β, must equal

$$\frac{1393.20 - 1500}{40} = \frac{-106.80}{40}$$

$$= -2.67$$

According to Table **C**,

$$F(-2.67) = .0038 \qquad \text{(approx.)}$$

and if H_1 is correct, therefore,

$$p(\bar{x} < 1393.20) = .0038$$

or

$$p(\bar{x} < 1393.20 \,|\, H_1 \text{ correct}) = .0038$$

In the present example, then

$$\beta = .0038$$

By definition β is the probability that the observed value of the test statistic will *not* fall in the critical region when H_1 is correct. In our anthropology example, therefore, β would be represented as the area corresponding to all values of $\overline{X}$ less than $\bar{x}_c$ under the distribution specified in the alternative hypothesis. Thus far, however, our discussion of Type II errors has been restricted to the case where $\alpha = .01$ and the critical value z_α therefore equals 2.33, with a corresponding mean value $\bar{x}_c = 1393.20$. For some other rejection rule, the critical value z_α and the corresponding sample mean $\bar{x}_c$ would assume other values. If, for example, the experimeter set his critical value z_α equal to 1.64, $\bar{x}_c$ would equal

$$40(1.64) + 1300 = 1365.60$$

Similarly, a critical value $z_\alpha = 3.08$ corresponds to a sample mean $\bar{x}_c = 1423.20$.

The effects on β of establishing rejection rules with critical values of 1.64, 2.33, and 3.08 are illustrated in Figure 11.1. In all three pairs, the distribution of $\overline{X}$ specified in the test hypothesis H_0 is represented by the solid curve. The shaded portion of each solid curve corresponds to all values *greater than* $\bar{x}_c$ and therefore represents the significance level α. The dotted curve in each pair represents the distribution of $\overline{X}$ specified in the alternative hypothesis H_1, and the hatched portion of this curve, which corresponds to all values *less than* $\bar{x}_c$, must therefore represent β.

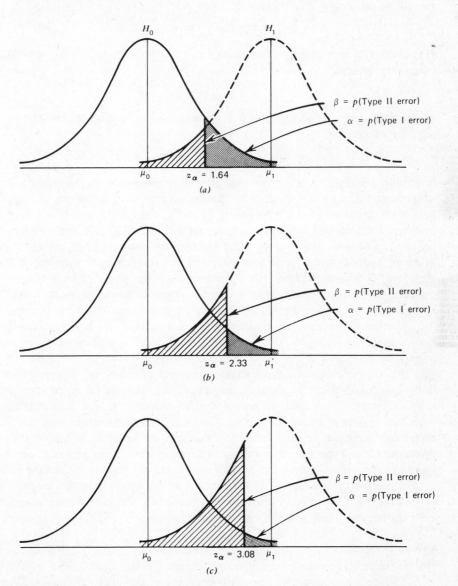

Figure 11.1   Relationship of critical value to probability of Type II error.

We see in Figure 11.1 that β increases as z_α becomes progressively larger. This should not be difficult to understand. We know that in the present example

$$\bar{x}_c = 40(z_\alpha) + 1300$$

Thus, any increase in z_α must be accompanied by a corresponding increase in $\bar{x}_c$, Furthermore, we know that according to our rejection rule the alternative hypothesis will be rejected if the observed sample mean $\bar{x}$ is less than $\bar{x}_c$. That is. H_1 will be rejected if $\bar{x}$ falls in the interval

$$\{\bar{x}\,|\,\bar{x} < \bar{x}_c\}$$

However, if $\bar{x}_c$ increases as a function of z_α , any increase in z_α must necessarily extend the interval

$$\{\bar{x}\,|\,\bar{x} < \bar{x}_c\}$$

upward, thereby including values of *increasing* probability under H_1. As z_α increases, therefore,

$$p(\bar{x} < \bar{x}_c\,|\ H_1 \text{ correct}) = \beta$$

must also increase.

In contrast, we know from prior discussion that as z_α increases, the significance level α *decreases*. In the present example, then, we see that α and β are inversely related, that any manipulation of the rejection rule that reduces the probability of one type of error must increase the probability of the other type of error. Indeed, this is generally true, and in establishing a significance level, therefore, it is not sufficient that the experimenter consider only the consequences that would result from a Type I error. Instead, he must weigh the consequences of a Type I error against the consequences of a Type II error.

These considerations are mathematically formalized in a branch of statistics known as *decision theory*. The fundamentals of decision theory were developed largely by economists, and most decision theory procedures therefore require that the consequences of any conclusion, correct or incorrect, be expressible in dollars, kilowatt-hours of power consumption, man-days of labor, or some other economic unit. When such an assignment of values is indeed meaningful, the investigator may compute the *expected value* of every conclusion that may arise under any particular rejection rule. For example, we know by definition that the *probability* of a Type I error is the significance level α. If, in addition, we let C_I represent the *cost* of such an error, then over many repetitions of the experiment, every rejection (correct as well as incorrect) of H_0 would incur an *average* loss equal to $\alpha(C_I)$. This quantity must therefore represent the *expected loss* associated with the rejection rule when H_0 is correct. That is,

$$\alpha(C_I) = E(\text{loss}\,|\,H_0 \text{ correct})$$

for the rejection rule under consideration. Once the significance level of the test is established, moreover, we may also compute β, the probability of a Type II error. And, if the cost of a Type II error is represented as C_{II}, the average loss associated with every decision to reject H_1 must be $\beta(C_{II})$. For this particular rejection rule, then,

$$\beta(C_{II}) = E(\text{loss} \mid H_1 \text{ correct})$$

The larger of these two values is, of course, the maximum loss expected under the rejection rule in question. In this regard, one decision theory strategy is to compute the expected losses for every rejection rule under consideration and to select the rule with the *smallest* maximum expected loss. To illustrate this *minimax loss* criterion, let us assign some arbitrary values to the possible errors that our anthropologist might make. It will be recalled that his test hypothesis is based on a theory that is widely accepted among other anthropologists. In this regard, Gallileo, Freud, and even Einstein discovered that the intellectual community is very intolerant of those who challenge scientific orthodoxy. We shall therefore suppose that if he rejects this hypothesis *incorrectly* he will be professionally discredited, and the agency that is sponsoring his research will not renew his $10,000 grant. The cost of a Type I error would therefore be

$$C_I = \$10,000$$

His alternative hypothesis, however, is based on a somewhat revolutionary theory, and if it is confirmed, a well-known publisher has agreed to advance him $20,000 for the rights to his findings. (He requested $650,000, but his editor informed him that such sums were available only for exclusive biographies of famous industrialists, such as Scrooge McDuck.) If he *incorrectly* rejects H_1, therefore, he stands to lose $20,000. Thus,

$$C_{II} = \$20,000$$

Now, let us imagine that he is considering rejection rules defined by significance levels of .01, .05, and .10. For the sake of illustration, we shall assume that he collects N skulls and that based on this sample size, the values of β for the three rules are .09, .04, and .02 (Table 11.1).

Table 11.1 Error Probabilities for Three Rejection Rules

Rule	p(Type I Error)	p(Type II Error)
I	$\alpha = .01$	$\beta = .09$
II	$\alpha = .05$	$\beta = .04$
III	$\alpha = .10$	$\beta = .02$

On this basis, we may compute the expected losses for each of his three rejection rules (Table 11.2).

The maximum loss expected under each rejection rule is indicated with an asterisk (*), and we see that the smallest of the three is $800, generated by Rule II. By the minimax loss criterion, therefore, our experimenter should select the

Table 11.2 Expected Losses for Three Rejection Rules

Rule	Expected Loss Given That	
	H_0 Correct	H_1 Correct
I	.01($10,000) = $ 100	.09($20,000) = $1,800*
II	.05($10,000) = $ 500	.04($20,000) = $ 800*
III	.10($10,000) = $1,000*	.02($20,000) = $ 400

second rule, in which the level of significance is set equal to .05 and the probability of a Type II error is .04.

The reader should understand that minimax loss is not the only criterion prescribed by decision theory. Indeed, it is one of the more conservative strategies, because it does not consider the payoffs that may be associated with *correct* conclusions. Whether the criterion considers just the cost or both costs and payoffs, however, such consequences must be assigned numerical values. Unfortunately, this requirement is unrealistic in most scientific work, and the applicability of formal decision theory is therefore restricted largely to economic decision making. Even though a scientist can seldom assign an *exact* cost to the results of an incorrect conclusion, however, he may often be able to state which of his two possible errors would be the *more* serious. This is particularly true in *applied* research. Suppose, for example, that a pharmaceutical company has developed a new tranquilizer and is conducting a study to determine whether or not the drug may be safely administered to pregnant women. In this regard, the company has two hypotheses:

H_A: Drug T is safe for pregnant women
H_B: Drug T causes birth defects

If H_A is rejected, the company is required by law to warn consumers that the drug is contraindicated for pregnant women. If the hypothesis is rejected incorrectly, therefore, they stand to lose some sales, but such a loss could probably be minimized by recommending another of their own products as an alternative. If, in contrast, the company incorrectly rejects H_B, the error could have tragic consequences for hundreds or thousands of children and their parents. Under these sorts of circumstances, the customary statistical practice is to make H_B the test hypothesis and to choose a very conservative significance level (say, $\alpha = $ .005), letting β fall where it may.

Even in *basic* research, as opposed to *applied* research, the scientist can occasionally assess the relative costs of his two potential errors. Sometimes, for example, a scientist will design a "crucial experiment" much like our artificial anthropology study, in which two competing theories generate hypotheses for which the *same* body of data will unequivocally disconfirm one or the other. In this regard, scientists are generally a conservative lot, and if one of the two theories is an "orthodox" theory and the other a "maverick" theory, it is customarily considered more important to avoid incorrect rejection of the ortho-dox theory. (Unfortunately, the more important the implications of the two theories, the greater the reluctance of the scientific community to abandon established thought.) Generally, however, competing theories, especially in the behavioral sciences, are rather like Napoleon's army and Nelson's navy. Until Waterloo there was simply no common ground on which Continental and British forces could be tested against one another. Thus, basic research produces relatively few crucial experiments, and the consequences of errors are generally in terms of time, thought, and personal effort. When such intangibles lie in the balance of our conclusions, the best rejection rule would be one in which the two errors are equally probable. The basic principles involved in establishing such rules are introduced in the next section, but actual computational procedures are deferred until later chapters.

3. Data Collection: Experimental Design and Sample Size

As we have indicated on several occasions, inferential statistics finds applica-tion in a wide variety of scientific disciplines. Thus, it is generally immaterial whether a particular random variable represents biological, anthropological, psychological, economic, or geographic observations. We shall see in later chapters, however, that it is sometimes necessary to classify data in terms of the *research methods* by which they were acquired. Traditional research method-ologists, for example, distinguish between the results of *experiments* and the results of *measurements*. The defining characteristic of an experiment is that the investigator manipulates some independent variable and observes the effects of his manipulations on some dependent variable. Thus, the psychologist who com-pares the maze-running performance of rats that have been deprived of food for 12 hours with a group that has been fed ad lib or the psychiatrist who administers different concentrations of vitamin B6 to two groups of austistic children and then compares their performance on a verbal learning task is conducting an experi-ment.

The investigator conducting a measurement study, in contrast, does not establish group differences by experimental manipulation. Instead, he *selects* his groups on the basis of some attribute that his subjects *already* possess. The anthropologist in our example, for instance, decided to examine skulls found in the Scandinavian Penninsula. He certainly exercised no control over the geo-graphic location of his potential specimens and was therefore conducting a measurement study. Similarly, a statistics teacher who tests the hypothesis that men perform better on statistics tests than women has no experimental control over his criterion variable; the sex of any subject is established long before he or she participates in the research. Strictly speaking, therefore, the teacher, like the anthropologist, is conducting a measurement study, not an experiment.

In recent years, *computer simulation* has emerged as another important source of data. At present, this method is becoming popular among urban sociologists and human ecologists who are using simulation techniques to formulate and test hypotheses about the long-term effects of resource depeletion, proposed transportation systems, technological developments, and social services.

a. Experimental Design. Whatever his method of data collection, however, the scientist is always obliged to assume that his observations reflect the actual state of nature. Remember, data exist in the real world, but theories and models exist only in the mind of the experimenter. And unless we are studying the thought processes involved in theory building, we are primarily concerned with the real world and not the mind of the experimenter. If we are to place such absolute reliance on our data, however, we must exert every effort to ensure that they are as representative as possible of the population(s) about which we hope to draw a conclusion. In this regard, a branch of statistics known as experimental design prescribes a number of techniques intended to minimize the sampling error to which all data are subject. In general, these techniques are more appropriate to data obtained by experiments than data that result from measurement. We shall therefore digress from our anthropology example and examine a study to which these considerations are more applicable. Suppose in this regard, that a psychologist in interested in the effects of tetrahydrocannabinol (THC), the active chemical agent in marijuana, on color perception. He therefore proposes to administer the drug orally to a group of subjects who will then perform a series of color discrimination tasks.

Experimental and Control Groups. If the experimenter simply administered THC to a single group and measured their discrimination performance, there would be no way to compare his results with the performance of individuals who were *not* under the influence of THC, and he could therefore draw no conclusions about the effects of the drug. Ordinarily, therefore, the experimenter would test at least *two* groups, one of which received the drug and the other of which received either a placebo or nothing at all. In all other respects, of course, the experiment would be conducted under identical conditions for both groups. The group that is subjected to the experimental treatment is called the experimental group, and the group that receives no treatment (or no treatment that is relevant to the experimental task) is called the control group.

Randomization. Let us further suppose that the experimenter has evidence indicating that under the influence of THC, experienced marijuana smokers differ significantly from naive subjects in the performance of many perceptual tasks. It is therefore important that users and nonusers be divided as equally as possible between his two groups. Because marijuana is illegal, however, he is reluctant to inquire about his subjects' use of the drug, and without such information he could not control this factor by simply assigning equal numbers of users and nonusers to each group. If, however, his assignment of subjects to the two groups were *independent* of the subjects' past use of marijuana, he could depend on *random chance* to achieve the even distribution he desires. It would not be advisable, therefore, for the experimenter to ask his subjects to volunteer for one or the other group; such a procedure would virtually guarantee a disproportionate number of marijuana users in his experimental group. Indeed, *any* systematic criterion by which he assigns subjects to his two groups might be somehow related to marijuana use and might therefore introduce spurious differences between his two groups.

His safest strategy, then, would be to make group assignments entirely at random. He could, for example, ask each subject to toss a coin and (appropriately) assign all "heads" to the experimental group and all "tails" to the control group. Or, he might pair each subject's name with a number drawn from a table of random digits and assign odd-numbered subjects to one group and even-numbered subjects to the other group. As defined in Chapter Four, two groups thus selected would constitute *random samples* drawn from the same parent population, and he could therefore assume that the population distribution of any random variable would be reflected with equal fidelity in both groups. Thus, the proportion of marijuana users in the experimental group should not be significantly different from the proportion in the control group. Differences in past experience with marijuana may still contribute to variability in color discrimination, but they will contribute only to performance differences *within* the groups, not to differences *between groups*. Thus, randomization serves to convert possible *systematic* variability into *random* variability.

Matching. Experience with marijuana is not, of course, the only variable that might conceivably influence group performance on his perceptual discrimination task. A subject's age and sex, for example, might effect either his response to the drug or his performance on the task irrespective of the drug. The experimenter has access to these variables, however, and can therefore balance their effects of by matching rather than by relying on randomization. The possible influence of sex, for example, could be eliminated by selecting the same ratio of men to women for both groups. He could use a similar, although more cumbersome, method to assure that the distribution of age was also identical for both groups.

With enough subjects, he might even be able to match his groups individual by individual instead of on the basis of distribution properties. That is, for each 19-year-old man in the experimental groups, he would assign a 19-year-old man to the control group; for each 27-year old woman in one group, he would assign a 27-year-old woman to the other group, and so forth.

Like randomization, matching serves to eliminate the influence of variables other than the experimental treatment that might introduce systematic differences between experimental subjects and control subjects. The essential distinction is that matching depends on systematic group assignment rather than the laws of probability to assure that such variables are identically distributed in the two groups.

Subjects as Own Controls. A more sophisticated method of avoiding spurious group differences is to use the *same* subjects in both the control condition and in the experimental condition. This use of subjects as their own controls assures that all variables are identical in both experimental and control subjects and therefore eliminates possible sources of systematic variability that the experimenter might not anticipate or be able to control by matching. Unfortunately, it introduces a new set of problems. Performance on any perceptual discrimination task, for example, is likely to improve with practice. Thus, if the control trials follow the experimental trials and if the performance is significantly better in the control condition, this result may as easily be attributed to practice as to the detrimental effects of THC. To eliminate the possibility of such practice effects, therefore, the subjects would be randomly separated into two groups. One group would perform the task first in the experimental condition and then in the control condition. For the second group, this order would be reversed. A *counterbalanced* design of this sort, however, requires that the entire experiment be conducted twice. The investigator must therefore be certain that *both* sets of experimental trials and both sets of control trials are administered under identical conditions. This would always include such obvious precautions as giving the same instructions, following the same procedures, and using the same apparatus. In some experiments, however, it might also include control of such subtle variables as the time of day that the experiment is administered, the sex of the experimenter, and even the temperature of the room.

b. Sample Size. The principles of experimental design emphasize consideration of *how* data are to be collected. In addition, however, the scientist must also consider *how much* data should be collected, because the *number* of observations in his sample(s) will bear heavily on both the results and meaningfulness of his statistical test. At the end of our discussion of rejection rules, the reader was probably left with the impression that β is determined entirely by the critical value of the test statistic and ultimately, therefore, by the significance level α. As we shall soon discover, however, the probability of a Type II error also depends on the variability of the test statistic. Furthermore, the variability of *every* test statistic developed in Part Three is a function of sample size. For some test statistics, the interrelationships among sample size, variability, and β are obscured by mathematical transformations. For tests based on the sample mean, however, the relationships are quite clear, and our anthropology example is therefore an exceptionally good vehicle for introducing these considerations.

By now, the reader should be well aware that the standard error of the mean $\sigma_{\bar{x}}$ is equal to

$$\frac{\sigma}{\sqrt{N}}$$

In this regard, both hypotheses in our anthropology example specified that $\sigma = 120$ cubic centimeters. Furthermore, much of our discussion of test statistics and rejection rules was based on a hypothetical sample of nine observations. Under either hypothesis, therefore, we have thus far assumed implicitly that $\sigma_{\bar{x}}$ equals

$$\frac{120}{\sqrt{9}} = \frac{120}{3}$$

$$= 40$$

It should be obvious, however, that for any sample size *other than nine*, the standard error of the mean would assume a different value, as would the results of any calculation performed in the last section in which $\sigma_{\bar{x}}$ was a factor.

Suppose, for example, that our anthropologist were able to find only four skulls. In this circumstance $\sigma_{\bar{x}}$ would equal

$$\frac{120}{\sqrt{4}} = \frac{120}{2}$$

$$= 60$$

Under the test hypothesis, then, the standardized value of any observed mean $\bar{x}$ would be

$$z_{\bar{x}} = \frac{\bar{x} - 1300}{60}$$

for any standardized value z, conversely, the corresponding sample mean would equal

$$z(60) + 1300$$

Suppose, now, that the anthropologist has decided in the interests of scientific conservatism that he should not reject the (orthodox) test hypothesis if his probability of committing a Type I error is greater than one in a hundred. Thus, he sets $\alpha = .01$. Furthermore, μ_1 is greater than μ_0, and his critical region should therefore include values in the *upper* extreme of $Z_{\bar{x}}$. On the basis of these considerations, he defines his critical region as

$$\{z \mid z \geq 2.33\}$$

Thus, his critical value is

$$z_\alpha = 2.33$$

and the corresponding value of the sample mean $\bar{x}_c$ must equal

$$2.33(60) + 1300 = 139.80 + 1300$$

$$= 1439.80$$

which is considerably larger than $\bar{x}_c = 1393.20$, which we calculated earlier on the basis of $N = 9$ observations.

Recall, in this regard, that for the present example p(Type II error) is equal to the cumulative probability of $\bar{x}_c$ under the distribution specified in the alternative hypothesis. That is,

$$\beta = F(z_\beta)$$

where z_β is computed by standardizing $\bar{x}_c$ with $\mu = 1500$ and $\sigma_{\bar{x}} = 120/\sqrt{N}$. For a sample of four observations, then,

$$z_\beta = \frac{1439.80 - 1500}{120/\sqrt{4}}$$

$$= \frac{1439.80 - 1500}{60}$$

$$= -1.00$$

and according to Table **C**

$$F(-1.00) = .16 \qquad \text{(approx.)}$$

We see, therefore, that for $\alpha = .01$ and a sample of four observations, the probability of a Type II error is .16, in contrast to $\beta = .0038$, which we calculated on the basis of $N = 9$.

To further explore the relationships among N, $\sigma_{\bar{x}}$, and β, let us imagine that our anthropologist was extremely fortunate and stumbled on an ice-age burial site that yielded 25 skulls. Thus, $N = 25$, and under the test hypothesis

$$\sigma_{\bar{x}} = \frac{120}{\sqrt{25}}$$

$$= 24$$

For $z_\alpha = 2.33$, therefore,

$$\bar{x}_c = 2.33(24) + 1300$$

$$= 55.92 + 1300$$

$$= 1355.92$$

and

$$z_\beta = \frac{1355.92 - 1500}{120/\sqrt{25}}$$

$$= \frac{1355.92 - 1500}{24}$$

$$= -6.00$$

Thus, for $N = 25$,

$$\beta = F(-6.00)$$

Table **C** has no entries for z-values less than -3.0, but according to Table C of Bailey (1971), the cumulative probability of -6.00 cannot exceed .0000. Hence

$$\beta = F(z = -6.00)$$

$$= .0000 \ldots$$

In the last few paragraphs we have considered the probability of a Type II error associated with a significance level $\alpha = .01$ for three different sample sizes, 4, 9, and 25. These results are summarized in Figure 11.2. As before, the solid

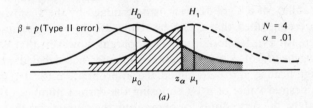

(a)

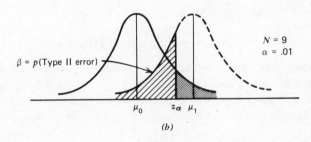

(b)

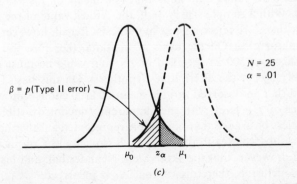

(c)

Figure 11.2   Relationship of sample size to probability of Type II error.

curve represents the distribution of $\overline{X}$ specified in the test hypothesis, and the shaded portion of each solid curve is the probability of a Type I error, $\alpha = .01$. The dotted curve in each pair represents the distribution of $\overline{X}$ specified in the alternative hypothesis, and the hatched portion of this curve represents β. It should be apparent from this illustration that for a fixed level of significance, the probability of a Type II error *decreases* as the sample size *increases*. As indicated earlier, this is because $\sigma_{\overline{X}}$ decreases with increased sample size, and in this regard, the reader will observe that our three distributions of $\overline{X}$ become progressively narrower as N increases.

In our discussion of rejection rules we demonstrated that for any *particular* sample size, the probability of a Type II error is determined by the significance level, and we therefore concluded that β can be reduced only by changing the rejection rule and increasing α. It should now be apparent, however, that β can be manipulated *independently* of α. Once the experimenter has decided on an acceptable level of significance and determined the appropriate critical region, he may establish *any* desired value of β by choosing the correct number of observations. Unfortunately, the procedures for computing the sample size that will yield a specific value of β are not the same for all test statistics. We shall therefore develop these procedures as we introduce various test statistics throughout the remainder of Part Three.

4. Conclusions: Rejecting Hypotheses versus Accepting Hypotheses

Throughout this chapter, the reader must have occasionally wondered why statistical conclusions were always discussed in terms of *rejecting* hypotheses and never in terms of *accepting* hypotheses. The basis for this convention becomes apparent if we again consider our anthropologist's two simple hypotheses,

$$H_0: \mu = \mu_0 = 1300$$

$$H_1: \mu = \mu_1 = 1500$$

and define some arbitrary critical value, such as $z_\alpha = 2.33$. Furthermore, let us assume that he collects nine skulls and finds that the mean of his sample is 1400 cubic centimeters. With a sample size $N = 9$, the critical value of the mean is 1393.20, and he must therefore reject his test hypothesis. Recall, however, that $\overline{X}$ is an unbiased estimator of μ. On the basis of his data, therefore the most likely value of μ is 1400, not 1500 as specified in H_1, and while he must reject H_0 he cannot reasonably accept the alternative hypothesis. On this basis, however, it might be argued that an observed sample mean of 1500 cubic centimeters *would* justify acceptance of H_1. Even with such apparently incontrovertible evidence, it could nonetheless be argued that his sample represented another heretofore unclassified human subspecies with the same skull capacity as Cro-magnon. It could *not* be claimed, however, that the skulls were Neanderthal, and his only completely defensible conclusion, therefore, is that H_0 is incorrect.

In general, then, disconfirmatory evidence is unequivocal, but confirmatory data are always subject to myriad interpretations. Thus, scientists are generally reluctant to accept statistical hypotheses and are frequently, therefore, in the curious position of testing the very hypotheses they hope to reject. This means, of course, that the greater the probability of rejecting H_0 incorrectly, the greater the likelihood that the scientist's data will yield the conclusion he desires. To avoid biasing the test in favor of his own expectations, therefore, the consciencious investigator seldom permits α to exceed .05 even when a Type I error would have no substantive consequences. Unfortunately, we shall see in the last section of this chapter that simply establishing a conservative level of significance is no real guarantee of an equitable statistical test.

C. Evaluation of the Statistical Test

1. The Concept of Power

We know from our discussion of statistical errors that

$$\beta = p(\text{rejecting } H_1 | H_1 \text{ correct})$$

That is, β is the probability that our statistical test will *fail* to detect that the population parameter under study is equal to θ_1, the value specified in the alternative hypothesis, and not θ_0, the parameter value specified in the test hypothesis. In this regard, it should be understood that the test hypothesis and alternative hypothesis typically represent mutually exclusive states of nature and that most rejection rules therefore stipulate that one of the two must be rejected. Thus, if we do *not* reject H_1, we *must* reject H_0. Assuming that H_1 is in fact correct, therefore, the probability of rejecting the *test* hypothesis must equal $1 - \beta$. That is,

$$1 - \beta = p(\text{rejecting } H_0 | H_1 \text{ correct})$$

Thus, $1 - \beta$ is the probability that our test *will* detect that our population parameter is equal to θ_1. This value is therefore a measure of the sensitivity of our test and, like the sensitivity of an optical microscope, is defined as *power*.

It should be apparent that if power $= 1 - \beta$, anything that *reduces* β will *increase* the power of the test. In our introduction to decision theory, for instance, we established that β decreases as α increases. It must therefore be true that a test may be made more powerful by increasing the level of significance. In our discussion of data collection we further established that β may be reduced by increasing sample size. Conversely, statistical tests based on large samples are more powerful than tests based on small samples.

2. The Power of a Test Against a Composite Alternative Hypothesis

In our anthropology example H_1 was a simple hypothesis. That is, the alternative hypothesis specified a single value for the mean of Scandinavian skull capacities,

$\mu_1 = 1500$. Once the significance level and sample size were established, therefore, the value of

$$z_\beta = \frac{\bar{x}_c - 1500}{\sigma_{\bar{x}}}$$

was completely determined, as was

$$\beta = F(z_\beta)$$

For a simple hypothesis, then, β assumes a single, fixed value. When H_1 is a *composite* hypothesis, however, β will assume a different value for every possible parameter value θ_1 that satisfies the conditions specified in the alternative. Suppose, for example, that our anthropologist had proposed the following alternative hypothesis:

$$H_1: \mu > \mu_0 = 1300$$

In this situation, μ_1 may assume any value greater than 1300 cubic centimeters, and

$$z_\beta = \frac{\bar{x}_c - \mu_1}{\sigma_{\bar{x}}}$$

and its cumulative probability $F(z_\beta) = \beta$ could be meaningfully computed for every such value. This is demonstrated in the Figure 11.3. The solid curve in each pair represents the distribution of $\overline{X}$ under the test hypothesis, and for purposes of illustration, we have assumed in all cases $\alpha = .01$ and $N = 9$. Thus, $z_\alpha = 2.33$ and $\bar{x}_c = 1393.20$. The three dotted curves represent the distributions of $\overline{X}$ for $\mu_1 = 1360$, $\mu_1 = 1420$, and $\mu_1 = 1480$, all of which satisfy the composite alternative hypothesis $\mu_1 > 1300$.

The hatched portion of each dotted curve represents the value of β for the particular alternative value μ_1 under consideration, and all area not identified by hatching must therefore represent $1 - \beta$, the power of the test against μ_1. It should be apparent that as μ_1 becomes larger, the probability of a Type II error, β, decreases and, conversely, the power increases. This latter result may be illustrated more comprehensively by plotting the power of the anthropologist's statistical test as a function of mean values μ_1 that satisfy the alternative hypothesis. In the *power function* graphed in Figure 11.4, power is represented on the vertical axis and, like any probability, extends from 0 to 1.0. The horizontal axis includes the set of values $\{\mu_1 | \mu_1 > 1300\}$, which the population mean may assume under the alternative hypothesis.

At the beginning of this section, we indicated that the power of a statistical test against a simple alternative hypothesis may be increased by raising the level of significance. This also holds true for a composite alternative hypothesis. To illustrate this, we have plotted the power functions for three tests of our fictitious anthropology data based on $N = 9$ and significance levels of .001, .01, and .05 (Figure 11.5). It is apparent that as α increases, the function is displaced to the left and that the power against any *particular* alternative mean μ_1 is therefore always greatest for the test with the largest significance level.

254

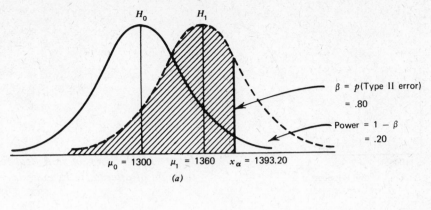

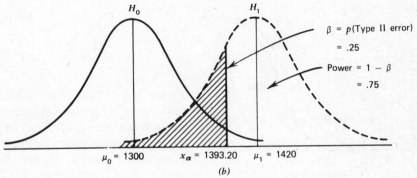

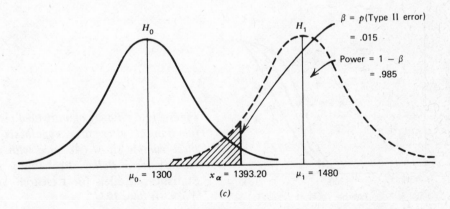

Figure 11.3 Power of statistical test against three alternative values of μ.

We also indicated earlier that the power of a statistical test is increased by increasing the sample size N. Figure 11.6 presents power functions for tests of the same hypothesis based on hypothetical samples of 4, 9, and 25 observations. In all cases, α is assumed equal to .01 and, for a fixed level of significance, we see that the test based on the largest N is uniformly the most powerful of the three.

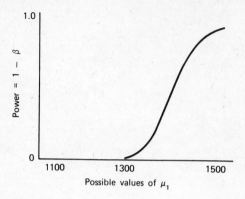

Figure 11.4   Power of statistical test as function of alternative hypothesis.

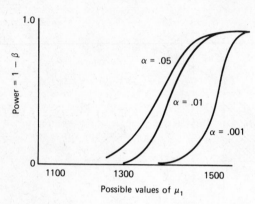

Figure 11.5   Power of statistical test as function of alternative hypothesis for three levels of significance. (Adapted with permission from Bailey, Probability and Statistics: Models for Research, *New York: Wiley, 1971.)*

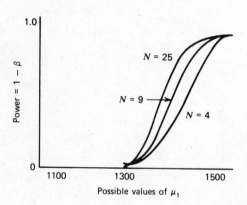

Figure 11.6   Power of statistical test as function of alternative hypothesis for three sample sizes. (Adapted with permission from Bailey, Probability and Statistics: Models for Research, *New York: Wiley, 1971.)*

In all of the functions illustrated above, it may be seen that the power of our test against possible mean values less than 1300 cubic centimeters is negligible. (Indeed, the reader should satisfy himself that the power of any of our tests against such a mean must be less than α.) This is because the composite alternative hypothesis under consideration was *directional*; it specified that if H_1 is correct, the population mean should be *larger* than some critical value, and our critical region was therefore confined to the upper tail of the curve. Frequently,

Hypothesis Testing

however, a scientist is unable to specify the direction of his alternative, and in such instances H_0 must be rejected if the observed test statistic is *either* much larger *or* much smaller than its expected value under the test hypothesis. If, for example, the alternative in our anthropology example were

$$H_1\colon \mu \neq \mu_0 = 1300$$

the experimenter would want a test by which he would be as likely to reject H_0 if his population mean were actually equal to, say, 1100 cubic centimeters as he would if the population mean were 1500 cubic centimeters. That is, he would want a test that was *equally powerful* against both such alternative values.

The appropriate rejection rule for such a test would therefore require *two* critical regions, one of which included Z-values in the upper tail of the distribution and the other of which included values in the lower tail. It should be noted, however, that if each critical region were defined by an interval Δz_α, where

$$p(z_{\bar{x}} \, \epsilon \, \Delta z_\alpha \,|\, H_0 \text{ correct}) = \alpha$$

the total probability of a Type I error would necessarily be 2α. Ordinarily, therefore, the desired probability of a Type I error is apportioned equally to the two critical regions, so that under the distribution specified in H_0 the probability associated with each interval is $\alpha/2$. Thus, if α were equal to .01, the anthropologist would want a critical region with a probability of .005 in each tail of the normal curve. In the upper tail, his critical value would therefore equal $z_{\alpha/2}$ where

$$p(z_{\bar{x}} \geq z_{\alpha/2} \,|\, H_0 \text{ correct}) = .005$$

This implies, of course, that if the test hypothesis is correct,

$$p(z_{\bar{x}} \leq z_{\alpha/2}) = .995$$

By definition this is the cumulative probability of $z_{\alpha/2}$ and from Table **C**, the Z-value with a cumulative probability must nearly equal to .995 is 2.57, for which

$$F(2.57) = .9949$$

If H_0 is correct, therefore,

$$p(z_{\bar{x}} \geq 2.57) = .005 \qquad \text{(approx.)}$$

Because the normal curve is symmetrical, moreover, it must also be true that

$$p(z \leq -2.57) = .005$$

This may be confirmed by referring again to Table **C**, which shows that

$$F(-2.57) = .0051$$

Thus

$$p(z_{\bar{x}} \leq -2.57 \,|\, H_0 \text{ correct}) = .005 \qquad \text{(approx.)}$$

The appropriate two-tailed rejection rule for $\alpha = .01$ would therefore require that H_0 be rejected if the observed test statistic $z_{\bar{x}}$ were either greater than 2.57 or less than -2.57.

In Figure 11.7, the power function for this test is superimposed on the power function for a one-tailed test with the same level of significance ($\alpha = .01$) and the same number of observations ($N = 9$).

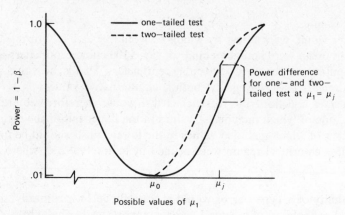

Figure 11.7   Power of one- and two-tailed tests as function of alternative hypothesis. (Adapted with permission from Bailey, Probability and Statistics: Models for Research, *New York: Wiley, 1971.*

Two important conclusions can be drawn from Figure 11.7. First, the two-tailed test does indeed provide power against alternative values less than 1300. Moreover, because the two critical values are of equal absolute value, the function is symmetrical around μ_0 and is therefore equally powerful against alternatives that are equidistant from 1300. However, it must also be noted that for any particular value of μ_1 *greater* than 1300, the one-tailed test is always more powerful. If the experimenter has sufficient information to formulate a directional alternative, therefore, his statistical test will be uniformly more powerful than a nondirectional test based on the same values of α and N. Indeed, it is generally true that the power of a statistical test is directly related to the number of assumptions an experimenter can make about his data.

3. Scientific versus Statistical Significance

Throughout our anthropology example we have discussed rejection rules by which H_0 is rejected if the observed test statistic $z_{\bar{x}}$ exceeds some critical value z_α. It must be emphasized, however, that statistical significance does not derive simply from the numerical magnitude of the test statistic. It depends, rather, on the difference between the observed value and the value to be expected if H_0 is correct. Thus, when we say that a particular result is statistically significant, we mean that the observed value of the test statistic is *significantly different* from its expected value under the test hypothesis. This was not always clear in our anthropology example because the statistic was generally expressed in standard form, and its expected value was therefore equal to zero. To further develop

258

this notion of a significant difference, let us therefore consider our rejection rule in terms of the unstandardized sample mean $\overline{X}$. If we define our critical value as $\bar{x}_c$, we know by definition that H_0 will be rejected if the observed sample mean $\bar{x}_o$ exceeds $\bar{x}_c$. That is, if

$$\bar{x}_o \geq \bar{x}_c$$

However, we also know that a positive constant may be subtracted from both sides of an inequality, so our rejection rule is unchanged if we say that H_0 will be abandoned if

$$\bar{x}_o - \mu_0 \geq \bar{x}_c - \mu_0$$

We have already established that if H_0 is correct, $E(\overline{X}) = \mu_0$, and in our present example, therefore $\bar{x}_o$ will be considered *significantly different* from its expectation under H_0 if and only if this difference exceeds $\bar{x}_c - \mu_0$.

In this regard, recall from our discussion of sample size that as N increases, $\bar{x}_c$ becomes smaller. Thus, any increase in sample size must also reduce the difference $\bar{x}_c - \mu_0$ required to reject the test hypothesis. Recall further that increasing N increases the power of the statistical test. It must therefore be true that the smaller the difference $\bar{x}_c - \mu_0$ the more powerful the test. Indeed, power is frequently interpreted as an index of how *small* a difference between the observed statistic t_o and its expected value under the test hypothesis $E(T | H_0 \text{ correct})$ may be regarded as statistically significant. Consequently, a scientist will often collect as many observations as possible in the belief that he is maximizing the sensitivity of his statistical test. The fallacy here is that while his test may become highly sensitive to small departures from θ_0, it may become *insensitive* to even large departures from θ_1, thereby making it unlikely that an *incorrect* alternative hypothesis will be rejected. To illustrate, let us suppose that our anthropologist had, as indicated in our original example, formulated a simple alternative hypothesis

$$H_1: \mu_{\text{S.P.}} = \mu_1 = 1500$$

and that he had stumbled on a veritable goldmine of ice-age remains that yielded him a sample of 100 skulls. His standard error of the mean would therefore be

$$\sigma_{\bar{x}} = \frac{120}{\sqrt{100}}$$

$$= \frac{120}{10}$$

$$= 12$$

If he sets $\alpha = .01$, then, $z_\alpha = 2.33$, and his critical value of $\overline{X}$ would be

$$\bar{x}_c = 2.33(12) + 1300$$

$$= 27.96 + 1300$$

$$= 1327.96$$

This means that he could reject his test hypothesis if he observed a sample mean of 1327.96 cubic centimeters and be nonetheless certain that his probability of a Type I error did not exceed .01. Please note, however, that 1327.96 cubic centimeters is actually closer to μ_0 than to μ_1. The anthropologist is therefore in the dubious position of rejecting H_0 *in favor of* H_1 on the basis of data that are even *less* likely under H_1 than under H_0. Indeed, $\bar{x}_c$ is *so unlikely* under H_1 that the rejection rule makes it almost impossible to reject H_1. If the alternative hypothesis is not rejected, of course, the experimenter must reject H_0, and under the present rule, therefore, rejection of H_0 is virtually assured. When we recall that a scientist customarily tests the hypothesis he hopes to reject we see that such a rejection rule biases the test in favor of the experimenter's expectations despite his establishment of a conservative significance level. To avoid making the alternative hypothesis a self-fulfilling prophesy, therefore, the experimenter must base his test on a smaller sample, thereby moving $\bar{x}_c$ closer to μ_1. By definition this reduces the power of the test and, conversely, increases β. Determination of the proper sample size must therefore depend on the importance of avoiding a Type II error. Once this is decided, the experimenter can establish an appropriate value of β and compute the corresponding value of N.

We have just demonstrated that establishment of a conservative significance level does not by itself guarantee an equitable statistical test. Because conventional rejection rules generally stipulate that one of the hypotheses *must* be rejected, the important consideration is not the magnitude of α, but the *relative* magnitudes of α and β. If, as is generally the case in basic scientific research, neither type of error is more serious than the other, the experimenter should simply set β equal to α. There will, of course, be occasions when one of the errors is substantially more serious than the other, but this should be decided before data are collected and α and β set accordingly.

Thus, far, our discussion of scientific and statistical significance has assumed that the anthropologist had sufficient information to formulate a simple alternative hypothesis. Let us now consider a more realistic situation, in which the investigator knows only that Cro-magnon skulls are generally larger than Neanderthal skulls. Under these conditions, his alternative hypothesis would be

$$H_1: \mu > \mu_0 = 1300$$

We shall assume once more that $\alpha = .01$, but we shall now suppose that $N = 900$. (*Note.* It is extremely improbable that 900 members of any human subspecies ever inhabited any geographic region the size of the Scandinavian Penninsula during the last ice age, but I am trying to make my point with minimal recourse to mathematics, so please excuse this lapse from realism.) For a sample of 900 observations, the standard error of the mean $\sigma_{\bar{x}}$ is equal to

$$\frac{120}{\sqrt{900}}$$

$$= \frac{120}{30}$$

$$= 4.0$$

Furthermore, for $\alpha = .01$, $z_\alpha = 2.33$, and

$$\bar{x}_c = 2.33(4) + 1300$$

$$= 9.32 + 1300$$

$$= 1309.32$$

If we now suppose that the mean capacity of his 900 skulls is 1310 cubic centimeters then on the basis of his observations the most likely value of his population mean is 1310. Had he formulated the *simple* alternative we discussed in the preceding example,

$$H_1: \mu = \mu_1 = 1500$$

it would be readily apparent that the population mean is probably closer to μ_0 than μ_1. However, under the present *composite* alternative, any value other than 1300 satisfies H_1, and it would therefore seem perfectly acceptable to reject H_0 in favor of H_1. Indeed, the size of his sample has generated a critical value $\bar{x}_c = 1309.32$, which leaves no other choice.

At this point, however, we must turn from statistical considerations to scientific considerations and ask if 10 cubic centimeters is really a difference that makes a difference. That is, is the difference between $\mu = 1310$ (our best present estimate) and $\mu_0 = 1300$ sufficient to discredit the *underlying theoretical basis for* H_0 that Neanderthal man inhabited the Scandinavian Penninsula during the last ice age? If we stop to consider that μ_0 necessarily represents *only* the population mean of Neanderthal skulls *yet discovered*, indeed a relatively small collection of which few are completely intact, we might have some reservations about rejecting H_0. If, in addition, we realize that the measurements on which μ_0 is based were probably recorded over many years by many different people using a variety of anthropometric techniques with differing degrees of skill, we should probably conclude that for all practical of scientific purposes, 1310 is indistinguishable from 1300 *even though it is significantly different at the .01 level.* (Indeed, the *most* appropriate conclusion might well be that the population of Neanderthal skull capacities is actually 1310 cubic centimeters. Alas, disillusioned reader, this is the *real* way of science.) Thus, the test has yielded a result that is *statistically* significant but not *scientifically* significant.

In this example, $\bar{x}_c - \mu_0$ was so small that H_0 could be rejected in favor of alternative values that were scientifically indistinguishable from μ_0. This situation occurs when the scientist formulates a composite alternative hypothesis and overpowers his test by making too many observations. To avoid this bind, therefore, he must reduce his power by choosing a smaller sample. However, computation of an appropriate sample size requires establishment of β, and β assumes a fixed value only for a *specific* alternative μ_1. If his test is to be scientifically meaningful, therefore, the experimenter *cannot* formulate a composite alternative hypothesis. Indeed, we submit that a scientist should *never* formulate a composite alternative if there is any possibility that a trivial result will be found statistically significant. Even in the absence of an alternative model, he should be able to determine how much μ must depart from μ_0 in order to constitute a *scientifically significant* difference. Once this is established, he may formulate a *synthetic* simple alternative hypothesis in the form

$$H_1: = \mu = \mu_1 = \mu_0 + \delta$$

where δ (lowercase Greek *delta*) is the smallest difference that he is willing to consider of sufficient scientific importance to abandon the theory under test. Once he has established his alternative, he may proceed to determine his value of β and compute the corresponding sample size. He will then be certain that the power of his test will not exceed $1 - \beta$ for any alternative μ_1 that lies between μ_0 and $\mu_0 + \delta$.

The alert reader will have recognized that this final section, Scientific versus Statistical Significance, really introduces no new issues. Instead, it supplements the four major topics discussed in the Foundations section. In the preceding paragraph, for example, we rounded out our treatment of *formulation* by suggesting that the investigator always cast a simple alternative hypothesis, even if it is synthetically derived from scientific and practical considerations instead of from a statistical model.

Similarly, the problems presented in connection with power are germane to our earlier discussion of *decisions*. In that section we developed the notion of *statistical* significance and therefore emphasized the role of α in the selection of a rejection rule. In the present section we point out that *scientific* significance is largely dependent on β. We therefore argue that if a statistically significant result is to be scientifically significant the experimenter must establish a rejection rule that yields specified values of *both* α and β.

Such a rule can be established, of course, only by controlling sample size, which is one aspect of *data collection*.

Finally, these precautions will assure the scientist that he has not permitted his expectations to influence his *conclusions* by conducting a test that is biased in favor of rejecting the test hypothesis.

D. Problems, Review Questions, and Exercises

Suppose you have a friend who always seems to win a coin toss, as long as it is his "lucky coin" and he calls the toss "heads." You therefore suspect that the coin is biased, but because he lets you call the toss about 85 percent of the time, the bias is unimportant unless the coin lands "heads" more than 85 percent of the time. You therefore decide to test the fairness of his coin by tossing it 10 times.

1. Formulation
   a. What is the "theory" under test?
   b. What model is generated by the theory?
   c. What are your hypotheses?

2. Decisions

    a. What are the appropriate units of observation?

    b. What is an appropriate test statistic?

In the event that you find his coin to be biased, let us suppose that you have decided to replace it with an uncirculated $10 gold piece, which costs about $90. If your decision is incorrect, therefore, the error will cost you $90. On the other hand, if you decide incorrectly that his coin is fair, you estimate that you will lose about $25 on sundry bets during the next year. In this regard, suppose that you are considering four possible rejection rules:

        Rule 1 If $x = 10$, reject H_0

        Rule 2 If $x \geq 9$, reject H_0

        Rule 3 If $x \geq 8$, reject H_0

        Rule 4 If $x \geq 7$, reject H_0

    c. Which rule should be used according to the minimax loss criterion?

3. Conclusions

    a. If the coin lands "heads" seven times, what conclusion is required by your rejection rule?

    b. What is the probability of seven heads under H_0?

    c. What is the probability of seven heads under H_1?

    d. What do **3a** and **3b** tell you about accepting H_0?

chapter twelve

testing
hypotheses
about means

Most of the preceding chapter was based on a single example, which involved the somewhat tedious efforts of a fictitious anthropologist to test an hypothesis about the mean capacity of ice-age skulls found in the Scandinavian Penninsula. Indeed, the majority of statistical inferences found in the literature of the behavioral and biological sciences involve one or another test of population means, and several of the most common techniques for conducting such tests are introduced in this chapter.

Many of the tests that the reader will encounter in the literature, or, indeed, may be called on to perform himself, involve only a single population. As was the case in our anthropology example, the test hypothesis in such instances takes the form

$$H_0: \mu = \mu_0$$

where μ_0 is a value specified by some theoretical model. The principles and techniques involved in testing one mean can be generalized to problems involving two population means, where the test hypothesis takes the form

$$H_0: \mu_1 = \mu_2$$

In problems of this sort, the experimenter is really concerned about the *difference* between the two population means, and his model need not actually specify values for either μ_1 or μ_2. Finally, the experimenter may be concerned about the means of many populations. However, testing hypotheses of the form

$$H_0: \mu_1 = \mu_2 = \ldots = \mu_J$$

is the purview of analysis of variance and is deferred until Chapter Fifteen after some necessary concepts are introduced in Chapters Thirteen and Fourteen. The content of the present chapter is therefore confined to tests of one or two means.

In Chapter Eleven we discussed four major aspects of hypothesis testing: formulation, decisions, data collection, and conclusions. These four steps will provide a convenient framework for much of the present chapter, but certain emphases will be apparent. We shall, for example, have almost nothing to say about statistical *conclusions*. In contrast, we shall devote considerable attention to *formulation* of hypotheses, but formulation of theories and models is not discussed. The various treatments of *data collection* are also selective. They focus almost entirely on sample size determination, and experimental design considerations receive only one or two passing comments. Similarly, the only *decision* to be considered in detail is selection of a test statistic. In this regard, recall from Chapter Eleven that computation of the test statistic $Z_{\bar{x}}$ involves the standard error of the mean $\sigma_{\bar{x}} = \sigma/\sqrt{N}$. This presented no difficulty in our anthropology example, because the value of σ was specified in the model, but in many research situations the standard deviation of the random variable is unknown. An obvious solution, of course, is to use the square root of the variance estimate

$$\hat{\sigma} = \sqrt{\frac{\sum_{i}^{N} (x_i - \bar{x})^2}{N - 1}}$$

Unfortunately, the sampling distribution of $\hat{\sigma}$ is extremely variable for small values of N, and a satisfactory estimate therefore requires a relatively large sample ($N \geq 30$). If σ is unknown, therefore, limitations on sample size may require the experimenter to use some test statistic other than $z_{\bar{x}}$. The techniques involved in testing hypotheses about means when σ is unknown (and sample size precludes estimation) are very similar to those employed when σ is known, but the principles and assumptions that underlie the two procedures are quite different, and they are therefore treated separately.

As indicated above, the appropriate choice of a test statistic may depend on sample size. The importance of sample size in this connection has led most traditional textbooks to distinguish between tests of means based on large samples and those based on small samples. This convenient distinction, however, assumes that the investigator will either set his sample size arbitrarily or collect as many observations as possible. In contrast, it is our position that the appropriate sample size should in many situations be determined by the experimenter on the basis of scientific significance. Thus, we necessarily draw the further distinction between *available* sample size and *computed* sample size.

A. Inferences About One Population Mean μ

Thus far in the present chapter and throughout the preceding chapter, we have indicated a test hypothesis about one population mean with the notation

$$H_0: \mu = \mu_0$$

We shall continue to use this notation, but in other books, the reader may occasionally find the test hypothesis represented as

$$H_0: \mu - \mu_0 = 0$$

By either convention, however, μ_0 assumes a single value, and H_0 must therefore be a simple hypothesis.

In contrast, we know that the alternative hypothesis may be simple or composite, and if composite the alternative may be either directional or nondirectional. In Chapter Eleven, however, we suggested that an experimenter is often well advised to cast a simple alternative even if he has no alternative model. In this connection, we defined δ as the minimal difference between μ_0 and the true population mean μ that the scientist would consider of sufficient scientific importance to abandon H_0 and suggested that a synthetic simple alternative could always be formulated in terms of μ_0 and δ. Recall, however, that δ was discussed in the context of a directional hypothesis in which $\mu_1 > \mu_0$, and it was therefore implicitly assumed that

$$\delta = \mu_1 - \mu_0$$

Had our alternative hypothesis been cast in the other direction, however, such that $\mu_1 \leq \mu_0$, our value of δ as given in the preceding expression would have been negative, and our alternative mean

$$\mu_1 = \mu_0 - \delta$$

would have been *larger* than μ_0. To avoid such problems, we shall always assume that δ is an *absolute* value, such as $|\mu_0 - \mu_1|$.

Although the δ notation was introduced to facilitate representation of *synthetic* simple alternative hypotheses, it is equally applicable to situations in which the experimenter has a *ready-made* simple alternative. However, where μ_1 is specified by the model rather than by the experimenter, we may not assume that δ represents the *smallest* departure from μ_0 of scientific interest to the experimenter. Instead, we should think of δ as indicating the *degree* of scientific significance associated with the difference between μ_0 and μ_1. In this regard, it should be pointed out that the scientific importance of any such difference depends largely on the dispersion of the random variable under consideration. In the anthropology example, for instance, the two simple hypotheses were

$$H_0: \mu = 1300$$

$$H_1: \mu = 1500$$

and δ was therefore equal to 200 cubic centimeters. However, our problem also stipulated that $\sigma = 120$ cubic centimeters. In *standard units*, therefore,

$$\delta = \frac{200}{120}$$

$$= 1.67$$

If, in comparison, the standard deviation of our random variable had been equal to 400 cubic centimeters, then the same two hypotheses would have generated $\delta = 0.50$ standard units. It should be apparent that a *statistically* significant result for which $\delta = 0.50$ is of less *scientific* significance that a result for which $\delta = 1.67$. Throughout this chapter, therefore δ is expressed in standard units instead of natural units. For tests of one mean, then, δ is formally defined as the absolute standardized difference between μ_0 and μ_1

$$\delta = \frac{|\mu_0 - \mu_1|}{\sigma}$$

where σ is the population standard deviation of the random variable under consideration and μ_1 is the alternative value of the mean, either formulated synthetically by the experimenter or specified by some theoretical model.

Even though δ is defined in standard units, hypotheses formulated about population means express both μ_0 and μ_1 in natural units. For any simple alternative hypothesis, synthetic or ready made, the alternative value μ_1 is therefore expressed in terms of μ_0 and the *unstandardized* difference

$$|\mu_0 - \mu_1| = \delta\sigma$$

For a simple nondirectional alternative, then,

$$H_1: \mu = \mu_1 = \mu_0 \pm \delta\sigma$$

268

For directional alternative hypothesis in which $\mu_1 > \mu_0$

$$H_1: \mu = \mu_1 = \mu_0 + \delta\sigma$$

and for a directional alternative in which $\mu_1 < \mu_0$

$$H_1: \mu = \mu_1 = \mu_0 - \delta\sigma$$

1. Testing Hypotheses About μ when σ^2 Is Known

In a spirit of loyalty to a worthy, if tiresome, friend, we shall develop the general methods involved in testing

$$H_0: \mu = \mu_0$$

when σ^2 is known in the framework of our well-worn anthropology example.

a. Formulation. We shall assume as before that the anthropologist is basing his test on a model that specifies that the population distribution of Neanderthal skull capacities has a mean of 1300 cubic centimeters and a standard deviation of 120 cubic centimeters. Thus

$$H_0: \mu = 1300$$

For purposes of our present discussion, however, we shall assume that his alternative model is incomplete and indicates only that the expected capacity of Cro-magnon skulls is larger than that of Neanderthal skulls (and that $\sigma = 120$). We shall further suppose that in the experimenter's best judgment, H_0 should be abandoned only if the true population mean of Scandinavian Penninsula skulls is at least $\frac{3}{4}$ of a standard unit larger than μ_0. Thus, δ equals 0.75 standard units, and his synthetic alternative hypothesis is therefore

$$\begin{aligned} H_1: \mu &= \mu_0 + \delta\sigma \\ &= 1300 + .75(120) \\ &= 1300 + 90 \\ &= 1390 \end{aligned}$$

b. Decisions. Once again, the appropriate units of observation are Scandinavian Penninsula ice-age skulls.

Concerning the appropriate test statistic, the reader may have noticed that neither model specified the *form* of the distribution of skull capacities. In point of fact, however, most continuous biological characteristics are determined by the additive effects of many independent polygenes and are therefore normally distributed. Even though it is not indicated in the model, then, the anthropologist may safely assume that Scandinavian Penninsula skull capacities are normally distributed. If X is normally distributed, we know that $\overline{X}$ must also be normally distributed and the distribution of $\overline{X}$ is therefore completely specified. As in Chapter Ten, therefore, the appropriate test statistic is

$$Z_{\overline{X}} = \frac{\overline{X} - \mu_0'}{\sigma_{\overline{X}}} = \frac{\overline{X} - \mu_0}{\sigma/\sqrt{N}}$$

It must be emphasized that $\overline{X}$ was assumed normally distributed under the Central Limit Theorem *irrespective of sample size* only because the anthropologist was willing to assume a normal distribution for X. If X is not normally distributed, sample size becomes a crucial consideration in the application of the CLT to the distribution of $\overline{X}$. In this regard, we indicated in Chapter Ten that the distribution of $\overline{X}$ will be approximately normal for samples even as small as $N = 10$, provided that the distribution of X is unimodal and approximately symmetrical. As the distribution of X departs from this configuration, however, one must obtain larger samples to justify the assumption that $\overline{X}$ is normally distributed. Thus, selection of a test statistic must occasionally be deferred until sample size is determined.

Finally, we shall suppose that the anthropologist wishes a rejection rule for which the probability of a Type I error is .01, that is, for which $\alpha = .01$. Because $\mu_1 \geq \mu_0$ he must therefore select a critical value z_α such that under the test hypothesis

$$p(z_{\overline{X}} \geq z_\alpha) = .01$$

and from Table **C** the appropriate value is $z = 2.33$.

In this study, the test hypothesis represents an orthodox theory and the alternative a maverick theory. Thus, the anthropologist feels that avoidance of a Type II error is somewhat less crucial than avoidance of a Type I error and consequently sets $\beta = .05$. His critical value must therefore be chosen such that under the alternative hypothesis

$$p(z_{\overline{X}} \leq z_\beta) = .05$$

From Table **C**, then, $z_\beta = -1.64$.

c. **Data Collection.** Because the anthropologist is conducting a measurement study rather than a formal experiment, few of the cannons of experimental design are applicable. However, the assumption that $\overline{X}$ is normally distributed relies on the Central Limit Theorem, which applies to samples of N *independent* observations. From Chapters Four and Eleven we know that independence is always guaranteed by random sampling, and we shall therefore permit our anthropologist the luxury of assuming that the skulls he finds represent a random sample of all the skulls in his population.

Moreover, if he is to conduct a test for which $\beta = .05$ against his (synthetic) alternative $\mu_1 = 1390$, he must determine the appropriate *number* of skulls to include in his random sample. In this regard, we know that

$$z_\alpha = \frac{\bar{x}_c - \mu_0}{\sigma_{\bar{x}}}$$

and that

$$\sigma_{\bar{x}} = \frac{\sigma}{\sqrt{N}}$$

Thus

$$z_\alpha \frac{\sigma}{\sqrt{N}} = \bar{x}_c - \mu_0$$

and

$$\bar{x}_c = \mu_0 + z_\alpha \left(\sigma / \sqrt{N} \right)$$

Under the distribution specified (albeit synthetically) in H_1, the standardized value of $\bar{x}_c$ is

$$z_\beta = \frac{\bar{x}_c - \mu_1}{\sigma_{\bar{x}}} = \frac{\bar{x}_c - \mu_1}{\sigma / \sqrt{N}}$$

Thus

$$z_\beta \frac{\sigma}{\sqrt{N}} = \bar{x}_c - \mu_1$$

and

$$\bar{x}_c = \mu_1 + z_\beta \left(\sigma / \sqrt{N} \right)$$

The critical value of the sample mean $\bar{x}_c$ is, of course, the same whether we ultimately standardize it under H_0 or H_1. Thus, our two expressions for $\bar{x}_c$ must be equal. That is,

$$\mu_0 + z_\alpha \left(\sigma / \sqrt{N} \right) = \mu_1 + z_\beta \left(\sigma / \sqrt{N} \right)$$

Hence

$$\mu_0 - \mu_1 = z_\beta \left(\sigma / \sqrt{N} \right) - z_\alpha \left(\sigma / \sqrt{N} \right)$$

$$= \frac{\sigma}{\sqrt{N}} (z_\beta - z_\alpha)$$

or

$$\sqrt{N} (\mu_0 - \mu_1) = \sigma (z_\beta - z_\alpha)$$

and

$$\sqrt{N} = \frac{\sigma (z_\beta - z_\alpha)}{\mu_0 - \mu_1}$$

Thus

$$N = \left[\frac{\sigma(z_\beta - z_\alpha)}{\mu_0 - \mu_1} \right]^2$$

In the present example, this expression becomes

$$N = \left[\frac{120(-1.64 - 2.33)}{1300 - 1390} \right]^2$$

$$= \left[\frac{120(-3.97)}{-90} \right]^2$$

$$= \frac{16(15.7609)}{9}$$

$$= 28.019$$

Thus, if our test is to yield $\beta = .05$ against $\mu_1 = 1390$, our anthropologist must collect exactly 28.019 skulls. In this study as in any other, however, the number of observations can assume only an integer value, and $N = 28.019$ must therefore be rounded to the nearest whole number, that is, $N = 28$. This will, of course, slightly alter the value of β. In general it may be assumed that rounding *down* will increase β and rounding *up* will reduce β. If the principal danger lies in overpowering the test, therefore, it is always safer to round down.

d. Conclusions. In accordance with his computations, the anthropologist stays in the field until he has collected 26 skulls and finds the mean of his sample to be 1360 cubic centimeters. Thus, the value of his test statistic is

$$z_{\bar{x}} = \frac{\bar{x}_0 - \mu_0}{\sigma / \sqrt{N}}$$

$$= \frac{1360 - 1300}{120 / \sqrt{28}}$$

$$= \frac{60}{22.68}$$

$$= 2.65$$

which exceeds his critical value of 2.33, and he therefore rejects his test hypothesis.

In the present example, our sample size was sufficient to justify the assumption that $\bar{X}$ is normally distributed simply on the basis of the Central Limit Theorem. For samples of fewer than 25 observations, however, this assumption, as indicated above, depends on the distribution of X. Thus, when X cannot be assumed normal, the experimenter may be forced to use a larger sample size N' than his computations indicate to be appropriate. This will, of course, decrease the value of β against all possible values of the population mean. To best evaluate the effect of this increase of power on the scientific significance of his test, the experimenter should compute the *effective* alternative mean μ_1' against which β is *now* equal to the value originally specified. With sample size now equal to N'

$$\bar{x}_c = \mu_0 + z_\alpha \frac{\sigma}{\sqrt{N'}}$$

Similarly

$$z_\beta = \frac{\bar{x}_c - \mu_1'}{\sigma/\sqrt{N'}}$$

and

$$\mu_1' = \bar{x}_c - \frac{z_\beta \sigma}{\sqrt{N'}}$$

He may then compute

$$\delta' = \frac{|\mu_0 - \mu_1'|}{\sigma}$$

If this *effective* difference is scientifically trivial, the experimenter may be reluctant to increase his sample size enough to justify the assumption that $\bar{X}$ is normally distributed. For cases such as this, statisticians have developed a body of *nonparametric* statistical techniques that require very few assumptions about the distribution of the random variable. Space precludes discussion of these techniques in the present text, but *Statistics Without Fear* by Nathaniel J. Ehrlich includes an introduction to nonparameterics for which the first 10 chapters of this book provide suitable preparation.

2. Testing Hypotheses About μ When σ^2 Is Not Known

In problems where σ^2 is unknown, the importance of sample size is even greater than in situations where σ^2 is known, and we shall therefore treat procedures for large and small samples separately.

Although

$$\hat{\sigma} = \sqrt{\frac{\sum_{i}^{N} (x_i - \bar{x})^2}{N - 1}}$$

is *not* an unbiased estimator of σ, it is, as indicated in the introduction to the chapter, reasonably efficient for $N \geq 30$. Thus, if the experimenter has a large number of available observations he may use these data to compute $\hat{\sigma}$ and use his estimate in place of σ to define δ and to compute

$$N = \left[\frac{\sigma (z_\beta - z_\alpha)}{\mu_0 - \mu_1} \right]^2$$

Once the sample size N is thus computed, he may select a *random subsample* of N observations from the larger group on which he based his estimate $\hat{\sigma}$. If his computed N is still greater than 25, he may assume from the CLT that $\overline{X}$ is normally distributed and may therefore use the test statistic

$$Z_{\overline{X}} = \frac{\overline{X} - \mu_0}{\sigma_{\overline{X}}}$$

substituting

$$\hat{\sigma}_{\overline{X}} = \frac{\hat{\sigma}}{\sqrt{N}}$$

for $\sigma_{\overline{X}}$. If, however, $N < 25$, the assumption that $\overline{X}$ is normally distributed again depends on the distribution of X. If the distribution of X departs too radically from a unimodal, symmetrical configuration to justify a normal approximation of $\overline{X}$ for his computed N, he must either increase his sample size or rely on nonparametric techniques.

When the available sample is less than 30 the problem is even more demanding. Even if X is normally distributed (or approximately so) the investigator cannot use the test statistic

$$Z_{\overline{X}} = \frac{\overline{X} - \mu_0}{\sigma_{\overline{X}}}$$

because it is impossible to obtain a satisfactory estimate of σ^2. If, however, the N available observations are independent of one another, he can use the *Student's t distribution*.

An (almost) apocryphal story. In the early 1900's William Sealy Gosset, a brewer's mathematician, was confronted by the brewmaster with an interesting problem. Breweries, like wineries, employ tasters. Throughout each day, the taster for Gosset's brewery drew several "observations" from every production lot and rated each observation on several qualities (body, bitterness, aftertaste, etc.). For any particular quality, such as bitterness, these daily ratings were averaged and tested against a hypothetical mean μ_0 established by the brewmaster as a standard. Because σ^2 was unknown, however, the taster was required to make at least 30 observations on each lot in order to obtain a satisfactory estimate of $\sigma_{\overline{X}}$. By the end of the day, therefore, he has so inebriated that he could not discriminate the "truly fine pale" from the "green and runny," and it was therefore necessary to develop some test of means that would permit use of smaller samples. In 1908, therefore, Gosset derived the distribution of the quotient

$$\frac{\overline{X} - \mu}{S/\sqrt{N - 1}}$$

Because Gosset's contract with the brewery stipulated that all "inventions, patents, etc." become property of his employer, he surriptitiously published his results under the pen name "Student." This statistic is therefore known as

Student's taste statistic or (in any context other than this absurd fable) Student's t statistic.

When Student's t is first presented to any group of introductory students, one or two inevitably figure out that $S/\sqrt{N-1}$ is precisely equal to $\hat{\sigma}_{\bar{X}}$ and conclude that t is simply an approximation of

$$Z = \frac{\bar{X} - \mu}{\sigma_{\bar{X}}}$$

with $\hat{\sigma}_{\bar{X}}$ substituted for $\sigma_{\bar{X}}$. It must be understood at the outset that this is not the case. We know from Chapter Seven that Z is entirely a function of X and that the probability distribution of Z therefore depends entirely on the distribution of X. Similarly, the distribution of $Z_{\bar{X}}$ depends entirely on the distribution of $\bar{X}$. In the expression for Student's t, however, the numerator includes the random variable $\bar{X}$, and the denominator includes the random variable S. Thus, the distribution of

$$t = \frac{\bar{X} - \mu}{S/\sqrt{N-1}}$$

must depend on the *joint* distribution of *two* random variables, $\bar{X}$ and S. It should therefore be apparent that the derivation of the t-distribution proceeded along very different lines from the derivation of the standard normal distribution that we presented intuitively in Chapter Eight.

To illustrate the basic principles underlying Gosset's derivation, we shall present a similarly nonmathemetical development of a random variable that is very closely related to Student's t.

$$Q = \frac{\bar{X} - \mu}{S}$$

To begin, let us consider the numerator of Q, $\bar{X} - \mu$, which we define as the random variable D. That is,

$$D = \bar{X} - \mu$$

We indicated above that t is an appropriate test statistic only when X is normally distributed. It follows, therefore, that $\bar{X}: N(\mu, \sigma_{\bar{X}}^2)$. Furthermore, by the definition given above, any particular D-value d_i is found by subtracting the constant μ from the corresponding $\bar{X}$-value $\bar{x}_i$. Thus, the distributions of $\bar{X}$ and D will differ only in location. In this regard,

$$E(D) = E(\bar{X} - \mu)$$
$$= E(\bar{X}) - \mu$$

However, we have already established that $E(\bar{X}) = \mu$, and it therefore follows that

$$E(D) = \mu - \mu$$
$$= 0$$

and we may therefore conclude that D: $N(0, \sigma_{\bar{x}}^2)$. Like all normal random variables, of course, D may assume any real value.

The denominator of Q is the sample standard deviation

$$S = \sqrt{\dfrac{\sum_{i}^{N} (x_i - \bar{x})^2}{N}}$$

which is distributed as the square root of a χ^2 (chi square) random variable. Because χ^2 is developed in Chapter Thirteen, a detailed consideration of the probability distribution of S is not possible at this time, and it must suffice to note that the function is unimodal and for small values of N exhibits a distinct positive skew. Moreover, we indicated in Chapter Six that the sample standard deviation is always assumed to be the positive root of the sample variance so the distribution of S is defined only for positive values.

Now, let us us define a Cartesian plane such that the horizontal axis represents the value set of D and the vertical axis represents the value set of S (Figure 12.1).

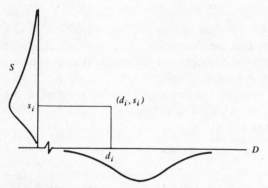

Figure 12.1   Marginal distributions of $D = \bar{X} - \mu$ *and* $S = \sqrt{\Sigma (x - \bar{x})^2/N}$ *and D,S plane.*

Any point in the plane illustrated in Figure 12.1 represents the joint occurrence of some value d_i with some value s_i. In this regard, two things must be noted. First, the Multiplication Theorem for independent events is as applicable to probability densities as to probabilities. Second, the sample mean and standard deviation for a normally distributed random variable are independent of one another. When *and only when* X is normally distributed, therefore, $D = f(\bar{X})$ is independent of S, and the probability density associated with any particular point (d_i, s_i) is equal to the product of the densities of d_i and s_i. Furthermore, any probability density thus computed could be graphically represented as the height of a line perpendicular to the D,S plane at (d_i, s_i). But since D and S are both continuous, the endpoints of all such lines would define a contoured surface, much like the correlation surface discussed in Chapter Nine. The probability density surface representing $\phi(D,S)$ is illustrated in Figure 12.2.

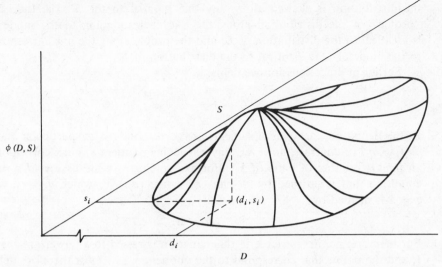

Figure 12.2   Joint distribution of (D,S) in three dimensions.

The geometric relationships among the marginal distributions of D and S and the joint distribution of D,S are illustrated in Figure 12.3. The joint distribution appears as a two-dimensional projection of the three-dimensional surface, but the contour lines indicate bands of equal density much like the contour lines on a topographical map represent zones of equal elevation. On a topographical map, the distance between adjacent contour lines indicates the steepness of a particular grade. Thus, if the bands representing elevations of 600 and 700 feet are a mile apart, the slope is relatively gradual, but if they are only 50 feet apart, one can anticipate a stiff hike or perhaps even a technical climb. These considerations tell us something of the form of our joint distribution. We see, for example, that the point of maximum density corresponds to the intersection of the marginal modes of D and S. It is also apparent that the joint distribution is symmetrical along any line that is parallel to the D axis, but that

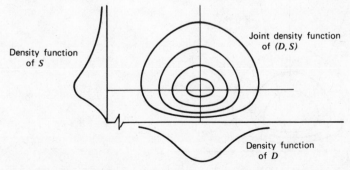

Figure 12.3   Marginal distributions of D and S and joint distribution of (D,S) in two dimensions.

the distribution is skewed along any line parallel to the S axis. Indeed, the "profile" of the distribution along the line corresponding to the mode of S is identical to the distribution of D, and the profile along the line corresponding to the mode of D is identical to the distribution of S.

Earlier in our discussion we defined

$$Q = \frac{\overline{X} - \mu}{S} = \frac{D}{S}$$

By this definition, each pair of values (d_i, s_i) corresponds to one particular quotient $d_i/s_i = q_j$. In contrast, however, any particular quotient q_j must correspond to an *infinite* number of pairs (d_i, s_i). For example, $q = 1$ means that $d/s = 1$, a condition that is satisfied by *all* pairs of values (d, s) for which $d_i = s_i$. Thus, $q = 1$ is defined by

$$D = S$$

Furthermore, neither variable in this equation is raised to a power greater than 1, and the points that correspond to the quotient $q = 1$ must therefore define a straight line. Similarly, $q = 2$ may be represented as the line defined by the equation

$$D = 2S$$

$q = 0.33$ may be represented by the line

$$3D = S$$

and so forth. In Figure 12.4, the joint distribution (D, S) has been superimposed on lines corresponding to several values of Q.

Suppose now that we slice into our joint density function along one of these lines, say $q = 1$, in a plane perpendicular to the (D, S) plane. The resulting cross section is illustrated in Figure 12.5. The area of this cross section is proportional to the total probability density associated with $q = 1$.

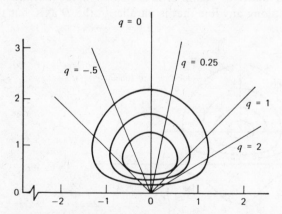

Figure 12.4   Loci of specific values q = d/s *in* D,S *plane superimposed on joint distribution of* (D,S).

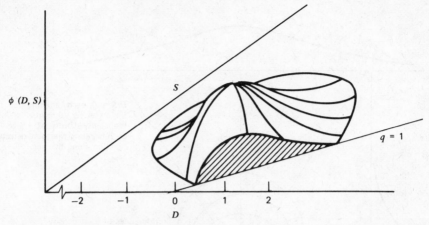

Figure 12.5 Probability density of q = *1 as area under curve defined by transecting joint distribution of* (D,S) *along* q = *1.*

In principle, the probability density of every *Q*-value could be determined by mathematical operations analogous to the graphic manipulations we have just conducted, thus generating the probability density function for *Q*. In Figure 12.6, the probability density $\phi(q_j)$ is represented as the height of the function where its baseline *Q* is intersected by the line $q = q_j$.

At the beginning of our discussion we defined Student's *t* as the quotient

$$t = \frac{\overline{X} - \mu}{S/\sqrt{N - 1}}$$

Thus

$$t = \frac{\overline{X} - \mu}{S(1/\sqrt{N - 1})}$$

$$= \frac{\overline{X} - \mu}{S} \sqrt{N - 1}$$

$$= Q \sqrt{N - 1}$$

Furthermore, if we consider only the positive root of $\sqrt{N - 1}$, then any particular *t*-value

$$t_j = q_j \sqrt{N - 1}$$

can occur when and only when q_j occurs. Thus

$$p(\Delta t) = p(t_a \leq t \leq t_b)$$

must equal

$$p(\Delta q) = p(q_a \leq q \leq q_b)$$

and it should therefore be apparent that the probability distribution of Student's *t* can be derived directly from the distribution of *Q*.

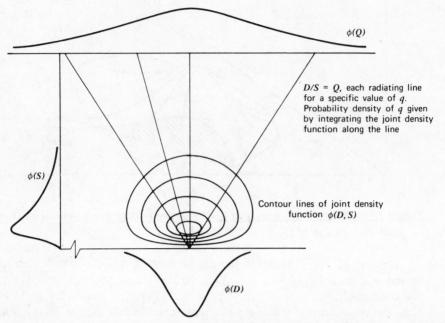

Figure 12.6   Density functions for D, S, (D,S) *and* Q. (*Adapted with permission from Bailey,* Probability and Statistics: Models for Research, *New York: Wiley, 1971.*

In Chapter Eight we established that the parameters of the normal curve are μ and σ^2. Moreover, we know from Chapter Seven that the values of μ and σ^2 can always be changed by performing certain transformations on the scale in which the random variable X is expressed. Thus, any normal random variable X may be transformed to the normal random variable Z, which has *standard* parameter values $\mu = 0$ and $\sigma^2 = 1$. In contrast, the only parameter for the distribution of Student's t is the number of *degrees of freedom*,

$$\nu = N - 1$$

where ν is the lower case Greek letter *nu* and N is the number of observations. Unlike the parameters of the normal distribution, then, ν is completely unrelated to the metric of the random variable under consideration, and even though values of t are expressed in standard units, there is no "standardized" t distribution. Strictly speaking, then, Student's t is not *a* distribution, but a *family* of distributions, each of which corresponds to a particular value of ν. Fortunately, ν can assume only integer values, and even a complete tabulation of probabilities for the t-statistic would therefore require a maximum of only 29 distributions (for $N \geq 30$ we could estimate σ^2 and use the Z-statistic). Generally, however, complete tabulations are seldom used. Instead, most t-tables present t values corresponding to *selected* cumulative probabilities for 30 or so values of ν.

The number of degrees of freedom associated with a particular t distribution is customarily indicated by a numerical subscript, but we shall depart from traditional notation and indicate the value of ν in parentheses, just as we have heretofore specified parameter values for the binomial and normal distributions. Thus, $t(1)$ is the distribution of Student's t for $\nu = 1$, the t distribution for $\nu = 2$ is $t(2)$, and so forth. When t appears *without* a parameter value indicated, it represents the *random variable*

$$t = \frac{\overline{X} - \mu}{S/\sqrt{N - 1}}$$

rather than a probability distribution, such as $t(N - 1)$. Concerning this random variable, we established earlier that

$$E(D) = E(\overline{X} - \mu) = 0$$

As $\overline{X} - \mu$ is the numerator of Student's t, it follows that

$$E(t) = 0$$

Furthermore, we state without proof that

$$V(t) = \frac{\nu}{\nu - 2}$$

or

$$V(t) = \frac{N - 1}{N - 3}$$

For large samples, therefore $V(t)$ approaches 1.0, and it should thus be apparent that despite the parametric difference between Student's t and Z, the two are closely related. In this regard we indicated at the beginning of our discussion that $S/\sqrt{N - 1} = \acute{\sigma}_{\overline{X}}$. This is easily demonstrated:

$$\frac{S}{\sqrt{N - 1}} = \sqrt{\frac{\sum_{i}^{N} (x_i - \bar{x})^2}{N(N - 1)}}$$

$$= \frac{\sqrt{\dfrac{\sum_{i}^{N} (x_i - \bar{x})^2}{N - 1}}}{\sqrt{N}}$$

$$= \frac{\acute{\sigma}}{\sqrt{N}}$$

$$= \acute{\sigma}_{\overline{X}}$$

Thus

$$t = \frac{\overline{X} - \mu}{\acute{\sigma}_{\overline{X}}}$$

Although $\hat{\sigma}$ is not an unbiased estimator of σ, it is consistent. As N increases, therefore, $\hat{\sigma}_{\bar{X}}$ approaches $\sigma_{\bar{X}}$, and t consequently approaches

$$\frac{\bar{X} - \mu}{\sigma_{\bar{X}}} = Z_{\bar{X}}$$

Thus, it should not be surprising that $\phi(t)$, like $\phi(Z)$, is unimodal, symmetical, and approaches the X-axis asymptotically. However, because $V(t)$ approaches 1.0 as a function of sample size, $\phi(t)$ for small values of $\nu = N - 1$ will exhibit more variability than $\phi(Z)$. These properties are apparent in Figure 12.7, which compares the t distributions for several values of ν with the normal distribution.

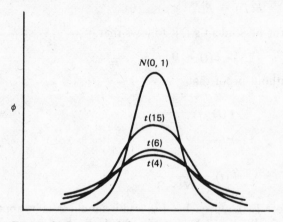

Figure 12.7   Probability density functions for three Student's t distributions and the standard normal random variable.

It is important to emphasize that the probability density function for Student's t is based on the assumption that $\bar{X}$ and S are independent. This assumption holds true only when X is normally distributed, and if the distribution of X departs severely from the normal, therefore, the experimenter should rely on nonparametric techniques. Slight departures from the normal produce relatively little error, however, the Student's t may therefore be used under a wide variety of circumstances. With the t-statistic, however, we encounter a number of problems defining δ and computing N to yield a specific value of β against $\mu_1 = \mu_0 \pm \delta\sigma$. The computational procedure we outlined earlier is predicated on the assumption that

$$\frac{\bar{X} - \mu_0}{\sigma_{\bar{X}}}$$

is normally distributed under either hypothesis. In contrast,

$$\frac{\bar{X} - \mu_0}{S/\sqrt{N - 1}}$$

is distributed as Student's t only if H_0 is correct. When H_1 is correct, this quotient is distributed as *noncentral t*. Instead of developing procedures to estimate δ and compute N under the noncentral t distribution, we shall simply assume that when Student's t is indicated, samples are so small that the experimenter is unlikely to overpower his test and reject H_0 on the basis of scientifically trivial differences. For tests requiring Student's t, therefore, formulation of a composite alternative is acceptable, if not exactly desirable.

To demonstrate the use of the t-statistic, let us return once again to our anthropologist, but let us now assume that neither model specifies a value for σ^2 and, in addition, that he is able to collect only 10 skulls.

a. Formulation. As before, we shall assume a simple test hypothesis

$$H_0: \mu = 1300$$

We shall also assume once more that his alternative model indicates only that $\mu_1 > \mu_0$. As indicated above, estimation of δ when σ^2 is unknown requires considerations that go beyond the scope of this book, and we shall therefore accept the necessity of formulating a composite alternative hypothesis

$$H_1: \mu > 1300$$

b. Decisions. Biological considerations again permit the experimenter to assume that skull capacities are normally distributed, and the experimenter may therefore assume that

$$\frac{\overline{X} - \mu_0}{S/\sqrt{N-1}}$$

is distributed as Student's t.

Because H_1 is a composite hypothesis β cannot be specified, and our only consideration in selecting a critical value t_α is that for $\nu = N - 1 = 9$

$$p(t_o \geq t_\alpha \,|\, H_0 \text{ correct}) = \alpha$$

If we again set $\alpha = .01$, the appropriate critical value t_α may be found in Table **D**. As indicated by the stubhead and column heads, each row represents the distribution of Student's t for a different value of ν, and each column represents a particular cumulative probability value. In the present example, therefore, our critical value must lie at the intersection of the column headed $F = .99$ and the row corresponding to $\nu = 9$. This value is 2.821.

c. Data Collection. As indicated above, the experimenter has 10 available observations. In lieu of a computed value of N, all 10 will be used in the computation of the test statistic.

d. Conclusions. Let us now suppose that his 10 skulls have a sample mean of 1445 cubic centimeters and a sample standard deviation of 150 cubic centimeters. With these values, his test statistic

$$\frac{\bar{x} - \mu_0}{S/\sqrt{N-1}} = \frac{1445 - 1300}{150/\sqrt{9}}$$

$$= \frac{145}{50}$$

$$= 2.9$$

This value exceeds $t_\alpha = 2.821$, and the anthropologist should therefore reject his test hypothesis.

In examining Table **D**, the student may have noted the absence of negative t-values. All t-distributions are symmetrical and if t_α is a *negative* value, therefore,

$$p(t \geq -t_\alpha) = \alpha$$

implies that

$$p(t \leq t_\alpha) = \alpha$$

Had the alternative hypothesis in our example specified $\mu_1 < \mu_0$, therefore, we would simply have defined our critical value as $t_\alpha = -2.821$ and rejected H_0 if our observed test statistic were less than this value.

Similarly, had our composite alternative been nondiredtional, the appropriate critical values would have been $\pm t_{\alpha/2}$ such that

$$p(t_o \geq t_{\alpha/2} | H_0 \text{ correct}) = \frac{\alpha}{2}$$

By this equation, it should be apparent that

$$F(t_{\alpha/2}) = 1 - \frac{\alpha}{2}$$

$$= .995$$

From Table **D**, the appropriate critical values would therefore be 3.250 and -3.250

B. Inferences About Two Population Means μ_1 and μ_2

All of the methods discussed in the preceding section assume that the model under test specifies a particular value μ_0 for the population mean. Frequently, however, the scientist will not be particularly concerned about the value of μ

in his population, but only with whether or not it differs from the mean of *another* population. Thus, a psychologist might administer a drug to one group of subjects and then compare their mean performance on some perceptual discrimination task with a group that was administered a placebo. Or, a college administrator might wish to compare the mean performances of black and white university students on a culture-free intelligence test. In such situations, the test hypothesis is usually represented as

$$H_0: \mu_1 = \mu_2$$

or

$$H_0: \mu_1 - \mu_2 = 0$$

When we test an hypothesis about two population means, therefore, we are really testing an hypothesis about a *difference*. In this regard, recall from Chapter Two that a difference is frequently indicated by the symbol Δ. To keep our notation as compact as possible, therefore, we shall define

$$\Delta = \mu_1 - \mu_2$$

and, our test hypothesis for the difference of two population means therefore becomes

$$H_0: \Delta = \Delta_0 = 0$$

In Chapter Eleven we suggested that the appropriate test statistic for tests of hypotheses about μ is $\overline{X}$. Similarly, for tests of hypotheses about $\mu_1 - \mu_2$, the appropriate test statistic is $\overline{X}_1 - \overline{X}_2$. In order to specify the distribution of this statistic, however, we must first consider the general case of the distribution of a difference $X_1 - X_2$. In this regard, we know from the Central Limit Theorem that the distribution of the sum of N independent random variables

$$Y = X_1 + X_2 + \ldots + X_N$$

becomes normal as N increases. Furthermore, we know that if $X_1, X_2, \ldots X_N$ are themselves normally distributed, then the size of N is immaterial. Thus, if X_1 and X_2 are normal random variables,

$$X_1 - X_2 = X_1 + (-X_2)$$

must also be normally distributed.

To complete our specification, then, we need only derive $\mu_{X_1-X_2}$ and $\sigma^2_{X_1-X_2}$. If we define $E(X_1) = \mu_1$ and $E(X_2) = \mu_2$

$$E(X_1 - X_2) = E(X_1) + E(-X_2)$$
$$= E(X_1) + E[-1(X_2)]$$
$$= E(X_1) + (-1)E(X_2)$$
$$= \mu_1 + (-1)\mu_2$$
$$= \mu_1 - \mu_2$$

Similarly, we shall define $V(X_1) = \sigma_1{}^2$ and $V(X_2) = \sigma_2{}^2$. Because we have defined X_1 and X_2 as independent of one another

$$V(X_1 - X_2) = V(X_1) + V(-X_2)$$
$$= V(X_1) + V[-1(X_2)]$$

By Rule 8 in Appendix II, a constant is squared when it is factored out of a variance. Thus

$$V(X_1 - X_2) = V(X_1) + (-1)^2 V(X_2)$$
$$= \sigma_1{}^2 + (-1)^2 \sigma_2{}^2$$
$$= \sigma_1{}^2 + (1)\sigma_2{}^2$$
$$= \sigma_1{}^2 + \sigma_2{}^2$$

Thus, if X_1 and X_2 are independent, normally distributed random variables with $E(X_1) = \mu_1$, $E(X_2) = \mu_2$, $V(X_1) = \sigma_1{}^2$, and $V(X_2) = \sigma_2{}^2$, then the difference $X_1 - X_2$ is normally distributed with

$$\mu_{X_1 - X_2} = \mu_1 - \mu_2$$

and

$$\sigma^2{}_{X_1 - X_2} = \sigma_1{}^2 + \sigma_2{}^2$$

In a test of hypotheses about the difference between two means, it is seldom that values are specified for either μ_1 or μ_2, and the alternative hypothesis is generally cast as a composite:

$$H_1: \mu_1 - \mu_2 \neq 0 \quad or \quad \Delta = \Delta_1 \neq 0$$
$$H_1: \mu_1 - \mu_2 > 0 \quad or \quad \Delta = \Delta_1 > 0$$
$$H_1: \mu_1 - \mu_2 < 0 \quad or \quad \Delta = \Delta_1 < 0$$

Whenever the availability of a large number of observations introduces the possibility of overpowering the test, however, we again recommend that the experimenter formulate a synthetic alternative hypothesis in terms of δ. In tests of two population means, δ represents the smallest value of $\mu_1 - \mu_2 = \Delta$ that the experimenter would *want* to detect as significantly different from $\Delta_0 = 0$. Once again, δ is assumed to be an absolute value and is expressed in standard metric. However, in this instance the most meaningful unit is the standard deviation of the *difference* $X_1 - X_2$. Thus

$$\delta = \frac{|\mu_1 - \mu_2|}{\sigma_{X_1 - X_2}} = \frac{|\mu_1 - \mu_2|}{\sqrt{\sigma_1{}^2 + \sigma_2{}^2}}$$

As in tests of one population mean, the value specified under the alternative hypothesis Δ_1 must be expressed in natural units, and simple alternatives are therefore given in terms of Δ_0 and the unstandardized difference

$$|\mu_1 - \mu_2| = \delta \sigma_{X_1 - X_2}$$

286

Thus

$$H_1: \Delta = \Delta_1 = \Delta_0 \pm \delta\sigma_{X_1-X_2}$$

$$H_1: \Delta = \Delta_1 = \Delta_0 + \delta\sigma_{X_1-X_2}$$

$$H_1: \Delta = \Delta_1 = \Delta_0 - \delta\sigma_{X_1-X_2}$$

1. Testing Hypotheses About μ_1 and μ_2 When $\sigma_1{}^2$ and $\sigma_2{}^2$ Are Known

To develop the methods involved in testing

$$H_0: \Delta = \Delta_0 = 0$$

when $\sigma_1{}^2$ and $\sigma_2{}^2$ are known, let us suppose that our anthropologist has obtained a significant result in his original study, concluded that Cro-magnon man displaced Neanderthaler from the Scandinavian Penninsula before the last ice age and has submitted his results to a professional journal. Let us further suppose, however, that the editor has raised the objection that all of the skulls in his sample were collected in the *southern* part of the penninsula. During the last ice age, the Straits of Denmark, which separate Denmark and Germany from the Scandinavian Penninsula, were frozen, and it is therefore possible that Cro-magnon, emmigrating from continental Europe, simply pushed Neanderthaler further north as the glaciers retreated. Thus, the presence of Cro-magnon skulls in the southern part of the penninsula may actually support the orthodox theory that migrations of Cro-magnon to the Scandinavian Penninsula began toward the *end* of the last ice age. In response the anthropologist has returned to the Scandinavian Penninsula to collect a sample of skulls from the north.

a. Formulation. The procedure under discussion concerns tests when $\sigma_1{}^2$ and $\sigma_2{}^2$ are known. We shall therefore assume once again that the standard deviation of skull capacities in the southern population, which he sampled first, is 120 cubic centimeters. That is, $\sigma_1 = 120$ cubic centimeters. Let us further assume that the standard deviation in his second, or northern, population is 110 cubic centimeters, that is, $\sigma_2 = 110$ cubic centimeters.

If the mean skull capacity of his northern sample is not significantly different from the capacity of his southern sample, our anthropologist could maintain his original conclusion that Cro-magnon was, in fact, the sole inhabitant of the Scandinavian Penninsula during the last ice age. Thus, his test hypothesis is

$$H_0: \mu_1 - \mu_2 = 0$$

or

$$H_0: \Delta = \Delta_0 = 0$$

His alternative hypothesis, of course, is that northern skulls are Neanderthal and therefore smaller than his southern skulls, in which case $\mu_1 > \mu_2$. Thus

$$H_1: \Delta = \Delta_1 > 0$$

However, he feels that any difference less than one standard unit would be scientifically inconclusive, so he sets $\delta = 1.0$. Thus

$$\Delta_1 = \Delta_0 + \delta\sigma_{X_1-X_2}$$

$$= 0 + 1.0\sqrt{\sigma_1^2 + \sigma_2^2}$$

$$= 0 + \sqrt{26,500}$$

$$= 162.79$$

The anthropologist is therefore prepared to test two simple hypotheses:

$$H_0: \Delta = 0$$

$$H_1: \Delta = 162.79$$

b. Decisions. After deciding to go to the northern part of the Scandinavian Penninsula and collect a second sample of skulls, he must next decide on an appropriate test statistic. We have already established that if X_1 and X_2 are normally distributed and independent, then

$$X_1 - X_2: N(\mu_1 - \mu_2, \sigma_1^2 + \sigma_2^2)$$

where $\mu_1 = E(X_1)$, $\mu_2 = E(X_2)$, $\sigma_1^2 = V(X_1)$, and $\sigma_2^2 = V(X_2)$. In the present example our experimenter knows that skulls are ordinarily normally distributed. Thus, mean skull capacities based on any sample size are also normally distributed, and it therefore follows that $\overline{X}_1 - \overline{X}_2$ is normally distributed with $\mu_{\overline{X}_1-\overline{X}_2}$ equal to $\mu_1 - \mu_2$ and $\sigma^2_{\overline{X}_1-\overline{X}_2}$ equal to $\sigma_{\overline{X}_1}^2 + \sigma_{\overline{X}_2}^2$. That is,

$$\overline{X}_1 - \overline{X}_2: N(\mu_1 - \mu_2, \sigma_{\overline{X}_1}^2 + \sigma_{\overline{X}_2}^2)$$

Like any normally distributed random variable, $\overline{X}_1 - \overline{X}_2$ may be expressed in standard form

$$Z_{\overline{X}_1-\overline{X}_2} = \frac{(\overline{X}_1 - \overline{X}_2) - \mu_{\overline{X}_1-\overline{X}_2}}{\sigma_{\overline{X}_1-\overline{X}_2}} = \frac{(\overline{X}_1 - \overline{X}_2) - (\mu_1 - \mu_2)}{\sqrt{\sigma_{\overline{X}_1}^2 + \sigma_{\overline{X}_2}^2}}$$

This, then, is his test statistic. Under the test hypothesis, of course,

$$\mu_1 - \mu_2 = \Delta_0 = 0$$

and the expression for the test statistic therefore reduces to

$$Z_{\overline{X}_1-\overline{X}_2} = \frac{\overline{X}_1 - \overline{X}_2}{\sqrt{\sigma_{\overline{X}_1}^2 + \sigma_{\overline{X}_2}^2}} = \frac{\overline{X}_1 - \overline{X}_2}{\sqrt{(\sigma_1^2/N_1) + (\sigma_2^2/N_2)}}$$

where N_1 is the number of observations drawn from the first population and N_2 is the number of observations on the second population.

In this example, the scientifically orthodox hypothesis is the *alternative* hypothesis, not the test hypothesis, and it is therefore more important to avoid a Type II error than a Type I error. Thus, the anthropologist sets $\alpha = .05$ and $\beta = .01$.

c. Data Collection. In establishing the distribution of $X_1 - X_2$ under the Central Limit Theorem, it was necessary to assume that X_1 and X_2 were independent. The assumption that $\overline{X}_1 - \overline{X}_2$ is normally distributed similarly requires that $\overline{X}_1$ and $\overline{X}_2$ be uncorrelated, and the two sample means must therefore represent *independent* sets of observations. This stipulation is clearly satisfied in our present example, but at the end of this section we shall briefly consider studies in which $\overline{X}_1$ and $\overline{X}_2$ are dependent.

In addition to the independence of his samples, the experimenter must also be concerned with the size of his samples, and computation of suitable sample sizes N_1 and N_2 to yield a specified value of β against a simple alternative $\Delta = \Delta_1$ involves essentially the same operations as in tests of one population mean. Under the test hypothesis the standardized value of the critical difference $(\bar{x}_1 - \bar{x}_2)_c$ is

$$z_\alpha = \frac{(\bar{x}_1 - \bar{x}_2)_c - \Delta_0}{\sqrt{\sigma_{\bar{x}_1}^{\,2} + \sigma_{\bar{x}_2}^{\,2}}}$$

Thus

$$z_\alpha \sqrt{\sigma_{\bar{x}_1}^{\,2} + \sigma_{\bar{x}_2}^{\,2}} = (\bar{x}_1 - \bar{x}_2)_c - \Delta_0$$

or

$$(\bar{x}_1 - \bar{x}_2)_c = z_\alpha \sqrt{\sigma_{\bar{x}_1}^{\,2} + \sigma_{\bar{x}_2}^{\,2}} + \Delta_0$$

$$= z_\alpha \sqrt{(\sigma_1^2/N_1) + (\sigma_2^2/N_2)} + \Delta_0$$

Under the alternative hypothesis, the standardized value of $(\bar{x}_1 - \bar{x}_2)_c$ is given by

$$z_\beta = \frac{(\bar{x}_1 - \bar{x}_2)_c - \Delta_1}{\sqrt{\sigma_{\bar{x}_1}^{\,2} + \sigma_{\bar{x}_2}^{\,2}}}$$

and

$$z_\beta \sqrt{\sigma_{\bar{x}_1}^{\,2} + \sigma_{\bar{x}_2}^{\,2}} = (\bar{x}_1 - \bar{x}_2)_c - \Delta_1$$

or

$$(\bar{x}_1 - \bar{x}_2)_c = z_\beta \sqrt{(\sigma_1^2/N_1) + (\sigma_2^2/N_2)} + \Delta_1$$

As before, our two expressions for $(\bar{x}_1 - \bar{x}_2)_c$ may be set equal to one another, but the simultaneous solution for N_1 and N_2 will be indeterminate. However, it may be shown that for a total of $N_1 + N_2$ observations, the power of the test against any specific alternative Δ_1 is maximized when the sample sizes are proportional to the population standard deviations. That is, when

$$\frac{N_1}{N_2} = \frac{\sigma_1}{\sigma_2}$$

If this stipulation is applied to our two simultaneous equations, a couple of scratch pages of elementary but tedious algebra will yield

$$N_1 = \sigma_1 (\sigma_1 + \sigma_2) \left(\frac{z_\beta - z_\alpha}{\Delta_0 - \Delta_1} \right)^2$$

and

$$N_2 = \sigma_2 (\sigma_1 + \sigma_2) \left(\frac{z_\beta - z_\alpha}{\Delta_0 - \Delta_1} \right)^2$$

In the present example, we know that $\sigma_1 = 120$ and $\sigma_2 = 110$. Furthermore, our hypothesized differences are $\Delta_0 = 0$ and $\Delta_1 = 162.79$. Finally, the error probabilities specified by the anthropologist are $\alpha = .05$ and $\beta = .01$. Because the difference specified under the alternative is *greater* than the difference specified in the test hypothesis, the rejection region must fall in the *upper* tail of the Z distribution; thus, $z_\alpha = 1.64$ and $z_\beta = -2.33$. Plugging these values into our formula for N_1

$$N_1 = 120(120 + 110) \left(\frac{-2.33 - 1.64}{0 - 162.79} \right)^2$$

$$= 120(120 + 110) \left(\frac{-3.97}{-162.79} \right)^2$$

$$= 120(230) \frac{15.76}{26,500}$$

$$= 16.41$$

$$= 16 \qquad \text{(approx.)}$$

Similarly

$$N_2 = 110(120 + 110) \left(\frac{-2.33 - 1.64}{0 - 162.79} \right)^2$$

but since we want

$$\frac{N_1}{N_2} = \frac{\sigma_1}{\sigma_2}$$

it must be true that

$$N_2 = \frac{N_1 \sigma_2}{\sigma_1}$$

$$= \frac{16(110)}{120}$$

$$= 14.67$$

$$15 \qquad \text{(approx.)}$$

d. Conclusions. On the basis of these computations, the anthropologist selects 16 skulls at random from the southern sample he collected on his first expenition and finds

$$\bar{x}_1 = 1420$$

In addition, he collects 15 skulls at his current dig in the north and finds

$$\bar{x}_2 = 1370$$

Thus, his test statistic

$$\frac{\bar{x}_1 - \bar{x}_2}{\sqrt{(\sigma_1{}^2/N_1) + (\sigma_2{}^2/N_2)}} = \frac{1420 - 1370}{\sqrt{900 + 806.67}}$$

$$= \frac{50}{41.31}$$

$$= 1.21$$

which is considerably less than his critical value $z_\alpha = 1.64$, and he must therefore reject his alternative hypothesis.

In this example, the computed sample sizes $N_1 = 16$ and $N_2 = 15$ were not large enough to assume $\overline{X}_1 - \overline{X}_2$ normally distributed solely on the basis of the Central Limit Theorem. Thus, had the anthropologist not been willing to assume that X_1 and X_2 were themselves normally distributed (or approximately so), he would have had to rely on nonparametric techniques to test his hypothesis.

Before we proceed to testing hypotheses when $\sigma_1{}^2$ and $\sigma_2{}^2$ are unknown, the reader is reminded that the procedures just discussed assumed independence of $\overline{X}_1$ and $\overline{X}_2$. In our example, there was no doubt that this condition was met, but there are many studies where each value of X_1 may be logically paired with a value of X_2. When we discussed experimental design in Chapter Eleven for, example, we suggested that one possibility in the THC experiment was to use subjects as their own controls. Thus, each subject s_i would have two scores, x_{i1} and x_{i2}. It should therefore be apparent that insofar as performance depends on individual difference variables (in addition to the experimental treatment), the two sets of scores X_1 and X_2 will be correlated, and the experimenter could not assume independence of $\overline{X}_1$ and $\overline{X}_2$. In this type of study the appropriate random variable is

$$D = X_1 - X_2$$

where the value associated with the ith pair of observations is

$$d_i = x_{i1} - x_{i2}$$

Thus, the experimenter is really testing an hypothesis about a *single* random variable:

$$H_0\colon \mu_D = \mu_0 = 0$$

and should therefore employ the methods developed in our discussion of inferences about one population mean.

2. Testing Hypotheses About μ_1 and μ_2 When σ_1^2 and σ_2^2 Are Not Known

When the experimenter has large samples ($N \geq 30$) available from both populations, he should compute the unbiased estimators $\hat{\sigma}_1^2$ and $\hat{\sigma}_2^2$ and use these values to estimate δ:

$$\hat{\delta} = \frac{|\mu_1 - \mu_2|}{\sqrt{\hat{\sigma}_1^2 + \hat{\sigma}_2^2}}$$

With $\hat{\delta}$ thus computed, he may define

$$\Delta_1 = \Delta_0 + \hat{\delta}\hat{\sigma}_{X_1 - X_2}$$

and proceed to compute N_1 and N_2, substituting $\hat{\sigma}_1$ and $\hat{\sigma}_2$ in the equations

$$N_1 = \sigma_1(\sigma_1 + \sigma_2)\left(\frac{z_\beta - z_\alpha}{\Delta_0 - \Delta_1}\right)^2$$

$$N_2 = \sigma_2(\sigma_1 + \sigma_2)\left(\frac{z_\beta - z_\alpha}{\Delta_0 - \Delta_1}\right)^2$$

He may then substitute $\hat{\sigma}_1^2$ and $\hat{\sigma}_2^2$ into the test statistic

$$Z_{\bar{X}_1 - \bar{X}_2} = \frac{\bar{X}_1 - \bar{X}_2}{\sqrt{(\sigma_1^2/N_1) + (\sigma_2^2/N_2)}}$$

and test this result against the appropriate critical value z_α.

In the preceding paragraph we indicated that if $N_1 \geq 30$ and $N_2 \geq 30$, the distribution of

$$\frac{(\bar{X}_1 - \bar{X}_2) - (\mu_1 - \mu_2)}{\hat{\sigma}_{\bar{X}_1 - \bar{X}_2}} = \frac{\bar{X}_1 - \bar{X}_2}{\sqrt{(\sigma_1^2/N_1) + (\sigma_2^2/N_2)}}$$

is approximated by the standard normal distribution. This is very similar to our earlier conclusion that for $N \geq 30$ the test statistic

$$\frac{\bar{X} - \mu}{\hat{\sigma}_{\bar{X}}} = \frac{\bar{X} - \mu}{\sqrt{\hat{\sigma}^2/N}}$$

may be considered a standard normal random variable. Recall, however, that for $N < 30$

$$\frac{\bar{X} - \mu}{\hat{\sigma}_{\bar{X}}}$$

is distributed as Student's t with $N - 1$ degrees of freedom. Similarly, if *either* $N_1 < 30$ or $N_2 < 30$, then

$$\frac{\bar{X}_1 - \bar{X}_2}{\sqrt{(\hat{\sigma}_1^2/N_1) + (\hat{\sigma}_2^2/N_2)}}$$

is distributed as Student's t with $N_1 + N_2 - 2$ degrees of freedom *provided that* the two unknown population variances $\sigma_1{}^2$ and $\sigma_2{}^2$ may be assumed equal. In this regard, we know that

$$\sigma^2{}_{\bar{X}_1 - \bar{X}_2} = \frac{\sigma_1{}^2}{N_1} + \frac{\sigma_2{}^2}{N_2}$$

If we assume that $\sigma_1{}^2 = \sigma_2{}^2 = \sigma^2$, this expression reduces to

$$\sigma^2{}_{\bar{X}_1 - \bar{X}_2} = \frac{\sigma^2}{N_1} + \frac{\sigma^2}{N_2}$$

$$= \sigma^2 \left(\frac{1}{N_1} + \frac{1}{N_2} \right)$$

$$= \sigma^2 \left(\frac{N_1 + N_2}{N_1 N_2} \right)$$

Similarly

$$\hat{\sigma}^2{}_{\bar{X}_1 - \bar{X}_2} = \hat{\sigma}^2 \left(\frac{N_1 + N_2}{N_1 N_2} \right)$$

Under our assumption of equal variances, of course, $\hat{\sigma}_1{}^2$ and $\hat{\sigma}_2{}^2$ are both estimates of the common variance σ^2, and either could therefore enter the preceding expression as $\hat{\sigma}^2$. It can be demonstrated, however, that neither $\hat{\sigma}_1{}^2$ nor $\hat{\sigma}_2{}^2$ provides as good an estimate of the common variance as may be obtained by pooling the variability observed in *both* samples:

$$\hat{\sigma}^2_{\text{pooled}} = \frac{\sum_i^{N_1} (x_i - \bar{x}_1)^2 + \sum_j^{N_2} (x_j - \bar{x}_2)^2}{N_1 + N_2 - 2} = \frac{(N_1 - 1)\hat{\sigma}_1{}^2 + (N_2 - 1)\hat{\sigma}_2{}^2}{N_1 + N_2 - 2}$$

Thus, when $\sigma_1{}^2$ is assumed equal to $\sigma_2{}^2$, the best estimate of $\sigma^2{}_{\bar{X}_1 - \bar{X}_2}$ becomes

$$\hat{\sigma}^2_{\text{pooled}} \left(\frac{N_1 + N_2}{N_1 N_2} \right)$$

and our test statistic

$$\frac{\bar{X}_1 - \bar{X}_2}{\sqrt{(\hat{\sigma}_1{}^2 / N_1) + (\hat{\sigma}_2{}^2 / N_2)}}$$

becomes

$$\frac{\bar{X}_1 - \bar{X}_2}{\sqrt{\hat{\sigma}^2_{\text{pooled}} \left(\dfrac{N_1 + N_2}{N_1 N_2} \right)}}$$

Because the population variances are unknown, the assumption that they are equal may be difficult to justify. Moreover, methods for testing the hypothesis that $\sigma_1^2 = \sigma_2^2$ (which we shall discuss in Chapter Thirteen) yield questionable results when applied to samples that are small enough to require a t-test. Fortunately, however, the assumption of equal variances may be violated with relatively little error if the two sample sizes N_1 and N_2 are approximately equal.

In our discussion of testing hypotheses when σ_1^2 and σ_2^2 are known, we solved a problem in which our computed sample sizes were 16 and 15, respectively. Let us now suppose that σ_1^2 and σ_2^2 are *not* known and that $N_1 = 16$ and $N_2 = 15$ are *available*, rather than computed, sample sizes. As before, we shall test the hypothesis.

$$H_0: \Delta = \Delta_0 = 0$$

With samples of 16 and 15, however, there is little danger of overpowering the test, so we shall settle for a composite alternative hypothesis

$$H_1: \Delta = \Delta_1 > 0$$

As before, we shall let $\alpha = .05$, but with a composite alternative we cannot specify a value for β. Furthermore, the samples are too small to compute $\hat{\sigma}_1^2$ and $\hat{\sigma}_2^2$ and substitute them for σ_1^2 and σ_2^2 in the Z-statistic. However, N_1 and N_2 are nearly equal, and we may therefore use the test statistic just discussed:

$$\frac{\overline{X}_1 - \overline{X}_2}{\sqrt{\hat{\sigma}_{\text{pooled}}^2 \left(\dfrac{N_1 + N_2}{N_1 N_2} \right)}}$$

As indicated above, this statistic is distributed as Student's t with $N_1 + N_2 - 2$ degrees of freedom, and for $\nu = 16 + 15 - 2 = 29$, Table **D** shows that

$$F(1.699) = .95$$

and for $\alpha = .05$, therefore,

$$t_\alpha = 1.699$$

To compute our test statistic we must first compute

$$\hat{\sigma}_{\text{pooled}}^2 = \frac{(N_1 - 1)\hat{\sigma}_1^2 + (N_2 - 1)\hat{\sigma}_2^2}{N_1 + N_2 - 2}$$

In this regard, we shall suppose that our computed variance estimates are

$$\hat{\sigma}_1^2 = (110)^2$$

and

$$\hat{\sigma}_2^2 = (140)^2$$

Thus

$$\hat{\sigma}^2_{pooled} = \frac{15(110)^2 + 14(140)^2}{16 + 15 - 2}$$

$$= \frac{455,900}{29}$$

$$= 15,720.69$$

Moreover, we shall assume that as before

$$\bar{x}_1 = 1420$$

and

$$\bar{x}_2 = 1370$$

Thus, our test statistic is equal to

$$\frac{1420 - 1370}{\sqrt{15,720.69 \left(\dfrac{16 + 15}{(16)(15)}\right)}} = \frac{50}{\sqrt{15,720.69 \left(\dfrac{31}{240}\right)}}$$

$$= \frac{50}{\sqrt{2,030.5206}}$$

$$= \frac{50}{45.06}$$

$$= 1.11$$

which is considerably less than the 1.699 required to reject H_0.

When σ_1^2 and σ_2^2 *cannot* be assumed equal and N_1 and N_2 are so different that violation of this assumption cannot be safely ignored,

$$\frac{\bar{X}_1 - \bar{X}_2}{\sqrt{(\hat{\sigma}_1^2/N_1) + (\hat{\sigma}_2^2/N_2)}}$$

is *still* distributed as Student's t, but

$$\nu = \frac{[(\hat{\sigma}_1^2/N_1) + (\hat{\sigma}_2^2/N_2)]^2}{\dfrac{(\hat{\sigma}_1^2/N_1)^2}{(N_1 + 1)} + \dfrac{(\hat{\sigma}_2^2/N_2)^2}{(N_2 + 1)}} - 2$$

and can be very tedious to compute without the aid of a desk calculator.

C. Problems, Review Questions, and Exercises

A social psychologist is investigating cultural bias in I.Q. tests. The test in question was developed using a population of primarily white, middle-class adults, and in this population it exhibits a mean $\mu = 100$ and a variance $\sigma^2 = 225$. The psychologist suspects that the test is more a test of cultural knowledge than a test of ability and that black subjects will therefore score lower than white subjects.

1. Formulation.
   a. What is the theory in question?
   b. What is the model under test?
   c. What are the hypotheses?
2. Decisions.
   a. What are the appropriate units of observation?
   b. What is the appropriate test statistic?
   c. If the experimenter sets $\alpha = .05$, what is his rejection rule?
3. Data collection: Suppose now that the psychologist administers his I.Q. test to a sample of 200 black subjects and that his sample exhibits the same distribution of age, income, and academic achievement as his white population. Suppose further that $\bar{x} = 98.25$.
4. Conclusions.
   a. Assuming that the variance for the black population is the same as for his white population, does his observed test statistic fall in the critical region?
   b. What does this conclusion imply for the theory under test?
5. Let us now suppose that his I.Q. test has a *standard error of measurement of* five points. This means that two I.Q. scores obtained by the same individual on different occasions may be expected to differ by five points. On this basis he decides that any difference $|\mu_B - \mu|$ less than five points is of no scientific importance.
   a. What is δ?
   b. Given δ, what is H_1?
   Suppose now that the experimenter feels that a Type II error is twice as serious as a Type I error. He therefore sets $\beta = .025$.
   c. What is z_β?
   d. What is the appropriate sample size of this test?
   e. What is the critical value of $\bar{X}$?
   f. If he draws the requisite number of observations and finds, as before, that $\bar{x} = 98$, what should he conclude?
   g. Is this a more reasonable conclusion than 4b? Why or why not?
6. Let us now add a note of realism to the problem and suppose (1) that the psychologist is unwilling to assume that $\sigma_B = 15$ and that (2) he has only 15 available I.Q. scores for black subjects.
   a. Given $\alpha = .05$ and $\hat{\sigma}_B = 18$, what will be the critical value of $\bar{X}$?
   b. Given the same sample size and level of significance, what would the critical value of $\bar{X}$ be if he *knew* that $\sigma_B = 18$?
   c. Comparing 6a and 6b, what can you conclude about the relative power of the *t*-test and *Z*-test?

7. In the midst of his study, the psychologist realizes that his white and black populations differ in one important particular. The test was initially developed in 1940. Past experience suggests that population I.Q.'s many change from generation to generation, so rather than test his black data against the parameter values established in 1940, he decides to test a contemporary group of white subjects and test the difference between his two groups.

   a. What is the test hypothesis?

   On the basis of 60 white subjects and 50 black subjects, he finds $\hat{\sigma}_W{}^2 = 169$ and $\hat{\sigma}_B{}^2 = 256$.

   b. If he decides that any difference $|\mu_W - \mu_B|$ less than half a standard unit $\sigma_{X_W - X_B}$ is unimportant, what is his alternative hypothesis?

   c. Assuming that he wants $\alpha = \beta = .05$, what are his two computed sample sizes N_W and N_B?

   d. Given the computed sample sizes, what should be concluded if $\bar{x}_W = 102.75$ and $\bar{x}_B = 97.50$?

8. Now assume that $\hat{\sigma}_W{}^2$, $\hat{\sigma}_B{}^2$, $\bar{x}_W$, and $\bar{x}_B$ are as given in problem **7**, but that $N_W = N_B = 12$. If $\alpha = .05$.

   a. What is the critical value of the appropriate test statistic?

   b. Is the difference significant under this test?

D. Summary Table

Distribution of X	Sample Size(s)	H_0	$H_1{}^a$	Test Statistic T	Distribution of T
			A. Testing Hypotheses about μ		
			1. σ^2 Known		
Normal	$N \geq 25$	$\mu = \mu_0$	$\mu = \mu_0 \pm \delta\sigma$	$\dfrac{\bar{X} - \mu_0}{\sigma_{\bar{X}}}$	$N(0, 1)$
Nonnormal	$N \geq 25$	Same	Same	Same	Same
Normal	$N < 25$	Same	Same[b]	Same	Same
Nonnormal	$N < 25$	—	—	Nonparametric	—
			2. σ^2 Not Known		
Normal	$N \geq 30$	$\mu = \mu_0$	$\mu = \mu_0 \pm \delta\sigma$	$\dfrac{\bar{X} - \mu_0}{\hat{\sigma}_{\bar{X}}}$	$N(0, 1)$
Nonnormal	$N \geq 30$	Same	Same	Same	Same
Normal	$N < 30$	Same	$\mu = \mu_1 \neq \mu_0$	Same	$t(N - 1)$
Nonnormal	$N < 30$	—	—	Nonparametric	—

D. Summary Table (continued)

Distribution of X	Sample Size(s)	H_0	H_1[a]	Test Statistic T	Distribution of T

B. Testing Hypotheses About μ_1 and μ_2
1. σ_1^2 and σ_2^2 Known

Distribution of X	Sample Size(s)	H_0	H_1[a]	Test Statistic T	Distribution of T
Normal	N_1 and $N_2 \geq 25$	$\Delta = \Delta_0$	$\Delta = \Delta_0 \pm \delta\sigma_{X_1-X_2}$	$\dfrac{\bar{X}_1 - \bar{X}_2}{\sqrt{\dfrac{\sigma_1^2}{N_1} + \dfrac{\sigma_2^2}{N_2}}}$	$N(0, 1)$
Nonnormal	N_1 and $N_2 \geq 25$	Same	Same	Same	Same
Normal	N_1 or $N_2 < 25$	Same	Same[b]	Same	Same
Nonnormal	N_1 or $N_2 < 25$	—	—	Nonparametric	—

2. σ_1^2 and σ_2^2 Not Known

Distribution of X	Sample Size(s)	H_0	H_1[a]	Test Statistic T	Distribution of T
Normal	N_1 and $N_2 \geq 30$	$\Delta = \Delta_0$	$\Delta = \Delta_0 \pm \delta\sigma_{X_1-X_2}$	$\dfrac{\bar{X}_1 - \bar{X}_2}{\sqrt{\dfrac{\hat{\sigma}_1^2}{N_1} + \dfrac{\hat{\sigma}_1^2}{N_2}}}$	$N(0, 1)$
Nonnormal	N_1 and $N_2 \geq 30$	Same	Same	Same	Same
Normal $\Big\{$	N_1 or $N_2 < 30$	Same	$\Delta = \Delta_1 \neq \Delta$	Same	$t(\nu)$[c]
	$N_1 = N_2 < 30$	Same	Same	$\dfrac{\bar{X}_1 - \bar{X}_2}{\sqrt{\hat{\sigma}^2{}_{\bar{X}_1-\bar{X}_2}{}^d}}$	$t(N_1 + N_2 - 2)$
Nonnormal	N_1 or $N_2 < 30$	—	—	Nonparametric	—

[a] Alternative hypotheses illustrated in summary table are nondirectional, but same statistical tests are applicable with directional alternative hypotheses.

[b] When tests are based on small samples, the experimenter may elect to formulate a composite alternative hypothesis. This procedure assumes that the available sample is small enough to provide an adequate δ and eliminates computation of N.

[c] $\nu = \dfrac{[(\hat{\sigma}_1^2/N_1) + (\hat{\sigma}_2^2/N_2)]^2}{\dfrac{(\hat{\sigma}_1^2/N_1)^2}{N_1 + 1} + \dfrac{(\hat{\sigma}_2^2/N_2)^2}{N_2 + 1}} - 2$

[d] $\hat{\sigma}^2{}_{\bar{X}_1-\bar{X}_2} = \hat{\sigma}^2_{\text{pooled}}\left(\dfrac{N_1 + N_2}{N_1 N_2}\right) = \dfrac{(N_1 - 1)\hat{\sigma}_1^2 + (N_2 - 1)\hat{\sigma}_2^2}{N_1 + N_2 - 2}\left(\dfrac{N_1 + N_2}{N_1 N_2}\right)$

298

chapter thirteen

testing hypotheses about variances

A. The x^2 Distribution

In the last chapter we found it necessary to introduce a new distribution, Student's t, in order to develop several tests of hypotheses about population means. Similarly, the methods involved in testing hypotheses about population variances require some understanding of the χ^2 (*chi* square) distribution. In this regard, recall from Chapter Eight that our development of the normal curve was based on the binomial probability function. In the present chapter, the normal curve provides a similar springboard for the development of the χ^2 distribution.

Let us define a normal random variable X with expectation μ and variance σ^2. Now, let us draw a random sample of one observation on this random variable, standardize our observed value x_1, and square it. Thus, we obtain

$$z_1{}^2 = \left(\frac{x_1 - \mu}{\sigma}\right)^2$$

By definition, x_1 represents our observed value, and both x_1 and $z_1{}^2$ must therefore be constants. We pointed out in Chapter Ten, however, that drawing a random sample is a type of random experiment. Before the sample is actually drawn, therefore, the observation may potentially assume any value in the value set of X and is therefore a random variable. If we stipulate that the experiment is yet to be conducted, the preceding expression becomes

$$Z_1{}^2 = \left(\frac{X_1 - \mu}{\sigma}\right)^2$$

The probability distribution of this random variable is defined as χ^2 with one degree of freedom. That is,

$$Z_1{}^2 : \chi^2(1)$$

Because Z^2 is a function of the standard normal random variable Z, we may draw a number of conclusions about the distribution properties of $\chi^2(1)$. We know, for example, that $\phi(Z)$ is defined for all Z-values from $-\infty$ to $+\infty$. By definition, however, the square of any real number is positive, and the probability density function $\phi(\chi^2)$ for $\chi^2(1)$ must therefore be defined only for positive values. In addition, we know that approximately 68 percent of the area under the standard normal curve falls between $z = -1$ and $z = +1$. That is,

$$p(-1.0 \leq z \leq 1.0) = .68 \qquad \text{(approx.)}$$

Because Z is a continuous variable, however, this probability is equivalent to

$$p(-1 \leq z \leq 0) + p(0 \leq z \leq 1)$$

In this regard, we know that $0^2 = 0$ and $1^2 = 1$, and it must therefore be true that all Z-values falling in the positive interval

$$\{z \mid 0 \leq z \leq 1\}$$

will yield Z^2-values in the interval

$$\{z^2 \mid 0 \leq z^2 \leq 1\}$$

We also know, however, that $(-1)^2 = 1$, and it must likewise be true, therefore, that Z-values in the *negative* interval

$$\{z \mid -1 \leq z \leq 0\}$$

will *also* yield Z^2-values in the interval

$$\{z^2 \mid 0 \leq z^2 \leq 1\}$$

Thus

$$p(0 \le z^2 \le 1) = p(-1 \le z \le 0) + p(0 \le z \le 1)$$

$$= p(-1 \le z \le 1) = .68 \qquad \text{(approx.)}$$

which means that approximately 68 percent of the area under $\phi(\chi^2)$ for $\chi^2(1)$ falls between 0 and $+1$. The distribution of $\chi^2(1)$ must therefore exhibit considerable positive skew with probability concentrated in the peak of the curve and relatively little probability in the shoulder of the curve (Figure 13.1).

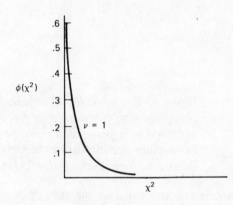

Figure 13.1 Probability distribution for chi-square with one degree of freedom. (Adapted with permission from Bailey. Probability and Statistics: Models for Research, New York: Wiley, 1971).

Let us now define a similar experiment in which our sample includes two independent observations x_1 and x_2. If these values are standardized and squared as before, we obtain

$$z_1{}^2 = \left(\frac{x_1 - \mu}{\sigma}\right)^2$$

and

$$z_2{}^2 = \left(\frac{x_2 - \mu}{\sigma}\right)^2$$

which may be added together to yield

$$z_1{}^2 + z_2{}^2 = \left(\frac{x_1 - \mu}{\sigma}\right)^2 + \left(\frac{x_2 - \mu}{\sigma}\right)^2$$

Prior to conducting our actual observations, however, this sum is a random variable

$$Z_1{}^2 + Z_2{}^2$$

which by definition is distributed as χ^2 with two degrees of freedom.

$$Z_1{}^2 + Z_2{}^2 : \chi^2(2)$$

Furthermore, we may infer a good deal about this probability distribution from what we have already established about $\chi^2(1)$. As before, we know that some 68 percent of the total probability associated with *either* Z_1^2 or Z_2^2 must lie between 0 and $+1$. It should therefore be apparent that under the distribution of $Z_1^2 + Z_2^2$ the probability associated with values *greater than* $+1$ is considerably larger than under the distribution of Z_1^2 considered alone. Hence, the $\chi^2(2)$ distribution is less skewed than the $\chi^2(1)$ distribution, with more area in the shoulder of the curve and less area in the peak (Figure 13.2).

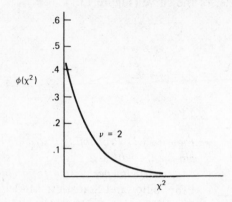

Figure 13-2 Probability distribution for chi-square with two degrees of freedom. (*Adapted with permission from Bailey,* Probability and Statistics: Models for Research, *New York: Wiley 1971*).

Finally, let us generalize our experiment and consider the sum of N independent, standard normal random variables squared:

$$\sum_i^N Z_i^2 = Z_1^2 + Z_2^2 + \ldots + Z_N^2$$

This sum is distributed as $\chi^2(N)$, and it now becomes apparent that the number of degrees of freedom ν associated with any χ^2 random variable is equal to the number of terms in the sum $Z_1^2 + Z_2^2 + \ldots + Z_N^2$. As was the case with Student's t, moreover, degrees of freedom is the only parameter of a χ^2 distribution, and as our examples have demonstrated, there is a different χ^2 distribution of every value of ν. Figure 13.3 illustrates χ^2 distributions for $\nu = 1, \nu = 2, \nu = 4$, and $\nu = 10$.

In Chapter Twelve we found it necessary to distinguish between the probability distribution $t(\nu)$ and the random variable t. We must similarly distinguish between the probability distribution $\chi^2(\nu)$ and the random variable χ^2, which is *equal to*

$$Z_1^2 + Z_2^2 + \ldots + Z_N^2$$

and is *distributed as* $\chi^2(N)$. In this regard, recall that the random variables Z_i^2 were defined as independent of one another. Thus, the random variable χ^2 satisfies the Central Limit Theorem, and as the number of variables Z_i^2 increases, therefore, the distribution of χ^2 must approach the normal. Indeed, it is apparent even in the four curves in Figure 13.3 that $\chi^2(\nu)$ loses its skew and becomes symmetrical as ν increases.

Testing Hypotheses About Variances

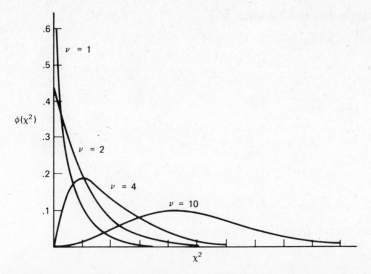

Figure 13.3 Several chi-square distributions. (Adapted with permission from Bailey, Probability and Statistics: Models for Research, *New York: Wiley, 1971).*

The distribution of χ^2 has another important property. Consider the two random variables

$$(Z_1^2 + Z_2^2 + \ldots + Z_m^2) : \chi^2(m)$$

and

$$(Z_1^2 + Z_2^2 + \ldots + Z_n^2) : \chi^2(n)$$

If these two sums are added together, the entire expression will include $m + n$ terms:

$$Z_1^2 + Z_2^2 + \ldots + Z_{m+n}^2$$

which by definition must be distributed as $\chi^2 (m + n)$. It must therefore be true in general that if

$$\chi_1^2 : \chi^2(\nu_1)$$

and

$$\chi_2^2 : \chi^2(\nu_2)$$

then

$$\chi_1^2 + \chi_2^2 : \chi^2(\nu_1 + \nu_2)$$

It may be demonstrated by similar argument that

$$\chi_1^2 - \chi_2^2 : \chi^2(\nu_1 - \nu_2)$$

1. The Expectation and Variance of χ^2

By definition we know that

$$\sum_i^N Z_i^2 = \chi^2$$

where $\chi^2 : \chi^2(N)$. Thus

$$E(\chi^2) = E\left(\sum_i^N Z_i^2\right)$$

$$= E(Z_1^2 + Z_2^2 + \ldots + Z_N^2)$$

$$= E(Z_1^2) + E(Z_2^2) + \ldots + E(Z_N^2)$$

In this regard, recall from Chapter Six that

$$V(X) = E(X^2) - [E(X)]^2$$

Thus

$$E(X^2) = V(X) + [E(X)]^2$$

and therefore

$$E(Z^2) = V(Z) + [E(Z)]^2$$

However, we know that the expected value of any standardized random variable is zero. Thus, $E(Z) = 0$, which means that

$$[E(Z)]^2 = 0$$

and

$$E(Z^2) = V(Z) = 1.0$$

Consequently

$$E(\chi^2) = E(Z_1^2) + E(Z_2^2) + \ldots + E(Z_N^2)$$

$$= \underbrace{1.0 + 1.0 + \ldots + 1.0}_{N \text{ times}}$$

$$= N$$

The expected value of a χ^2 random variable is therefore equal to the number of variables comprising the sum, which by definition equals its number of degrees of freedom. Thus, if $\chi^2 : \chi^2(\nu)$

$$E(\chi^2) = \nu$$

and we state without proof that

$$V(\chi^2) = 2\nu$$

B. Inferences About One Population Variance σ^2

In testing hypotheses about a population variance, our test hypothesis takes a familiar form:

$$H_0: \sigma^2 = \sigma_0{}^2$$

Similarly, the alternative hypothesis may be simple:

$$H_1: \sigma^2 = \sigma_1{}^2$$

or composite:

$$H_1: \sigma^2 \neq \sigma_0{}^2$$

$$H_1: \sigma^2 > \sigma_0{}^2$$

$$H_1: \sigma^2 < \sigma_0{}^2$$

Moreover, for problems in which the model specifies a composite alternative, there are methods for establishing a synthetic simple alternative, which takes the form

$$H_1: \sigma^2 = \sigma_1{}^2 = \delta^2 \sigma_0{}^2$$

and for computing sample size to yield a particular value of β against such an alternative. These methods are relatively simple and are presented in Chapters Sixteen and Seventeen of Bailey's (1971) *Probability and Statistics: Models for Research*. The principles that underlie these methods, however, lack the intuitive appeal of the methods discussed in Chapter Twelve, and we shall therefore confine ourselves in the present text to problems involving composite alternatives. Furthermore, it should be noted that while the chapter explicitly considers only directional alternative hypotheses, the tests that we develop are equally applicable to nondirectional hypotheses.

In the preceding chapter we saw that the test statistic for hypotheses about μ was always some function of $\overline{X}$. Similarly, the appropriate test statistic for testing hypotheses about σ^2 is a function of $\hat{\sigma}^2$:

$$\frac{\hat{\sigma}^2 (N - 1)}{\sigma_0{}^2}$$

The probability distribution for this statistic is not readily apparent from inspection, so it must be derived. To begin, we know that any constant may be added to and then subtracted from a mathematical expression without changing the value of that expression. Thus

$$x_i - \mu = x_i - \mu + \bar{x} - \bar{x}$$

or, rearranging terms,

$$x_i - \mu = (x_i - \bar{x}) + (\bar{x} - \mu)$$

Thus

$$(x_i - \mu)^2 = [(x_i - \bar{x}) + (\bar{x} - \mu)]^2$$

$$= (x_i - \bar{x})^2 + 2(x_i - \bar{x})(\bar{x} - \mu) + (\bar{x} - \mu)^2$$

and

$$\sum_i^N (x_i - \mu)^2 = \sum_i^N [(x_i - \bar{x})^2 + 2(x_i - \bar{x})(\bar{x} - \mu) + (\bar{x} - \mu)^2]$$

The difference $(\bar{x} - \mu)$ is a constant, so distributing our sum yields

$$\sum_i^N (x_i - \mu)^2 = \sum_i^N (x_i - \bar{x})^2 + 2(\bar{x} - \mu) \sum_i^N (x_i - \bar{x}) + N(\bar{x} - \mu)^2$$

Furthermore, we know from Chapter Six that the sum of deviations about the sample mean $(x_i - \bar{x})$ must always equal zero. This eliminates the middle term from our expression, thus reducing it to

$$\sum_i^N (x_i - \mu)^2 = \sum_i^N (x_i - \bar{x})^2 + N(\bar{x} - \mu)^2$$

To simplify our notation throughout the remainder of this derivation, we shall assume that all sums run from 1 to N. Except where required for clarification or emphasis, therefore, our summations will include neither the subscript i nor the limit N. With this in mind, we divide both sides of the preceding equation by σ^2 and obtain

$$\frac{\sum (x_i - \mu)^2}{\sigma^2} = \frac{\sum (x_i - \bar{x})^2}{\sigma^2} + \frac{N(\bar{x} - \mu)^2}{\sigma^2}$$

If we now assume that the observations on the random variable X are yet to be made, the constant values x_i and $\bar{x}$ become random variables X_i and $\bar{X}$ and, rearranging terms, our expression becomes

$$\frac{\sum (X_i - \bar{X})^2}{\sigma^2} = \frac{\sum (X_i - \mu)^2}{\sigma^2} - \frac{N(\bar{X} - \mu)^2}{\sigma^2}$$

Consider the terms of the right-hand side of the equation. By definition

$$\frac{\sum (X_i - \mu)^2}{\sigma^2} = \sum \left(\frac{X_i - \mu}{\sigma} \right)^2$$

$$= \sum_i^N Z_i^2$$

Furthermore

$$\frac{N(\overline{X} - \mu)^2}{\sigma^2} = \frac{(\overline{X} - \mu)^2}{\sigma^2/N}$$

$$= \frac{(\overline{X} - \mu)^2}{\sigma_{\overline{X}}^2}$$

$$= \left(\frac{\overline{X} - \mu}{\sigma_{\overline{X}}}\right)^2$$

$$= Z_{\overline{X}}^2$$

Thus

$$\frac{\sum (X_i - \overline{X})^2}{\sigma^2} = \sum_i^N Z_i^2 - Z_{\overline{X}}^2$$

However, we know that ΣZ_i^2 is a χ^2 random variable with N degrees of freedom and that Z^2 is a χ^2 random variable with one degree of freedom. Our equation may therefore be rewritten as

$$\frac{\sum (X_i - \overline{X})^2}{\sigma^2} = \chi_1^2 - \chi_2^2$$

where $\chi_1^2 : \chi^2(N)$ and $\chi_2^2 : \chi^2(1)$. Moreover, we established earlier that the difference between two χ^2 random variables is distributed as χ^2 with $\nu_1 - \nu_2$ degreees of freedom. Thus, $\chi_1^2 - \chi_2^2 : \chi^2(N - 1)$, and

$$\frac{\sum (X_i - \overline{X})^2}{\sigma^2} : \chi^2(N - 1)$$

By definition we know that for any particular sample, the unbiased estimate of the variance

$$\hat{\sigma}^2 = \frac{\sum (x_i - \bar{x})^2}{N - 1}$$

is a constant. From our discussion of sampling distributions in Chapter Ten, however, we also know that across all possible samples of N observations, $\hat{\sigma}^2$ is a random variable and in this context may therefore be expressed as

$$\hat{\sigma}^2 = \frac{\sum (X_i - \overline{X})^2}{N - 1}$$

Multiplying both sides of this expression by $N - 1$, we obtain

$$\hat{\sigma}^2 (N - 1) = \sum (X_i - \overline{X})^2$$

and dividing both sides by σ^2

$$\frac{\hat{\sigma}^2(N-1)}{\sigma^2} = \frac{\sum (X_i - \overline{X})^2}{\sigma^2}$$

We have already established that $\Sigma(X_i - \overline{X})^2/\sigma^2$ is distributed as χ^2 with $N - 1$ degrees of freedom. It must likewise be true, therefore, that

$$\frac{\hat{\sigma}^2(N-1)}{\sigma^2} : \chi^2(N-1)$$

Throughout our discussion of χ^2 we assumed that the variables X_i underlying the sum $Z_1^2 + Z_2^2 + \ldots + Z_N^2$ were normally distributed. Strictly speaking, therefore, the ratio $\hat{\sigma}^2(N - 1)/\sigma^2$ is distributed as $\chi^2(N - 1)$ *if and only if* the random variable under observation is normally distributed. In contrast to the rather liberal normality assumption underlying applications of Student's t, moreover, this assumption must be rigidly satisfied if the χ^2 distributions are to be used without serious inferential errors. In addition, we have stipulated that these N normally distributed random variables are independent of one another. We must further assume, therefore, that the observations that generate $\hat{\sigma}^2$ are independent of one another.

To see how the statistic $\hat{\sigma}^2(N - 1)/\sigma^2$ is applied, let us return to the saga of the frustrated anthropologist. In the last episode, our protagonist concluded once again that Cro-magnon was the only human inhabitant of the Scandinavian Penninsula during the last ice age. Suppose, however, that in a last-ditch effort to preserve scientific orthodoxy, the journal editor rejects the paper because the mean of his southern sample

$$\bar{x} = 1420$$

is actually closer to the mean skull capacity of *modern man* than to the mean capacity of Cro-magnon man. It is therefore possible, he argues, that these skulls are specimens of early modern man. In response to this obstinate pettifoggery, the anthropologist reviews the available literature to discover some distribution property for which an appropriate statistical test might distinguish a sample of Cro-magnon skulls from a sample of modern skulls.

1. Formulation

In the course of his review, he discovers that modern skull capacities are more variable than Cro-magnon skull capacities. Recall in this regard that the population variance of Cro-magnon skulls was given in Chapter Eleven as

$$\sigma^2 = (120)^2$$

In contrast, we shall suppose that the variance of skull capacities in modern man is

$$\sigma^2 = (180)^2$$

The anthropologist therefore decides to test the hypothesis that Scandinavian Penninsula ice-age skulls have a variance of $(180)^2$. That is,

$$H_0: \sigma^2 = (180)^2$$

Because Cro-magnon skulls are less variable than modern skulls, his alternative hypothesis is

$$H_1: \sigma^2 < (180)^2$$

2. Decisions

In Chapter Eleven we indicated that the distribution of an appropriate test statistic must be specified by the model under test and must be expressed in some form that permits comparison with entries in a conventional table of cumulative probabilities. In the present example

$$\frac{\hat{\sigma}^2(N-1)}{\sigma^2}$$

satisfies these requirements. First, all of the required values are known. Under the test hypothesis $\sigma^2 = \sigma_0^2 = (180)^2$, the estimate $\hat{\sigma}^2$ is computed from the data and N is sample size. Second, we know that skull capacities may be assumed normally distributed, and the statistic is therefore distributed as χ^2 with $N-1$ degrees of freedom.

The only remaining decision, then, is the selection of a rejection rule. Because the test hypothesis represents the scientifically orthodox position (at least insofar as the journal editor represents orthodoxy), he sets a rather conservative level of significance, $\alpha = .025$. Under the alternative hypothesis $\sigma^2 < \sigma_0^2$ and the critical region must therefore fall in the *lower* tail of the distribution of $\chi^2(N-1)$. Furthermore, it will be recalled from Chapter Twelve that the southern sample, which yielded the disputed mean of 1420 cubic centimeters, included 16 skulls. Our critical region must therefore be

$$\{\chi^2 | \chi^2 \leq \chi_\alpha^2\}$$

where if H_0 is correct,

$$p\left[\frac{\hat{\sigma}^2(N-1)}{\sigma_0^2} \leq \chi_\alpha^2\right] = .025$$

under the distribution of χ^2 for $\nu = N - 1 = 15$ degrees of freedom. According to Table **E**,

$$F(6.26) = .025$$

for $\chi^2(15)$. Thus

$$\chi_\alpha^2 = 6.26$$

3. Conclusions

Recall from Chapter Twelve that variance estimate $\hat{\sigma}^2$ for the sample under consideration is $(110)^2$. Thus, the value of the test statistic is

$$\frac{\hat{\sigma}^2(N-1)}{\sigma_0^2} = \frac{(110)^2\,(15)}{(180)^2}$$

$$= \frac{12,100\,(15)}{32,400}$$

$$= 5.60$$

The observed value 5.60 falls in the critical region $\{\chi^2 | \chi^2 \leq 6.26\}$, and the anthropologist may therefore conclude that the variance of his southern population is smaller than the variance of modern skull capacities.

C. Inferences About Two Population Variances σ_1^2 and σ_2^2

In this section of our chapter we develop methods for testing hypotheses of the form

$$H_0: \sigma_1^2 = \sigma_2^2$$

against any of the following alternatives:

$$H_1: \sigma_1^2 \neq \sigma_2^2$$

$$H_1: \sigma_1^2 > \sigma_2^2$$

$$H_1: \sigma_1^2 < \sigma_2^2$$

To develop the appropriate test statistic, let us consider two populations with variances σ_1^2 and σ_2^2, respectively. Let us further suppose that we draw a sample of N_1 observations from the first population and compute the unbiased estimator of the variance $\hat{\sigma}_1^2$ and that we draw a sample of N_2 observations from the second population and compute $\hat{\sigma}_2^2$. For the first sample we know that

$$\frac{\hat{\sigma}_1^2(N_1-1)}{\sigma_1^2} = \chi_1^2$$

where $\chi_1^2: \chi^2(N_1-1)$. Dividing both sides of this equation by $N-1$ we obtain

$$\frac{\hat{\sigma}_1^2}{\sigma_1^2} = \frac{\chi_1^2}{N_1-1}$$

For the sample drawn from the second population it must be similarly true that

$$\frac{\hat{\sigma}_2^2(N_2-1)}{\sigma_2^2} = \chi_2^2$$

where χ_2^2: $\chi^2(N_2 - 1)$ and that

$$\frac{\hat{\sigma}_2^2}{\sigma_2^2} = \frac{\chi_2^2}{N_2 - 1}$$

If we now divide by one another the two equations just derived,

$$\frac{\hat{\sigma}_1^2/\sigma_1^2}{\hat{\sigma}_2^2/\sigma_2^2} = \frac{\chi_1^2/(N_1 - 1)}{\chi_2^2/(N_2 - 1)}$$

This is a very important result, because the ratio of two independent *chi* square variables, χ_1^2 and χ_2^2, each divided by its own degrees of freedom ν_1 and ν_2, respectively, is distributed as a statistic known as Fisher's F. The F-distribution is derived from the joint distribution of the χ^2 variables entering the ratio, and it should not be surprising, therefore, that the probability function of F is defined only for positive values. Nor should it be surprising that every distribution of Fisher's F is specified by *two* parameters, ν_1 and ν_2, where ν_1 is the degrees of freedom associated with χ_1^2, and ν_2 is the degrees of freedom associated with χ_2^2. As before, we shall distinguish the random variable F from the probability distribution $F(\nu_1, \nu_2)$. Thus

$$\frac{\chi_1^2/\nu_1}{\chi_2^2/\nu_2} = F$$

where F: $F(\nu_1, \nu_2)$ and, likewise,

$$\frac{\hat{\sigma}_1^2/\sigma_2^2}{\hat{\sigma}_2^2/\sigma_2^2} = \frac{\chi_1^2/(N_1 - 1)}{\chi_2^2/(N_2 - 1)} = F$$

where F: $F(N_1 - 1, N_2 - 1)$ *if* $\hat{\sigma}_1^2$ and $\hat{\sigma}_2^2$ are each based on a sample of independent observations on a normally distributed random variable.

Cumulative probability tables for a large number of Fisher's F distributions are provided in Table **F**. To find an F-value corresponding to a particular cumulative probability **F**, the reader must first find the *section*, or block of rows, corresponding to ν_2. Within this section, the desired entry will be found at the intersection of the column headed ν_1 and the row corresponding to the cumulative probability in question.

Armed with this new statistical tool, let us suppose that our anthropologist decides to play it safe. He knows that the estimated variance based on his northern sample was considerably larger than the variance estimate for his southern sample (which he tested in our last example) and therefore even closer to the variance of modern skull capacities. To forestall any objections from the editor, therefore, he decides to test the hypothesis that the variances of his northern and southern parent populations are equal.

1. Formulation

Thus, his test hypothesis is

$$H_0: \sigma_1^2 = \sigma_2^2$$

and the appropriate alternative is

$$H_1: \sigma_1{}^2 < \sigma_2{}^2$$

2. Decisions

We have just ascertained that

$$\frac{\hat{\sigma}_1{}^2/\sigma_1{}^2}{\hat{\sigma}_2{}^2/\sigma_2{}^2}$$

is distributed as Fisher's F with $N_1 - 1$ and $N_2 - 1$ degrees of freedom, where N_1 is the size of the first sample and N_2 is the size of the second sample. If, however H_0 is correct, then $\sigma_1{}^2 = \sigma_2{}^2$ and our ratio reduces to

$$\frac{\hat{\sigma}_1{}^2}{\hat{\sigma}_2{}^2}$$

When *and only when* H_0 is correct and each sample consists of independent observations on a normally distributed random variable, this test statistic is distributed as $F(N_1 - 1, N_2 - 1)$. Moreover, the reader is reminded that Fisher's F is defined as a function of two *independent* χ^2 variables. Thus, if both samples are believed to represent the same population, the experimenter must ensure independence *between* samples as well as independence of observations *within* samples.

Because his alternative hypothesis is the orthodox hypothesis in this instance, he decides to minimize β by setting a rather liberal level of significance, $\alpha = .10$. Moreover, his alternative hypothesis specifies that $\sigma_1{}^2 < \sigma_2{}^2$, and his region of rejection must therefore fall in the lower tail of the distribution for $F(N_1 - 1, N_2 - 1)$. Thus, his critical value is F_α where

$$p[(\hat{\sigma}_1{}^2/\hat{\sigma}_2{}^2) \leq F_\alpha | H_0 \text{ correct}] = .10$$

and H_0 will be rejected if $\hat{\sigma}_1{}^2/\hat{\sigma}_2{}^2$ is equal to or less than F_α.

According to the instructions given above, our critical value $F_{.10}$ *should* be found in the row corresponding to $F = .10$ under the column headed $\nu_1 = 15$ in the section designated $\nu_2 = 14$. However, for larger values of ν the distribution of Fisher's F changes very slowly as ν increases, and it is therefore impractical to tabulate every distribution $F(\nu_1, \nu_2)$ for values of ν greater than 12. For this reason, the distribution $F(15,14)$ is not included in Table F, but we can obtain a satisfactory estimate of F_α from the distribution $F(15, 15)$. In this distribution we find that the cumulative probability of $F = .507$ is equal to .10. Thus

$$F_\alpha = .507 \qquad \text{(approx.)}$$

3. Conclusions

The two variance estimates, once more, are $\acute{\sigma}_1^2 = (110)^2$ and $\acute{\sigma}_2^2 = (140)^2$. Thus

$$\frac{\acute{\sigma}_1^2}{\acute{\sigma}_2^2} = \frac{(110)^2}{(140)^2} = \frac{12,100}{19,600}$$

$$= .617$$

which is substantially larger than his critical value .507. He must therefore reject his alternative hypothesis and conclude that $\acute{\sigma}_1^2$ and $\acute{\sigma}_2^2$ represent independent estimates of the same population variance σ^2.

D. Problems, Review Questions, and Exercises

1. A random variable is distributed as $\chi^2(14)$. The probability is .10 that a single observation of this random variable will exceed what value?

2. What are the expected value and variance of this variable?

3. What values in the distribution of $\chi^2(6)$ satisfy the following relationships?
   a. $p(\chi^2 \geq \chi_b^2) = .05$.
   b. $p(\chi^2 \leq \chi_a^2) = .05$.

4. What is the expected value of a variable distributed as $\chi^2(6)$?

5. Is the interval χ_a^2 to χ_b^2 symmetrical about the value computed in problem 4? If not, why?

6. Given the data in problem 6 of Chapter Twelve test the hypothesis ($\alpha = .05$) that $\sigma_B^2 = 225$.
   a. Is the value of your test statistic significant?
   b. How large would $\acute{\sigma}^2$ have to be in order for your result to be significant?
   c. Does this value seem surprisingly larger than σ_0^2?
   d. Compute the square root of the value computed in 6b and compare it with σ_0. Does this difference seem surprisingly large?

7. A certain variable is distributed as $F(6,6)$. What is the cumulative probability of $F = 1.78$?

8. If $F:F(11,12)$, what value F_α satisfies the relationship $p(F \geq F_\alpha) = .05$?

9. Given the data in problem 8 of Chapter Twelve, test the hypothesis that $\sigma_W^2 = \sigma_B^2$. Set $\alpha = .05$.
   a. What is your alternative hypothesis?
   b. Is the value of your test statistic significant?
   c. For approximately what value of α *would* your result be significant?
   d. What does this tell you about the psychologist's "tongue in cheek" assumption that $\sigma_W^2 = \sigma_B^2$?

E. Summary Table

H_0	$H_1{}^a$	Test Statistic T	Distribution of T
		A. Testing Hypotheses About σ^2	
$\sigma^2 = \sigma_0{}^2$	$\sigma^2 \neq \sigma_0{}^2$	$\dfrac{\hat{\sigma}^2(N-1)}{\sigma_0{}^2}$	$\chi^2(N-1)$

Assumptions: X normally distributed[b]

Observations of X independent

		B. Testing Hypotheses About $\sigma_1{}^2$ and $\sigma_2{}^2$	
$\sigma_1{}^2 = \sigma_2{}^2$	$\sigma_1{}^2 \neq \sigma_2{}^2$	$\dfrac{\sigma_1{}^2}{\sigma_2{}^2}$	$F(N_1 - 1, N_2 - 1)$

Assumptions: X_1 and X_2 normally distributed[b]

X_1 and X_2 independent of one another

Observations of both X_1 and X_2 independent

[a] Alternative hypotheses illustrated in summary table are nondirectional, but same statistical tests are applicable with directional alternative hypotheses.

[b] This assumption is strictly enforced. Violators will obtain meaningless results!

chapter fourteen

testing hypotheses about entire distributions: pearson's x^2

In Chapter Twelve we developed techniques specifically to test hypotheses about population means, and in Chapter Thirteen we developed tests of hypotheses about population variances. Thus, every statistical test we have considered to this point was developed to test an hypothesis about one particular distribution property. While it may therefore seem that such tests yield relatively little information about a distribution, it must be recalled that, barring application of the Central Limit Theorem, all of these tests require that the statistician assume the probability distribution of his random variable to be approximated by some *known* mathematical function. Thus, in order to submit the test statistic

$$\frac{\bar{x}_0 - \mu_0}{\hat{\sigma}_{\bar{x}}}$$

to the distribution of Student's t, it is necessary to assume that X is normally distributed. Similarly, the assumption that

$$\frac{\hat{\sigma}_1{}^2}{\hat{\sigma}_2{}^2}$$

is distributed as $F(N_1 - 1, N_2 - 1)$ requires that X_1 and X_2 both be normally distributed. If the investigator can make such assumptions, then he *already* knows everything about the distribution of the random variable *except* its parameter values. In Chapter Twelve, however, we

319

mentioned in passing that such assumptions are often unwarranted, and in such instances, testing an hypothesis about a single distribution property may not yield sufficient information to justify an inference about the theory underlying the test. Frequently, therefore, the scientist must formulate a test hypothesis that simultaneously considers many of the distribution properties that would be implicitly stipulated if the probability function were known or assumed.

A. Testing Hypotheses About One Distribution: Goodness of Fit Tests

Even if an experimenter cannot *assume* that the distribution of his random variable conforms to some specific probability function, he may nonetheless have reason to *hypothesize* that his random variable is so distributed. That is, he may wish to test the hypothesis that the distribution of his random variable "fits" some particular hypothetical distribution. In this circumstance the investigator requires a test statistic that is sensitive to any significant departure of his observed distribution from the distribution specified in the model. To develop the logic underlying one such test statistic, let us suppose that a statistics teacher with a class of 50 students has constructed a midterm examination for which the maximum score is 49 points. Suppose further that for various psychometric reasons he hopes that his test scores will be *uniformly distributed*. That is, he hopes that every possible test score x_i will be observed with the same frequency as every other test score x_j. He therefore formulates two hypotheses:

$$H_0: X \text{ distributed uniformly}$$

$$H_1: X \text{ not distributed uniformly}$$

If he groups his observations into C intervals of equal class width ΔX, then under the test hypothesis he should expect to obtain the same number of scores in each interval. Thus, for $C = 5$ intervals of 10 points each and a total enrollment of 50 students, the distribution of expected frequencies is given in Table 14.1, where Δx_j is the range of test scores defining the jth interval and $\mathbf{F}_j$, the entry in the jth cell of the table, is the *expected* frequency of observations in Δx_j.

Now let us suppose that he administers his examination and obtains the distribution of *observed* frequencies shown in Table 14.2.

Table 14.1 Expected Frequency Distribution Based on Hypothesis That Examination Scores Are Uniformly Distributed

Δx_j	0–9	10–19	20–29	30–39	40–49	Σ
$\mathbf{F}_j$	10	10	10	10	10	50

Table 14.2 Observed Frequency Distribution of Examination Scores

Δx_j	0–9	10–19	20–29	30–39	40–49	Σ
f_j	4	6	15	15	10	50

If H_0 is correct, any differences between the observed frequencies f_j and their corresponding expected frequencies F_j must be entirely attributable to random error, and such differences should therefore be relatively small. If the alternative hypothesis is correct, however, there should be substantial differences between the observed and expected frequencies. To determine how well his observed distribution "fits" his expected distribution, he therefore computes the difference between the expected and observed frequency in each cell:

$$f_1 - F_1 = 4 - 10 = -6$$

$$f_2 - F_2 = 6 - 10 = -4$$

$$f_3 - F_3 = 15 - 10 = 5$$

$$f_4 - F_4 = 15 - 10 = 5$$

$$f_5 - F_5 = 10 - 10 = 0$$

Now he is faced with the problem of determining whether these differences represent a significant departure from his expectations. To explore this problem, let us consider, say, the difference

$$f_3 - F_3 = 15 - 10 = 5$$

This same difference could, of course, have resulted from $f_j = 6$ and $F_j = 1$. Were these indeed our values, the observed frequency would be *six times* as large as the expected frequency, and simply on intuitive grounds we would probably agree that it represented a significant departure. Suppose, however, that $f_j = 10{,}000$ and $F_j = 9995$. Once again, $f_j - F_j = 5$, but in this instance the observed frequency is just fractionally larger than the expected frequency, and we would probably agree that the difference was relatively unimportant. From these examples, it should be apparent that any particular difference $f_j - F_j$ assumes meaning only *in comparison with* the expected frequency F_j. On this basis the teacher computes the ratio $(f_j - F_j)/F_j$ for each cell:

$$\frac{f_1 - F_1}{F_1} = \frac{4 - 10}{10} = \frac{-6}{10}$$

$$\frac{f_2 - F_2}{F_2} = \frac{6 - 10}{10} = \frac{-4}{10}$$

$$\frac{f_3 - F_3}{F_3} = \frac{15 - 10}{10} = \frac{5}{10}$$

$$\frac{f_4 - F_4}{F_4} = \frac{15 - 10}{10} = \frac{5}{10}$$

$$\frac{f_5 - F_5}{F_5} = \frac{10 - 10}{10} = \frac{0}{10}$$

He then reasons that the extent to which his *entire* distribution of observed frequencies departs from his distribution of expected frequencies should be reflected in the *sum* of these ratios, and he therefore computes

$$\sum_{j=1}^{5} \frac{f_j - F_j}{F_j}$$

In this particular problem, $F_1 = F_2 = \ldots = F_5 = 10$. The denominator F_j may therefore be factored out of the summation as a constant, and the above expression becomes

$$\frac{1}{10} \sum_{j=1}^{5} (f_j - F_j) = \frac{1}{10} (-6 - 4 + 5 + 5 + 0)$$

The factor in parentheses represents the sum of the differences $(f_j - F_j)$, and the reader will observe that this sum is equal to zero. Indeed, the sum

$$\sum_{j=1}^{C} (f_j - F_j)$$

will *always* equal zero. In our present example, the teacher has 50 students enrolled in his class. No matter how they are distributed, therefore, the sum of frequencies, *expected or observed*, must equal 50. And in general

$$\sum_{j=1}^{C} f_j = \sum_{j=1}^{C} F_j = N$$

Thus, if f_j is less than F_j in one cell, this deficit must be distributed as a surplus among the other cells. The numerical sum of the *negative* differences

$$\sum (f_j - F_j) | f_j < F_j$$

must therefore equal the numerical sum of the *positive* differences.

$$\sum (f_j - F_j) | f_j > F_j$$

yielding an overall sum of zero.

This zero-sum property was also encountered in Chapter Six when we attempted to measure dispersion by taking the sum of the deviations about the mean, and it will be recalled that the difficulty was overcome by *squaring* each difference. Our teacher employs this same tactic and obtains

$$\frac{(f_1 - F_1)^2}{F_1} = \frac{(4 - 10)^2}{10} = \frac{36}{10} = 3.60$$

$$\frac{(f_2 - F_2)^2}{F_2} = \frac{(6 - 10)^2}{10} = \frac{16}{10} = 1.60$$

$$\frac{(f_3 - F_3)^2}{F_3} = \frac{(15 - 10)^2}{10} = \frac{25}{10} = 2.50$$

$$\frac{(f_4 - F_4)^2}{F_4} = \frac{(15 - 10)^2}{10} = \frac{25}{10} = 2.50$$

$$\frac{(f_5 - F_5)^2}{F_5} = \frac{(10 - 10)^2}{10} = \frac{0}{10} = 0.00$$

If the expected and observed distributions are identical, of course, f_j will always equal F_j. Thus, the differences $f_j - F_j$ would equal zero, as would the squared differences $(f_j - F_j)^2$. When corresponding observed and expected frequencies are not identical, however, these *squared* differences must be positive. Consequently, each value $(f_j - F_j)^2/F_j$ must always assume a *nonnegative* value representing the departure from perfect fit in one particular cell, and the sum of these values

$$\sum_{j=1}^{5} \frac{(f_j - F_j)^2}{F_j} = 3.60 + 1.60 + 2.50 + 2.50 + 0 = 10.20$$

must therefore reflect the departure from perfect fit across the entire distribution. In 1900, Karl Pearson demonstrated that under certain experimental conditions this statistic is distributed approximately as χ^2, and it is therefore known as Pearson's χ^2. In the present text Pearson's χ_2 will be distinguished from the random variable $\chi^2 = \Sigma Z_i^2$ by the tilde ($\sim$), which is customarily used in mathematical notation to indicate an approximation. Thus

$$\tilde{\chi}^2 = \sum_{j=1}^{c} \frac{(f_j - F_j)^2}{F_j}$$

1. Degrees of Freedom as Unconstrained Cells and Yates' Correction

We know from Chapter Thirteen that the distribution of any χ^2 random variable is specified by its number of degrees of freedom. Thus far, however, degrees of freedom has remained a rather hazy notion, defined only implicitly as some function of sample size. But in the context of Pearson's χ^2, degrees of freedom assumes a very concrete meaning as the *number of cells* in the distribution of F_j that may assume *any* arbitrary positive value. In this regard, the expected frequencies in our present example were computed under the assumption that X was uniformly distributed. Had our teacher specified some other distribution for X, our expected frequencies might have been entirely different. Whatever the form of our expected distribution, however, we know that the sum of our frequencies F_j must always equal the number of observations. Theoretically, therefore, our teacher could have assigned any expected frequencies to four of his cells, just so long as the fifth cell brought this total to 50. Thus, one cell was *constrained* by the total, but the other $5 - 1 = 4$ cells were *free*. In general, then, Pearson's χ^2 is distributed as χ^2 with

$$\nu = \text{total number of cells} - \text{number of constrained cells}$$

degrees of freedom. If the expected frequency distribution consists of a single row of C cells, there is only one constraint, which is imposed by the total number of observations, and

$$\nu = C - 1$$

In our present example, therefore,

$$\tilde{\chi}^2 = \sum_{j=1}^{5} \frac{(f_j - F_j)^2}{F_j} : \chi^2(4)$$

In establishing a critical region for Pearson's χ^2, it must be remembered that χ^2 is equal to zero when the expected and observed frequency distributions are identical and that the value of the statistic increases as a function of their dissimilarity. Thus, the critical region in any test of Pearson's χ^2 is always in the *upper* tail of the appropriate χ^2 distribution. If we suppose that our teacher sets $\alpha = .05$, therefore, our critical value χ_α^2 has a cumulative probability of .950 in the distribution of $\chi^2(4)$. According to Table **E**, this value is

$$\chi_\alpha^2 = 9.49$$

Our observed test statistic $\chi^2 = 10.20$ exceeds χ_α^2, and our teacher must therefore conclude that his test scores are not uniformly distributed.

In closing our discussion of degrees of freedom, it is important to note that while Pearson's χ^2 is a discrete variable, all χ^2 distributions are continuous. The use of χ^2 to approximate the distribution of $\tilde{\chi}^2$ is therefore analogous to the normal approximation of the binomial discussed in Chapter Eight. Recall, in this regard, that determination of binomial probabilities from the normal curve involved the establishment of fictitious real limits 0.5 units beyond the apparent limits of the binomial interval in question. A similar correction for continuity may be applied to Pearson's χ^2, but for $\nu > 1$ the correction is cumbersome and makes little practical difference. For $\nu = 1$, however, use of $\chi^2(1)$ to approximate probabilities for Pearson's χ^2 can introduce substantial error, and it is therefore a good practice to adjust the computation of $\tilde{\chi}^2$ with *Yates' correction for continuity*:

$$\tilde{\chi}^2 = \sum_{j=1}^{2} \frac{(|f_j - \mathbf{F}_j| - .5)^2}{\mathbf{F}_j}$$

B. Testing Hypotheses About Two or More Distributions: Tests of Association

1. Tests of Homogeneity

In the last section we saw that Pearson's χ^2 may be used to determine whether or not a distribution of observed frequencies "fits" a distribution of expected frequencies based on some hypothesized distribution of the random variable. This is very similar to a test of one population mean where the investigator specifies an hypothesized mean value μ_0. As we found in Chapter Twelve, however, a scientist will often formulate the test hypothesis that two population means are equal without specifying values for μ_1 and μ_2. Similarly, Pearson's χ^2 may be used to test the hypothesis that a random variable is identically distributed in two (or more) populations without specifying the distribution of the random variable. In this regard, we know that if a random variable X is distributed identically in R populations, then by definition the probability associated with any particular interval $p(\Delta x_j)$ must be the same in every population. Formally, then, the investigator tests for equality or *homogeneity* of probability:

$$p(\Delta x_j)_1 = p(\Delta x_j)_2 = \ldots = p(\Delta x_j)_R = p(\Delta x_j)$$

where j runs from 1 to C, the number of intervals, and the subscripts $1, 2, \ldots R$ identify the populations in question.

By way of example, let us suppose that our teacher suspects that student performance on his midterm may be related to high school mathematics background. If he is correct, he intends to make two years of high school albegra the prerequisite for his course, but he does not want to add this requirement unless he can demonstrate significant differences between the midterm scores of students who have the algebra background and those who do not. He therefore formulates the following hypotheses:

$$H_0: p(\Delta x_j)_1 = p(\Delta x_j)_2$$

$$H_1: p(\Delta x_j)_1 \neq p(\Delta x_j)_2$$

where Population 1 is defined as all students with fewer than two years of high school algebra, Population 2 is defined as all students with two or more years of high school algebra, and X is the score on the midterm examination.

To test H_0 using Pearson's χ^2, we must first obtain estimates of the *marginal* probabilities $p(\Delta x_j)$ associated with our C intervals. In this regard, recall that the frequencies f_j associated with these intervals were compiled in the goodness of fit problem, and by definition each frequency may be divided by $N = 50$ to obtain the corresponding *relative* frequency (Table 14.3).

Table 14.3   Relative Frequency Distribution of Examination Scores

Δx_j	0–9	10–19	20–29	30–39	40–49	Σ
$f_j/50$	.08	.12	.30	.30	.20	1.00

Furthermore, recall from Chapter Five that for a binomial random variable

$$E(X) = Np$$

where N is the number of trials and p is the probability of success. If we multiply both sides of this expression by $1/N$ we obtain

$$\frac{1}{N} E(X) = p$$

and because $1/N$ is a constant, this becomes

$$E\left(\frac{X}{N}\right) = p$$

By definition, however, a binomial random variable is equal to the *frequency* of successes in N trials. Thus

$$E\left(\frac{f}{N}\right) = p$$

which means that for a binomial random variable, relative frequency is an unbiased estimator of probability:

$$\frac{f}{N} = \hat{p}$$

This conclusion also applies to *multinomial* variables, in which each experimental result is classified into one of $n > 2$ mutually exclusive categories. In the present example, therefore, the relative frequency of observations in any particular interval Δx_j is an unbiased estimate of the probability associated with that interval in the population represented by our 50 students.

$$\frac{f_j}{50} = \hat{p}(\Delta x_j)$$

If H_0 is correct, furthermore, we know that for any particular interval, $p(\Delta x_j)_1$ and $p(\Delta x_j)_2$ are equal to the common value $p(\Delta x_j)$. Thus

$$\frac{f_j}{N}$$

should be an unbiased estimate of $p(\Delta x_j)$ for both of our populations. That is,

$$\frac{f_j}{N} = \hat{p}(\Delta x_j) = \hat{p}(\Delta x_j)_1 = \hat{p}(\Delta x_j)_2$$

Moreover, if

$$\frac{f_j}{N} = \hat{p}(\Delta x_j)$$

then

$$f_j = N\hat{p}(\Delta x_j)$$

Thus, if we make N_1 observations on Population 1 and N_2 observations on Population 2 *and* if H_0 is correct, then our expected frequencies in Δx_j are

$$N_1 \hat{p}(\Delta x_j)_1 = N_1 \hat{p}(\Delta x_j)$$

and

$$N_2 \hat{p}(\Delta x_j)_2 = N_2 \hat{p}(\Delta x_j)$$

Let us suppose, therefore, that our instructor submits an inquiry to the office of admission and finds that 15 of his students have completed fewer than two years of high school algebra and that the other 35 have completed two or more years. If H_0 is correct, therefore, he should expect

$$N_1 \hat{p}(\Delta x_j) = 15 \left(\frac{f_j}{50} \right)$$

students with fewer than two years of algebra to obtain test scores in Δx_j, and he should expect

$$N_2 \hat{p}(\Delta x_j) = 35 \left(\frac{f_j}{50} \right)$$

students with two or more years to obtain scores in this interval. The entire set of such computations is presented in Table 14.4. and working out the indicated multiplications we obtain the values given in Table 14.5.

In the goodness of fit problem, each expected frequency was identified by a subscript corresponding to its column, but in this problem, each frequency must be identified by two subscripts, one of which indicates its column and one of which indicates its row. Thus, the cell entry in Table 14.5 for subjects with fewer than two years of algebra scoring from 30 to 39 points on the test would be identified as F_{41}, indicating the fourth column of the first row. The generalized notation is F_{jk} where j can take any value from 1 to C, the number of columns, and k can take any value from 1 to R, the number of rows.

Now let us suppose that he separates the midterms of his two groups and finds the observed frequencies in Table 14.6.

Table 14.4 Joint Distribution of Row Totals Multiplied by Column Probability Estimates (Relative Frequencies)

Δx_i	0–9	10–19	20–29	30–39	40–49	Σ
Population						
1 (0–1 year)	15(.08)	15(.12)	15(.30)	15(.30)	15(.20)	15
1 (2—years)	35(.08)	35(.12)	35(.30)	35(.30)	35(.20)	35
$f_i/N = \hat{p}(\Delta x_i)$	.08	.12	.30	.30	.20	1.00

Table 14.5 Expected Joint Frequency Distribution Based on Hypothesis of Homogeneity

Δx_i	0–9	10–19	20–29	30–39	40–49	Σ
Population	Expected Frequencies					
1 (0–1 year)	1.2	1.8	4.5	4.5	3.0	15
2 (2—years)	2.8	4.2	10.5	10.5	7.0	35
Σ	4	6	15	15	10	50

Table 14.6 Observed Joint Frequency Distribution of Examination Scores by Algebra Background

Δx_i	0–9	10–19	20–29	30–39	40–49	Σ
Population	Observed Frequencies					
1 (0–1 year)	1	4	7	2	1	15
2 (2—years)	3	2	8	13	9	35
Σ	4	6	15	15	10	50

Once again we compute $\tilde{\chi}^2$ by summing the squared differences between observed and expected frequencies in corresponding cells with each square divided by the expected frequency. We have already established, however, that each frequency is now identified by double subscripts, and under the notation conventions discussed in Appendix I, computation of our sum is therefore indicated with double summation signs:

$$\sum_{k}^{R} \sum_{j}^{C} \frac{(f_{jk} - F_{jk})^2}{F_{jk}}$$

Quite simply, this notation tells us to compute the sum

$$\sum_{j=1}^{C} \frac{(f_{jk} - F_{jk})^2}{F_{jk}}$$

for each row and then to take the sum of the R row totals. Following this procedure, the row totals are

$$\frac{(f_{11} - F_{11})^2}{F_{11}} + \frac{(f_{21} - F_{21})^2}{F_{21}} + \frac{(f_{31} - F_{31})^2}{F_{31}} + \frac{(f_{41} - F_{41})^2}{F_{41}} + \frac{(f_{51} - F_{51})^2}{F_{51}} =$$

$$\frac{(1 - 1.2)^2}{1.2} + \frac{(4 - 1.8)^2}{1.8} + \frac{(7 - 4.5)^2}{4.5} + \frac{(2 - 4.5)^2}{4.5} + \frac{(1 - 3.0)^2}{3.0} = 6.83$$

and

$$\frac{(f_{12} - F_{12})^2}{F_{12}} + \frac{(f_{22} - F_{22})^2}{F_{22}} + \frac{(f_{32} - F_{32})^2}{F_{32}} + \frac{(f_{42} - F_{42})^2}{F_{42}} + \frac{(f_{52} - F_{52})^2}{F_{52}} =$$

$$\frac{(3 - 2.8)^2}{2.8} + \frac{(2 - 4.2)^2}{4.2} + \frac{(8 - 10.5)^2}{10.5} + \frac{(13 - 10.5)^2}{10.5} + \frac{(9 - 7.0)^2}{7.0} = 2.93$$

Pearson's χ^2 is therefore equal to

$$6.83 + 2.93 = 9.76$$

In order to test this result, of course, we must determine the appropriate number of degrees of freedom. As before, this parameter is equal to the number of unconstrained cells and may be computed by the general formula

$$\nu = \text{total number of cells} - \text{number of constrained cells}$$

In the present example we have a total of $RC = 2 \times 5 = 10$ cells, and computation of ν requires only that we determine how many of these cells are constrained. In this regard, let us suppose that we have computed the first four expected frequencies in the first row according to the multiplications indicated earlier

$$F_{11} = N_1 \hat{p}(\Delta x_1) = 15(.08) = 1.2$$

$$F_{21} = N_1 \hat{p}(\Delta x_2) = 15(.12) = 1.8$$

$$F_{31} = N_1 \hat{p}(\Delta x_3) = 15(.30) = 4.5$$

$$F_{41} = N_1 \hat{p}(\Delta x_4) = 15(.30) = 4.5$$

It is not necessary to compute the expected frequency of the last cell, because we know that there are 15 students with fewer than 2 years of algebra and that the sum of the expected frequencies in the first row must therefore equal 15. Thus, the expected frequency of the last cell must equal

$$15 - (F_{11} + F_{21} + F_{31} + F_{41}) = 15 - (1.2 + 1.8 + 4.5 + 4.5)$$

$$= 15 - 12 = 3.0$$

Furthermore, we know from our earlier distribution of observed frequencies that a total of 10 students obtained scores in this last interval $40 - 49$. However they are distributed between our two groups, therefore, we must have a total of 10 observations in the last column. If, as indicated above, we expect three students with fewer than two years of algebra to obtain test scores in this interval, then we must also expect seven students with two or more years. By this reasoning, *every* cell in the second row is constrained, and we therefore have a total of six constrained cells. Thus

$$\nu = 10 - 6$$

$$= 4$$

As the reader has probably concluded, there is one constraint associated with every row total and one constraint associated with every column total. In the present example this yields $R + C = 2 + 5 = 7$ constraints. However, the last cell to be filled is always constrained *twice*. It is constrained once by its row total and once by its column total. Thus, the number of constrained cells is always *one less* than the number of rows plus the number of columns, and therefore

$$\nu = RC - (R + C - 1)$$

$$= RC - R - C + 1$$

$$= (R - 1)(C - 1)$$

In the present example, of course, our formula produces the same result as our manual solution

$$\nu = (2 - 1)(5 - 1) = 4$$

We know from our first example that for $\alpha = .05$, the critical value in the $\chi^2(4)$ distribution is $\chi_\alpha^2 = 9.49$. The value of our Pearson χ^2 test statistic exceeds this, and our teacher should therefore reject the hypothesis that statistics performance is unrelated to algebra background (which tells you something about my prejudices).

2. Tests of Independence

In the preceding example, the teacher was concerned with two variables, performance on the midterm examination and years of high school algebra. Furthermore, the second variable clearly defined two populations, students with fewer than two years of algebra and students with two or more years. Suppose, however, that our statistics instructor believed that performance on his midterm might be related to performance on the mathematics portion of the Scholastic Aptitude Test. In this situation, neither variable defines a meaningful set of populations, and it would therefore be more appropriate to conceptualize his hypothetical association as a relationship between two variables jointly distributed in the *same* population.

As before, midterm scores may be grouped into C intervals Δx_j, and in addition, the scores obtained by his students on the mathematics portion of the SAT may be similarly grouped into R intervals Δy_k. In the algebra example, each student fell simultaneously into one X-interval Δx_j and one of the two populations. In the present example, each student falls similarly into one X-interval Δx_j and one Y-interval Δy_k. And as before, the instructor may record the joint frequency f_{jk} associated with each combination of his two variables (Table 14.7).

Recall now from Chapter One that if two events E_1 and E_2 are independent, then

$$p(E_1E_2) = p(E_1)p(E_2)$$

Thus, if midterm performance is, as hypothesized, independent of SAT score, then the probability that any student selected at random from the population under investigation will fall simultaneously in Δx_j and Δy_k must be given by

$$p(\Delta x_j \Delta y_k) = p(\Delta x_j)p(\Delta y_k)$$

Formally, this equation defines the test hypothesis in a χ^2 test of independence.

To test this hypothesis we must, as before, generate a set of expected frequencies $\mathbf{F}_{jk}$ corresponding to the observed frequencies f_{jk} discussed above.

Table 14.7 Layout for Observed Joint Frequency Distribution of Examination Scores by SAT Scores

Midterm Scores	Δx_1 . . . Δx_j . . . Δx_C		
SAT Scores	Observed Frequencies		
Δy_1	f_{11}	f_{j1}	f_{C1}
. . .	. . .	. . .	. . .
Δy_k	f_{1k}	f_{jk}	f_{Ck}
. . .	. . .	. . .	. . .
Δy_R	f_{1R}	f_{jR}	f_{CR}

Once again, we estimate the multinominal probability associated with Δx_j by the relative frequency of observations in that interval:

$$\hat{p}(\Delta x_j) = \frac{f_j}{N}$$

Similarly, we estimate the probability associated with Δy_k by its relative frequency:

$$\hat{p}(\Delta y_k) = \frac{f_k}{N}$$

And under the test hypothesis, the probability associated with cell jk is therefore estimated as

$$\hat{p}(\Delta x_j \Delta y_k) = \hat{p}(\Delta x_j)\hat{p}(\Delta y_k)$$

In this regard, we again note that if

$$\frac{f}{N} = p$$

then

$$f = N\hat{p}$$

To obtain the expected frequency of observations in any cell, we therefore multiply the total number of observations by our estimate of the probability expected for that cell under H_0:

$$\mathbf{F}_{jk} = N\hat{p}(\Delta x_j \Delta y_k)$$
$$= N\hat{p}(\Delta x_j)\hat{p}(\Delta y_k)$$

We see, then, that the expected frequencies derive directly from the condition of independence expressed the test hypothesis:

$$p(\Delta x_j \Delta y_k) = p(\Delta x_j)p(\Delta y_k)$$

Thus, the rationale underlying our computations of $\mathbf{F}_{jk}$ in the test of independence is quite different from the reasoning that generates $\mathbf{F}_{jk}$ in a test of homogeneity. The reader will be relieved to discover, however, that irrespective of these differences, the test statistic is identical:

$$\tilde{\chi}^2 = \sum_k^R \sum_j^C \frac{(f_{jk} - \mathbf{F}_{jk})^2}{\mathbf{F}_{jk}}$$

Furthermore, recall that $\hat{p}(\Delta y_k)$ is equal to f_k/N. Thus

$$\mathbf{F}_{jk} = N\hat{p}(\Delta x_j)\hat{p}(\Delta y_k)$$
$$= N\hat{p}(\Delta x_j)\frac{f_k}{N}$$
$$= \hat{p}(\Delta x_j)f_k$$

where the frequency f_k is the total number of observations in row k. Using the notation established earlier, our expression may therefore be rewritten as

$$\mathbf{F}_{jk} = N_k \hat{p}(\Delta x_j)$$

which is identical to the computation for $\mathbf{F}_{jk}$ is a test of homogeneity. While in principle the expected frequencies in χ^2 tests of homogeneity and independence derive from different experimental assumptions, therefore, in practice they are equivalent.

When $\nu = 1$ in either test of association it is somewhat easier to calculate Pearson's χ^2 using

$$\tilde{\chi}^2 = \frac{N(ad - bc)^2}{(a + b)(c + d)(a + c)(b + d)}$$

where $a = f_{11}$, $b = f_{21}$, $c = f_{12}$, and $d = f_{22}$. However, it is customary in this situation to apply Yates' correction for continuity, which is incorporated into the following computational formula:

$$\tilde{\chi}^2 = \frac{N(|\, ad - bc\,| - N/2)^2}{(a + b)(c + d)(a + c)(b + d)}$$

C. Assumptions Underlying Pearson's χ^2

1. Distribution Properties of X: "Nonparametrics" and Other Misnomers

Earlier in this chapter we pointed out that the test statistics developed in Chapters Twelve and Thirteen are appropriate to hypotheses about particular distribution properties. In this regard, the reader may have noted that every test statistic developed in Chapter Twelve was a function of the estimator $\bar{X}$ and that every test statistic developed in Chapter Thirteen involved the estimator $\hat{\sigma}^2$. In contrast, Pearson's χ^2 will detect differences in location, dispersion, skew, or any other descriptive property, and, concomitantly, its mathematical expression includes no obvious estimator of any particular parameter. Consequently, such statistics are often characterized as "estimation-free" or "nonparametric." At least in the case of Pearson's χ^2, however, these labels are manifestly inappropriate, because the marginal relative frequencies f_j/N and f_k/N are used to estimate, respectively, the multinomial probabilities $p(\Delta x_j)$ and $p(\Delta y_k)$.

Perhaps the feature of Pearson's χ^2 that most clearly distinguishes it from our other test statistics is in the *assumptions* that must be met if the test is to yield meaningful results. In order to assume that the distribution of Pearson's χ^2 is approximated by $\chi^2(\nu)$, it must be true that:

i.   Every observation falls into one and only one class Δx_j.
ii. The N observations are independent of one another.
iii. The sample size N is large.

This last requirement is the most difficult to pin down precisely and is generally expressed as a rule of thumb: If $\nu = 1$, no expected frequency F_{jk} should be less than 10; If $\nu > 1$, no expected frequency F_{jk} should be less than 5. (We should note, however, that the frequency requirement for $\nu = 1$ may be relaxed a bit if Yates' correction is applied.) Unlike the other tests we have considered, therefore, the assumptions underlying Pearson's χ^2 are primarily concerned with data collection rather than the probability distribution of the random variable. Some writers have therefore suggested that test statistics like χ^2 should be designated as "distribution-free." This label is also misleading, however, because the investigator must always be able to specify the probability distribution of his *test statistic*, even if he cannot specify the distribution of his random variable. In the long run, therefore, it may be more useful to abandon existing nomenclature and identify tests like Pearson's χ^2 as *minimal assumption* techniques.

2. Scaling Properties of *X:* Levels of Measurement

In Chapter Seven we stated that two measurements could be meaningfully compared only if they were expressed in the same units of measurement and taken from the same reference point. It should be pointed out, however, that these stipulations define a particular *level of measurement* that is not necessarily implied by the simple operation of assigning a numerical value to an observation. To gain some understanding of levels of measurement, let us consider a grand prix in which five cars have qualified. The starting lineup includes one Ferrari, one Porsche, one Honda, one Cooper-Maserati, and one Lotus. To facilitate identification of the cars for officials and spectators, the cars are assigned identification numbers as follows:

Car	Number
Cooper-Maserati	1
Ferrari	2
Honda	3
Porsche	4
Lotus	5

In this example, the numbers assigned to our units of observation (cars) are simply arbitrary symbols that serve to distinguish them from one another, and it should therefore be apparent that numbers used in this fashion have no quantitative implications. Designating the Honda as entry number 3, for example, in no way implies that it possesses less of some attribute than entries 4 and 5 or that it possesses two more units of some attribute than entry number 1. Indeed, the number serves merely as a shorthand name, and any such array of numbers is therefore called a *nominal* scale.

Random variables are expressed in terms of nominal scales when observations must be classified in a set of *qualitative* categories that cannot be meaningfully ordered according to any *quantitative* property or attribute. Examples of nominally scaled variables in the behavioral and biological sciences therefore include sex, political voting preference, blood type, agreement or disagreement with a statement on an opinion questionnaire, presence or absence of a particular biochemical condition, and so forth.

Let us now suppose that the race is over and that the cars finished in the following order:

Car	Position
Ferrari	1
Lotus	2
Porsche	3
Honda	4
Cooper-Maserati	5

In this instance, the numbers do represent some quantitative attribute, for example, time. Assuming that all of the cars finished, we know that car number 1 spent less total time on the course than any of the others, that car number 2 spent less time than car number 3 but more time than car number 1, and so forth. However, this scale does not tell us how *many* more minutes car number 2 spent than car number 1. That is, we cannot make quantitative statements about the *difference* between the amount of this attribute possessed by two observations. Thus, even though the difference in *position* between the Porsche and the Maserati

$$5 - 3 = 2$$

is the same as the difference in position between the Porsche and the Ferrari

$$3 - 1 = 2$$

we cannot infer that the corresponding time differences were equal. The scale in this example provides information only about the order in which the observations may be arrayed and is therefore an example of an *ordinal* scale.

Even with well-developed measurement techniques, ordinal scaling is often the highest level of measurement that can be legitimately assumed in the behavioral sciences. In an intelligence study, for example, a psychologist may be reasonably certain that an I.Q. of 145 indicates more of some attribute (which he loosely calls "intelligence") than an I.Q. of 140. However, he also knows that intelligence measurement breaks down at the extremes. Thus, he would be hard put to assume that the difference in intelligence between persons with I.Q. scores of 140 and 145 is the same as the difference between persons scoring, say, 100 and 105.

Finally, let us return to our racing example and suppose that the timekeeper has posted the official times, to the nearest minute, recorded for each entry:

Car	Time in Minutes
Ferrari	67
Lotus	72
Porsche	75
Honda	81
Cooper-Maserati	84

Using this scale, we can both array the observations in a meaningful quantitative order and we can make inferences about the differences between observations. We can, for example, say that the difference between the Maserati and the Honda

$$84 \text{ minutes} - 81 \text{ minutes} = 3 \text{ minutes}$$

is the same as the difference between the Porsche and the Lotus

$$75 \text{ minutes} - 72 \text{ minutes} = 3 \text{ minutes}$$

Furthermore, we can say that this difference is half as large as the difference between the Honda and the Porche

$$81 \text{ minutes} - 75 \text{ minutes} = 6 \text{ minutes}$$

That is, we can perform a number of meaningful arithmetic operations on the intervals defined by our observations. Scales that have these properties are therefore defined as *interval scales*.

It should be apparent from this last example that the stipulations laid down in Chapter Seven refer explicitly to interval scales. Indeed, all of the random variables that we have discussed in our presentation of inferential techniques have possessed interval properties. In addition to requiring assumptions concerning the probability distribution of X, therefore, the test statistics developed in Chapters Twelve and Thirteen implicitly assume the random variable to represent an interval scale of measurement. In contrast, Pearson's χ^2 techniques may be applied to data at any level of measurement. In this regard, a significant test of independence is frequently interpreted as indicating *covariation* of X and Y. Indeed, such tests are often conducted when data fail to meet the rigid distribution and scaling assumptions required for a meaningful test of the Pearson correlation ρ. In this light, it seems even more appropriate to adopt the designation "minimal assumption" in place of such terms as "nonparametric" or "distribution-free," which refer *only* to assumptions about distribution properties and *not* scaling properties.

D. Problems, Review Questions, and Exercises

1. One of the most important applications of Pearson's χ^2 is the test for goodness of fit with the normal distribution. To understand this test, suppose that you intend to draw and standardize 50 observations on a random variable that you believe to be normally distributed. Assume, too, that the standard normal random variable runs from -3.00 to $+3.00$. (This assumption is not necessary; it simply avoids openended intervals.)
   a. If you wish to define 10 intervals of equal width, what will your real limits be?

b. Multiply the probability associated with every such interval by $N = 50$ to determine the expected frequency in each cell.

c. Do your expected frequencies satisfy the assumption that $F_j \geq 5$?

d. What sample size would be necessary to satisfy this requirement?

From the frequencies computed in **1b** it should be apparent that the difficulty considered in **1c** is attributable to the very low probabilities in the tails of the normal curve. One way to avoid this problem is to establish intervals of equal *probability* rather than intervals of equal *width*.

e. To establish the limits for 10 equally probable intervals, use Table **C** to find Z-values with cumulative probabilities of .10, .20, .30, . . . 1.00.

f. Assuming 50 observations, tabulate the expected frequencies for intervals with limits defined by the Z-values in **1e**.

From Table **B**, draw 50 two-digit numbers for which the first digit is 0, 1, or 2. Divide every even number you draw by $+10$ and every odd number by -10.

g. Assuming that these numbers are your standardized observations, construct their observed frequency distribution for the intervals in **1f**.

h. Compute $\tilde{\chi}^2$ based on the expected frequencies in **1f** and the observed frequencies in **1g**.

i. What is ν for this test?

j. Is your test statistic significant for $\alpha = .001$. Why do you suppose this is so?

2. After collecting student course evaluations for several semesters, a statistics teacher begins to suspect that he is a more effective instructor for juniors and seniors than for freshmen and sophomores. His course is rated A, B, C, D, or F by the students, and separating the evaluations of upper and lower division students, he obtains the following distribution:

	A	B	C	D	F
Lower division	3	8	15	5	1
Upper division	7	12	9	2	2

a. Generate an appropriate expected frequency distribution for a Pearson's χ^2 test of homogeneity.

b. Compute $\tilde{\chi}^2$. Is your value of $\tilde{\chi}^2$ significant at the .05 level?

3. One of the examples in Chapter Fourteen violates an important assumption underlying Pearson's χ^2.

a. What is the violation and in which table does it appear?

b. How can the violation be corrected?

c. Compute Pearson's χ^2 based on your answer to **3b**.

d. Is your result significant at $\alpha = .025$?

4. If your answer to **3d** is "Yes," some sources would argue that the result is indeed significant at $\alpha/2 = .0125$ because the alternative hypothesis was implicitly *directional*. Can you explain this assertion?

E. Summary Table

H_0	H_1	Test Statistic T	Distribution of T

A. Goodness of Fit Test for a Single Population Distribution

$X:P(X)^a$	H_0 incorrect	$\sum_{j=1}^{C} \dfrac{(f_j - F_j)^{2}}{F_j}\,^b$	$\chi^2 (C - 1)$

Assumptions: Classes Δx_j mutually exclusive
$\qquad\qquad\quad N$ observations independent
$\qquad\qquad\quad$ Sample size N large

B. Tests of Association
1. Homogeneity

$p(\Delta x_j)_k = p(\Delta x_j)$	H_0 incorrect	$\sum_{k}^{R} \sum_{j}^{C} \dfrac{(f_{jk} - F_{jk})^{2}}{F_{jk}}\,^c$	$\chi^2([R - 1][C - 1])$

Assumptions: Same as for goodness of fit tests

2. Independence

$p(\Delta x_j \Delta y_k) =$ $\quad p(\Delta x_j)p(\Delta y_k)$	Same	Same ^{c}	Same

Assumptions: Same

[a] $P(X)$ is any probability function or probability density function specified by the theoretical model under test.

[b] When $C = 2$, a better approximation to $\chi^2(1)$ is obtained by incorporating Yates' correction for continuity:

$$\tilde{\chi}^2 = \sum_{j=1}^{2} \frac{(|f_j - F_j| - .5)^2}{F_j}$$

[c] When $C = R = 2$, a better approximation to $\chi^2(1)$ is obtained by incorporating Yates' correction for continuity:

$$\tilde{\chi}^2 = \frac{N(|ad - bc| - N/2)^2}{(a + b)(c + d)(a + c)(b + d)}$$

where

$$a = f_{11}$$
$$b = f_{21}$$
$$c = f_{12}$$
$$d = f_{22}$$

testing hypotheses about many population means: introduction to analysis of variance

A. Overview

In Chapter Twelve we developed techniques for testing hypotheses about one population mean and about two population means. Frequently, however, the scientist will be concerned with testing hypotheses about differences among the means of three or more populations. In this final chapter we therefore introduce the general strategy and methods involved in testing the hypothesis

$$H_0: \mu_1 = \mu_2 = \ldots = \mu_J = \mu$$

Paradoxically, these methods focus on dispersion rather than location, and the general rubric for such tests is therefore called *analysis of variance*.

Before we introduce the formal mathematical bases of analysis of variance, it will be helpful to consider a typical analysis of variance problem and to intuitively develop the basic logic underlying this extremely versatile and elegant statistical tool. Suppose, therefore, that we wish to determine whether or not three consciousness-altering drugs differ in their effects on performance of a particular learning task. To begin, we assign subjects at random to three groups with N_1 subjects in the first group, N_2 subjects in the second group, and N_3 subjects in the third group. We then administer a quantity of ethanol to each of the subjects in Group 1, a quantity of tetrahydrocannibinol to every member of Group 2, and a dose of dextroamphetamine sulfate to the subjects in Group 3. Finally, we present the learning task to all $N_1 + N_2 + N_3 = N$ subjects and record their scores. This procedure would yield the array of data shown in Table 15.1.

Table 15.1 Arrangement, Identification, and Description of Three Groups of Observations

Group 1 (Ethanol)	Group 2 (THC)	Group 3 (Amphetamine)	
x_{11}	x_{12}	x_{13}	
x_{21}	x_{22}	x_{23}	
. . .	. . .	. . .	
$x_{N_1 1}$	$x_{N_2 2}$	$x_{N_3 3}$	
Σx_1	Σx_2	Σx_3	$\Sigma\Sigma x$
$\bar{x}_1$	$\bar{x}_2$	$\bar{x}_3$	$\bar{x}$

In this display, each score has two subscripts. As indicated in Appendix I, the second subscript j indicates the group to which each subject belongs, and the first subscript i distinguishes among members of the same group. Thus, x_{11} is the score obtained by the first subject in Group 1, x_{21} is the score obtained by the second subject in Group 1, and so forth. Immediately below our two-dimensional array of scores, we have the individual group sums Σx_j and the grand sum $\Sigma\Sigma x$ based on all $N_1 + N_2 + N_3 = N$ subjects. Finally, our bottom line gives the mean score for each group $\bar{x}_j$ and the grand mean $\bar{x}$ computed across all N subjects.

1. The Test Hypothesis

In this experiment (as in any inference problem) we are concerned with performance differences among our subjects only insofar as these differences may be generalized to the populations from which the subjects were drawn. In selecting three groups therefore, we implicitly define three populations:

> Population 1. All persons who *might* have performed the experimental task under the influence of ethanol.
>
> Population 2. All persons who *might* have performed the experimental task under the influence of THC.
>
> Population 3. All persons who *might* have performed the experimental task under the influence of dextroamphetamine sulfate.

We assume, of course, that all persons in any population or group of populations will exhibit variability in performing our experimental task. As indicated on several occasions, some of this variability will derive from *random* differences in skill, interest in the task at hand, attention, experience with similar tasks, and so forth. Such variation is called *error* because it is not under experimental control. In addition, some of the variation may be attributable to the drugs that were administered to the subjects. In contrast to the random differences attributable to error, any such *treatment effects* should produce *systematic* differences among subjects who receive different drugs, which by definition implies systematic differences among our three experimental populations. In this regard, we know that performance typical of a given population can best be summarized by its population mean μ. Thus, any systematic differences attributable to drug treatments should be reflected in differences among our three population means. If, however, our treatments produce *no* systematic differences, our population means should be equal to one another and therefore equal to the mean of the *grand population* comprised of all persons who might have performed our experimental task under *any* of the three conditions. Thus, the appropriate test hypothesis in our example is

$$H_0: \mu_1 = \mu_2 = \mu_3 = \mu$$

where the common value μ is the grand population mean, and the alternative hypothesis is

$$H_1: \mu_j \neq \mu \text{ for at least two of the populations}$$

2. The Test Statistic

We suggested earlier that each treatment group represents a sample drawn from one of our three conceptual populations. Any differences among our population means should therefore be reflected in differences among our group means, which may be described by their variance $\sigma_{\bar{X}}^2$. If, in this regard, we assume that each group includes N_j observations, then we know from Chapter Ten that

$$\sigma_{\bar{X}}^2 = \frac{\sigma^2}{N_j}$$

In the general expression $\sigma_{\bar{X}}{}^2 = \sigma^2/N$, the parameter σ^2 is the variance of the the population from which the experimenter draws his sample(s). In the present context, then, σ^2 is the variance of the grand population and therefore taps *all* sources of performance variability. We must thus consider σ^2 to be composed of *treatment* variance plus *error* variance, both of which must consequently be reflected in $\sigma_{\bar{X}}{}^2$. If, however, we assume the test hypothesis to be correct, our drugs produce no systematic performance differences, and all variability in the grand population must be attributable to random error. Under this assumption, therefore, the grand population variance σ^2 is equal to the error variance, which we designate $\sigma_e{}^2$, and

$$\sigma_{\bar{X}}{}^2 = \frac{\sigma_e{}^2}{N_j}$$

By the preceding argument, of course, $\sigma_{\bar{X}}{}^2$ is entirely a function of error variance *only* when H_0 is correct. In contrast, we know by definition that all members of Population j receive the same drug. Thus, variance within each group, unlike variation among group means, must be entirely attributable to randon error *whether or not H_0 is correct.* The unbiased estimate of variance for Population j

$$\hat{\sigma}_j{}^2 = \frac{\sum_{i}^{N_j} (x_{ij} - \bar{x}_j)^2}{N_j - 1}$$

is therefore an estimate of the *error* variance in Population j. If, in addition, we recognize that each treatment group constitutes a sample drawn from the grand population, then $\hat{\sigma}_j{}^2$ is also an estimate of error variance in the grand population $\sigma_e{}^2$. Because we have more than one group, however, our best estimate of $\sigma_e{}^2$ is obtained by *pooling* $\hat{\sigma}_1{}^2$, $\hat{\sigma}_2{}^2$, and $\hat{\sigma}_3{}^2$ as discussed in Chapter Twelve. Even if H_0 is incorrect, therefore,

$$\hat{\sigma}^2_{\text{pooled}} = \hat{\sigma}_e{}^2$$

If H_0 is correct, however, we have already established that

$$\sigma_{\bar{X}}{}^2 = \frac{\sigma_e{}^2}{N_j}$$

Similarly

$$\hat{\sigma}_{\bar{X}}{}^2 = \frac{\hat{\sigma}_e{}^2}{N_j}$$

and

$$N_j \hat{\sigma}_{\bar{X}}{}^2 = \sigma_e{}^2$$

When *and only when* the test hypothesis is correct, therefore, $\hat{\sigma}^2_{\text{pooled}}$ and $N_j \hat{\sigma}_{\bar{X}}{}^2$ are *both* estimates of $\sigma_e{}^2$. In this regard, we know from Chapter Thirteen that if

$\hat{\sigma}_1{}^2$ and $\hat{\sigma}_2{}^2$ are estimates of the same variance, then $\hat{\sigma}_1{}^2/\hat{\sigma}_2{}^2$ is distributed as $F(N_1 - 1, N_2 - 1)$. We may thus test H_0 by submitting the ratio

$$\frac{N_j \hat{\sigma}_{\bar{x}}{}^2}{\hat{\sigma}^2_{\text{pooled}}}$$

to the appropriate distribution of Fisher's F.

3. Assumptions Underlying Analysis of Variance

In Chapter Thirteen we pointed out that Fisher's F is derived from the joint distribution of two independent χ^2 random variables and that the test statistic

$$\frac{\hat{\sigma}_1{}^2}{\hat{\sigma}_2{}^2}$$

can be distributed as $F(N_1 - 1, N_2 - 1)$ only if $\hat{\sigma}_1{}^2$ and $\hat{\sigma}_2{}^2$ are based on independent samples, each of which is comprised of independent observations. Even if the analysis of variance test hypothesis is correct, therefore, the ratio

$$\frac{N_j \hat{\sigma}_{\bar{x}}{}^2}{\hat{\sigma}^2_{\text{pooled}}}$$

will be distributed as Fisher's F only if our N observations x_{ij} are independent of one another both within groups and across groups. This assumption can be justified, of course, only if subjects are assigned completely at random to the various treatment groups.

The second assumption underlying analysis of variance is also related to the requirement that $\hat{\sigma}_1{}^2$ and $\hat{\sigma}_2{}^2$ be independent of one another. In this regard we have indicated that $\hat{\sigma}^2_{\text{pooled}}$ is based on *variance* estimates $\hat{\sigma}_j{}^2$ and that $N_j \hat{\sigma}_{\bar{x}}{}^2$ is based on variation among the J sample *means* $\bar{x}_j$. Furthermore, recall from our development of Student's t in Chapter Twelve that sample means and variances are independent of one another only when X is normally distributed. Thus, if our ratio is indeed to be distributed as Fisher's F, we must assume that X is normally distributed in all of our J populations. Fortunately, this assumption is not difficult to justify if one keeps in mind that all variability within any population is attributable to error. Error, remember, includes such uncontrolled sources of variability as individual differences in personality, skills that may be relevant to the experimental task, transient motivational states, and so forth. If we are willing to assume that these influences are operating independently of one another and that each contributes some positive or negative increment to every score, then the total error contribution to any observation becomes a sum of N independent random variables, and under the Central Limit Theorem we may conclude that such a sum is normally distributed. Even when this conception of error is not appropriate, the experimenter may nonetheless capitalize on the CLT if all of the sample sizes N_j are relatively large.

Finally, if our pooled estimate of variance is indeed to estimate the error variance of the grand population, we must assume that all of our experimental population variances are identical:

$$\sigma_1{}^2 = \sigma_2{}^2 = \ldots = \sigma_J{}^2 = \sigma_e{}^2$$

Because it is virtually impossible to identify all sources of error, however, this assumption is often difficult to justify. Fortunately, it may be violated without serious risk if the number of observations in each group is the same. That is, if

$$N_1 = N_2 = \ldots = N_J$$

B. Components of Variability

In the Overview to this chapter, we developed several important points. First, we argued that a test hypothesis concerning differential effects of J experimental treatments should be formulated in terms of differences among the means of our J experimental populations μ_j. Moreover, we suggested that any such differences should be reflected in variability among the means $\bar{x}_j$ of our J treatment groups. Pursuing this line of reasoning, however, we discovered that such variability reflects *both* treatment effects and random error *unless H_0 is correct*, in which case variability among sample means may be attributed entirely to random error. In contrast, we pointed out that variability among observations in the *same* treatment group is entirely attributable to error whether H_0 is *correct or incorrect*. Given certain assumptions we therefore concluded that the hypothesis of equal population means could be tested by testing a variance estimate based on differences *among* treatment groups against an estimate based on differences *within* treatment groups.

The logic of analysis of variance really requires no further elaboration, and in the present section we therefore focus entirely on mathematical considerations. First, we demonstrate formally that total variability can be expressed as variability within groups plus variability among groups. Second, we transform these two components into variance estimates that may be submitted to an F test.

1. Variability as Sums of Squares

In the past we have frequently expressed statistical variability as some function of the sum of squared deviations about the mean

$$\sum_{i}^{N} (x_i - \bar{x})^2$$

This *sum of squares (SS)* appears in both the sample standard deviation s^2 and in the unbiased estimator of the population variance $\hat{\sigma}^2$. It should not be surprising, therefore, that the total variability in any collection of N scores x_{ij} obtained under J experimental conditions is expressed in terms of the sum of squared deviations from the grand mean $\bar{x}$. In analysis of variance, however, each score is identified by double subscripts, and by the conventions established in Appendix I, the notation for total sums of squares (SST) requires double summation signs:

$$SST = \sum_{j}^{J} \sum_{i}^{N_j} (x_{ij} - \bar{x})^2$$

$$= \sum_{i}^{N_1} (x_{i1} - \bar{x})^2 + \sum_{i}^{N_2} (x_{i2} - \bar{x})^2 + \ldots + \sum_{i}^{N_J} (x_{iJ} - \bar{x})^2$$

It is also indicated in Appendix I that double summations are often expressed without limits. Thus

$$\sum_{j}^{J} \sum_{i}^{N_j} (x_{ij} - \bar{x})^2 = \sum_{j} \sum_{i} (x_{ij} - \bar{x})^2$$

where

$$\sum_{i}^{N_j} = \sum_{i}$$

is based entirely on observations in Group j and is therefore called the *within-group* sum, and

$$\sum_{j}^{J} = \sum_{j}$$

is computed over all J groups and is thus called the *across-group* sum.

   With these notation conventions established, we may proceed to decompose the total sum of squares into two components, one of which reflects differences within treatment groups, and one of which reflects differences among treatment groups. To begin, we know that any constant may be added to and then subtracted from a value without changing that value, and it must therefore be true that

$$x_{ij} - \bar{x} = x_{ij} - \bar{x} + \bar{x}_j - \bar{x}_j$$

or

$$x_{ij} - \bar{x} = (x_{ij} - \bar{x}_j) + (\bar{x}_j - \bar{x})$$

Thus

$$\sum_{j} \sum_{i} (x_{ij} - \bar{x})^2 = \sum_{j} \sum_{i} [(x_{ij} - \bar{x}_j) + (\bar{x}_j - \bar{x})]^2$$

$$= \sum_{j} \sum_{i} [(x_{ij} - \bar{x}_j) + 2(x_{ij} - \bar{x}_j)(\bar{x}_j - \bar{x}) + (\bar{x}_j - \bar{x})^2]$$

$$= \sum_{j} \sum_{i} (x_{ij} - \bar{x}_j)^2 + 2 \sum_{j} \sum_{i} (x_{ij} - \bar{x}_j)(\bar{x}_j - x) + \sum_{j} \sum_{i} (\bar{x}_j - \bar{x})^2$$

Consider the second term

$$2 \sum_j \sum_i (x_{ij} - \bar{x}_j)(\bar{x}_j - \bar{x})$$

Although $\bar{x}$ is a constant across all groups, $\bar{x}_j$ varies from group to group. Within any given group, however, $\bar{x}_j$ is also a constant. Thus $(\bar{x}_j - \bar{x})$ is a constant within any group and by our algebra of summations may therefore be factored out of the within-group sum $\sum_i$. The second term may consequently be rearranged as follows:

$$2 \sum_j (\bar{x}_j - \bar{x}) \sum_i (x_{ij} - \bar{x}_j)$$

However, $\sum_i (x_{ij} - \bar{x}_j)$ is the sum of deviations from the group mean and is therefore equal to zero, which means that the entire second term is equal to zero and therefore falls out of our expression. Thus

$$\sum_j \sum_i (x_{ij} - \bar{x})^2 = \sum_j \sum_i (x_{ij} - \bar{x}_j)^2 + \sum_j \sum_i (\bar{x}_j - \bar{x})^2$$

Because $(\bar{x}_j - \bar{x})$ is a constant within Group j, however, the sum

$$\sum_i^{N_J} (\bar{x}_j - \bar{x})^2 = N_j(\bar{x}_j - \bar{x})^2$$

and the last term

$$\sum_j \sum_i (\bar{x}_j - \bar{x})^2 = \sum_i^{J} N_j(\bar{x}_j - \bar{x})^2$$

Thus

$$\sum_j \sum_i (x_{ij} - \bar{x})^2 = \sum_j \sum_i (x_{ij} - \bar{x})^2 + \sum_j^{J} N_j(\bar{x}_j - \bar{x})^2$$

If we now consider the terms on the right-hand side of the expression, it will be noted that

$$\sum_j \sum_i (x_{ij} - \bar{x}_j)^2$$

is based entirely on differences of observations from their own group means. Hence, this sum of squared deviations reflects only within-group variability and is therefore defined as the *sum of squares within groups*, or *SSW*. In contrast, the term

$$\sum_j^{J} N_j (\bar{x}_j - \bar{x})^2$$

is based entirely on differences of group means from the grand mean. This sum of squared deviations therefore reflects across-group or among-group variability and is consequently defined as the *sum of squares among groups*, or *SSA*. Thus

$$SST = SSW + SSA$$

Decomposing the total sum of squares into the sum of squares within groups plus the sum of squares among groups is known as *partitioning* the sum of squares.

a. Computational Formulas for the Sums of Squares. The expressions for sums of squares we have just derived are the definitional formulas and clearly demonstrate that the total variability in a collection of N observations may indeed be partitioned into components that reflect within-group variability and among-group variability. To actually calculate the sums of squares, however, it is far more convenient to use the computational formulas derived in Appendix III:

$$SST = \sum_j \sum_i x_{ij}{}^2 - N\bar{x}^2$$

$$SSW = \sum_j \sum_i x_{ij}{}^2 - \sum_j^J N_j \bar{x}_j{}^2$$

$$SSA = \sum_j^J N_j \bar{x}_j - N\bar{x}^2$$

2. Mean Squares and the *F*-Ratio

In addition to facilitating calculation, the computational formulas derived above permit us to develop a number of theoretical points with minimal recourse to extraneous mathematics. First, it will be recalled that in our development of Pearson's χ^2 we defined degrees of freedom as the number of cells that can assume arbitrary frequencies. More generally, the number of degrees of freedom associated with any statistic is the number of *observations* that may vary without constraint. If, in this regard, we examine the computational formula for *SSW*:

$$SSW = \sum_j \sum_i x_{ij}{}^2 - \sum_j^J N_j \bar{x}_j{}^2$$

we see that the first term is a random variable defined by $N_1 + \ldots + N_J = N$ observations. Under the assumptions discussed earlier in this chapter, all of these observations are independent of one another, and each is therefore free to assume any value in the value set of X. Thus, the first term $\Sigma\Sigma x_{ij}{}^2$ must have N degrees of freedom. Examining the second term, we see that it is based on the J sample means $\bar{x}_1, \ldots \bar{x}_J$. Once again, our assumptions stipulate that our treatment groups be independent of one another, and each mean is therefore free to assume any value in the value set of $\overline{X}$. The second term thus has J degrees of freedom, and by the additive property of degrees of freedom discussed in Chapter Thirteen, SSW must therefore have a total of $N - J$ degrees of freedom.

The degrees of freedom associated with SSA may also be determined by inspection of our computational formula.

$$SSA = \sum_{j}^{N} N_j \bar{x}_j^2 - N\bar{x}^2$$

We have already established that the first term is a random variable with J degrees of freedom. In contrast, the second term $N\bar{x}^2$ is the product of two constants and must likewise be a constant. For any value of N that the investigator may abritrarily select, therefore, the value of the other factor $\bar{x}^2$ is completely determined. Thus, $N\bar{x}^2$ has only one degree of freedom, and SSA must have a total of $J - 1$ degrees of freedom.

To determine the degrees of freedom for SST, we simply note that

$$SST = SSW + SSA$$

and that SST must have a total of

$$(N - J) + (J - 1) = N - 1$$

degrees of freedom. In this connection, our definitional formula for SST was given as

$$\sum_{j} \sum_{i} (x_{ij} - \bar{x})^2$$

If there is no need to identify the treatment group to which each subject belongs, this may be rewritten as

$$SST = \sum_{i}^{N} (x_i - \bar{x})^2$$

Knowing that SST has $N - 1$ degrees of freedom, it now becomes apparent that unbiased estimator of the population mean

$$\hat{\sigma}^2 = \frac{\sum_{i}^{N} (x_i - \bar{x})^2}{N - 1}$$

is the total sum of squares divided by its degrees of freedom. Any such sum of squares divided by its own degrees of freedom is defined as a *mean squares*.

a. **Expected Value of the Mean Squares Within Groups.** Earlier in this section we established that the sum of squares within groups has $N - J$ degrees of freedom, and by our definition of mean squares, therefore,

$$\frac{SSW}{N - J} = \frac{\sum_{j} \sum_{i} (x_{ij} - \bar{x}_j)^2}{N - J}$$

becomes the *mean squares within groups* (*MSW*). Furthermore, we may expand the right-hand side of this equation and obtain

$$\frac{\sum_{i}^{N_1} (x_{i1} - \bar{x}_1)^2 + \sum_{i}^{N_2} (x_{i2} - \bar{x}_2)^2 + \ldots + \sum_{i}^{N_J} (x_{iJ} - \bar{x}_J)^2}{N_1 + N_2 + \ldots + N_J - J}$$

which the reader might conceivably recognize as a generalization of the formula given in Chapter Twelve for the pooled estimate of variance. Thus

$$MSW = \hat{\sigma}^2_{\text{pooled}}$$

In this regard, we argued in the chapter overview that $\hat{\sigma}^2_{\text{pooled}}$ is an unbiased estimate of the error variance, and by the definition of unbiasedness given in Chapter Ten, it should therefore follow that

$$E(MSW) = \sigma_e^2$$

To prove this, we remind the reader that the sum of squares within groups may be expressed as

$$SSW = \sum \sum x_{ij}^2 - \sum_{j}^{J} N_j \bar{x}_j^2$$

To simplify our notation, however, we shall let the first term

$$\sum_{j} \sum_{i} x_{ij}^2 = A$$

and the second term

$$\sum_{j}^{J} N_j \bar{x}_j^2 = B$$

in which case

$$SSW = A - B$$

and *MSW* becomes

$$\frac{A - B}{N - J}$$

Thus

$$E(MSW) = E\left(\frac{A - B}{N - J}\right)$$

Because both N and J are constants, however,

$$E(MSW) = \frac{1}{N - J} E(A - B)$$

$$= \frac{1}{N - J} [E(A) - E(B)]$$

In this regard, it is proved in Appendix III that

$$E(A) = N\sigma_e{}^2 + \sum_j^J N_j\mu_j{}^2$$

and

$$E(B) = J\sigma_e{}^2 + \sum_j^J N_j\mu_j$$

Thus

$$E(MSW) = \frac{1}{N-J}[E(A) - E(B)]$$

$$= \frac{1}{N-J}\left(N\sigma_e{}^2 + \sum_j^J N_j\mu_j{}^2 - J\sigma_e{}^2 - \sum_j^J N_j\mu_j{}^2\right)$$

$$= \frac{1}{N-J}(N\sigma_e{}^2 - J\sigma_e{}^2)$$

$$= \frac{1}{N-J}(N-J)\sigma_e{}^2$$

$$= \sigma_e{}^2$$

b. Expected Value of the Mean Squares Among Groups. When we introduced the notion of mean squares we determined that the sum of squares among groups has $J - 1$ degrees of freedom. We may therefore define the *mean squares among groups* (*MSA*) as

$$\frac{SSA}{J-1} = \frac{\sum_j^J N_j(\bar{x}_j - \bar{x})^2}{J-1}$$

If all our N_j treatment groups include the same number of observations, then N_j is a constant and may be factored out of the summation:

$$MSA = \frac{N_j\sum_j^J (\bar{x}_j - \bar{x})^2}{J-1}$$

It may be easily demonstrated, however, that the grand mean $\bar{x}$ is equal to the mean of the sample means $\bar{x}_j$, and our expression therefore becomes

$$MSA = N_j\hat{\sigma}_{\bar{x}}{}^2$$

In this connection, we argued earlier that $N_j \hat{\sigma}_{\bar{x}}^2$ estimates the grand population variance and therefore taps both treatment effects (if any) and random error. If this intuitive conclusion was correct, the expected value of MSA should equal σ_e^2 plus a bias term that reflects differences among the population means μ_j. By the computational formula given earlier

$$SSA = \sum_j^J N_j \bar{x}_j^2 - N\bar{x}^2$$

As in the preceding derivation, however, we shall let

$$\sum_j^J N_j \bar{x}_j^2 = B$$

and in addition, we shall define

$$N\bar{x}^2 = C$$

Thus

$$SSA = B - C$$

and

$$MSA = \frac{B - C}{J - 1}$$

Proceeding as before, then,

$$E(MSA) = \frac{1}{J - 1} [E(B) - E(C)]$$

From Appendix III,

$$E(B) = J\sigma_e^2 + \sum_j^J N_j \mu_j^2$$

and

$$E(C) = \sigma_e^2 + N\mu^2$$

Thus

$$E(MSA) = \frac{1}{J - 1} \left(\sum_j^J N_j \mu_j^2 + J\sigma_e^2 - N\mu^2 - \sigma_e^2 \right)$$

$$= \frac{1}{J - 1} \left[\sum_j^J N_j \mu_j^2 - N\mu^2 + (J - 1)\sigma_e^2 \right]$$

$$= \frac{\sum_j^J N_j \mu_j^2 - N\mu^2}{J - 1} + \sigma_e^2$$

The derivations in this section have generated two very important conclusions:

$$E(MSW) = \sigma_e^2$$

$$E(MSA) = \sigma_e^2 + \frac{\sum_j^J N_j \mu_j^2 - N\mu^2}{J-1}$$

If the test hypothesis is correct, however, and

$$\mu_1 = \mu_2 = \ldots = \mu_J = \mu$$

then the bias term in $E(MSA)$ reduces to zero:

$$\frac{\sum_j^J N_j \mu_j^2 - N\mu^2}{J-1} = \frac{\sum_j^J N_j \mu^2 - N\mu^2}{J-1}$$

$$= \frac{\mu^2 \sum_j^J N_j - N\mu^2}{J-1}$$

$$= \frac{N\mu^2 - N\mu^2}{J-1}$$

$$= 0$$

and

$$E(MSW) = E(MSA) = \sigma_e^2$$

That is, when H_0 is correct, both MSW and MSA are unbiased estimates of σ_e^2, and the ratio

$$\frac{MSA}{MSW} = \frac{SSA/J-1}{SSW/N-J}$$

will therefore be distributed as Fisher's F with $J-1$ and $N-J$ degrees of freedom. When H_0 is incorrect, however, the expected value of MSA is larger than σ_e^2, and we may therefore test

$$H_0: \mu_1 = \mu_2 = \ldots = \mu_J = \mu$$

by submitting the ratio

$$\frac{MSA}{MSW}$$

to the distribution of $F(J-1, N-J)$ and establishing a critical value in the *upper* tail of the curve.

C. The ANOVA Summary Table

To get some feel for this admittedly complex welter of statistical notation, let us consider a problem from biology. Until recently animals were considered members of the same species if they produced fertile offspring. This criterion breaks down, however, when we realize that such apparently distinct species as the bobcat and domestic cat produce viable offspring, as do matings of dogs and wolves. The taxonomy problem becomes even more complex when the biologist is called on to determine whether two organisms are separate subspecies or simply local variants of the same species. In this regard, some biologists recognize as many as 26 distinct subspecies of timberwolf. Theoretically, such classifications are based on differences in size, color, and body conformation. Whatever the morphological characteristics involved, however, the criterion is whether the groups in question display more variation across groups than within groups, and as such, the problem of distinguishing among subspecies is clearly in the domain of analysis of variance. Let us suppose, therefore, that we have three mating pairs of timberwolves, captured respectively in Ontario, Minnesota, and Alaska. Let us further suppose that the three bitches have delivered litters of pups that were removed from their mothers at birth and that have been reared on the same commercial diet for 10 days. To explore the possibility that the three pairs represent three subspecies, we propose to test the hypothesis that the litters are not different in average 10-day body weight. Thus, our random variable is weight measured in ounces at 10 days of age, and our hypotheses are

$$H_0: \mu_1 = \mu_2 = \mu_3 = \mu$$

$$H_1: \mu_j \neq \mu \text{ for at least two of the subspecies}$$

The data for this problem appear in Table 15.2.

Table 15.2 Body Weight at 10 Days of Age for Three Litters of Timber Wolf Pups

	Group 1 ($N = 5$)	Group 2 ($N = 4$)	Group 3 ($N = 5$)	
	22	25	29	
	24	26	31	
	24	29	32	
	27	32	33	
	28		35	
Σ	125	112	160	397
$\overline{X}$	25	28	32	28.36

If H_0 is correct, then with a total of $N = 14$ observations among $J = 3$ groups, the ratio

$$\frac{MSA}{MSW}$$

should be distributed as $F(2, 11)$. Thus, if we set $\alpha = .05$, the critical value of our test statistic is found in Table **F** to be 3.98. To determine the observed value of our test statistic we first compute the basic quantities A, B, and C discussed above:

$$A = \sum_j \sum_i x_{ij}^2 = 11{,}455.00$$

$$B = \sum_j^J N_j \bar{x}_j^2 = 5(25)^2 + 4(28)^2 + 5(32)^2$$

$$= 3125 + 3136 + 5120$$

$$= 11{,}381.00$$

$$C = N\bar{x}^2 = 14(28.36)^2$$

$$= 11{,}260.05$$

Thus

$$SSW - A - B = 11{,}455.00 - 11{,}381.00$$

$$= 74$$

and

$$MSW = \frac{SSW}{N - J}$$

$$= \frac{74}{11}$$

$$= 6.73$$

Similarly

$$SSA = B - C = 11{,}381.00 - 11{,}260.05$$

$$= 120.95$$

and

$$MSA = \frac{SSA}{J - 1}$$

$$= \frac{120.95}{2}$$

$$= 60.48$$

Our F-ratio is therefore

$$\frac{MSA}{MSW} = \frac{60.48}{6.73}$$

$$= 8.99$$

This exceeds our critical F-value, and we must therefore conclude that the three populations of wolves from which we drew our samples do not have identical mean body weights.

To facilitate the representation of data in analysis of variance, the important computations are displayed in the analysis of variance (ANOVA) summary table (Table 15.3).

Table 15.3 Analysis of Variance (ANOVA) Summary Table

Source	SS	v	MS	F
Treatments (across groups)	120.95	2	60.48	8.99
Error (Within groups)	74.00	11	6.73	
Totals	197.21	13		

D. More Complex Analyses of Variance

In the two examples discussed in this chapter, the experimenter considered only one independent variable, or factor. In the drug study, the factor under investigation was the influence of three drugs, and in the wolf study, the independent variable was site of capture. Thus, this chapter introduced only *one-way* analysis of variance. It must be understood, however, that analysis of variance is also applicable to problems involving two or more factors. Suppose, for example, that a statistics teacher is trying to determine which of two textbooks produces better student performance on examinations and whether assigning problem sets provides better test preparation than having students critique statistical analyses in published journal articles. He therefore divides his class into two random groups, assigning one book to each group. Each of the two groups is again divided at random such that half the students using each book are assigned weekly problem sets and half are assigned articles to critique. At the end of the semester he records final examination scores in the table illustrated in Figure 15.1. This experiment constitutes a typical *two-way* analysis of variance and permits the experimenter to test one hypothesis about mean differences produced by the

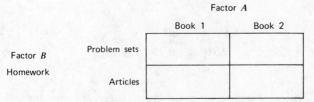

Figure 15.1   Layout of data for two-way analysis of variance.

books and another hypothesis about mean differences produced by the home-work assignments. To this point, two-way analysis of variances involves simply a generalization of the techniques developed in the present chapter. With two factors, however, the experimenter will test a third hypothesis concerning their *interaction*. To illustrate the concept of interaction, consider the row means, column means, and cell means in our fictitious experiment (Table 15.4).

Table 15.4   Row Means, Column Means, and Cell Means in Typical Two-Factor Analysis of Variance Design

		Factor A		
		Book 1	Book 2	Row Means
Factor B	Problem sets	$\bar{x}_{11} = 30$	$\bar{x}_{21} = 50$	$\bar{x}_{B_1} = 40$
	Articles	$\bar{x}_{12} = 50$	$\bar{x}_{22} = 30$	$\bar{x}_{B_2} = 40$
	Column means	$\bar{x}_{A_1} = 40$	$\bar{x}_{A_2} = 40$	

It is apparent from the row means that the homework assignments produce no systematic *main* effects, and it is likewise apparent from the column means that the textbooks produce no systematic main effects. Nevertheless, we see that when Book 1 is supplemented with articles or Book 2 is supplemented with problem sets, student performance is better than with the other book-homework combinations. If these performance differences are significant, we would say that the two variables exhibit a significant interaction.

Another property shared by the examples discussed in this chapter is that in no instance did the experimenter wish to generalize beyond the treatments actually involved in the study. In the drug experiment, for example, we were concerned only with the drugs administered to the subjects, and in the wolf example, our conclusion applied only to the subspecies actually observed. When the treatments exhaust all experimental categories of interest to the experimenter we have a *fixed effects* analysis of variance, and it should be apparent that such an analysis is generally applicable only to nominal, or qualitative, independent variables. Frequently, however, the experimental factor about which the investigator wishes to draw an inference will be a quantitative variable (ordinal or nominal) defined by a large, if not infinite, value set. This would have been the case, for example, had we considered three different doses of the *same* drug in our learning experiment and generalized our results to *all possible* doses. In such an experiment, the *levels* of the independent variable that are actually represented in the treatment effects constitute a random sample of all possible treatments, and we therefore have what is called a *random effects* analysis of variance. Although the assumptions, test hypothesis, and expected mean squares in the random effects situation differ slightly from those in the fixed effects analysis, there are no computational differences unless the study involves two or more factors.

E. Problems, Review Questions, And Exercises

1. What source of variability is tapped by MSW? Support your answer both with a plausibility argument based on the definitional formula for MSW and by giving the expected value of MSW.
2. What source of variability is tapped by MSA? Support your answer both with a plausibility argument based on the definitional formula for MSA and by giving the expected value of MSA.
3. Why is the ratio MSA/MSW submitted to a Fisher's F distribution?
4. In the drug study discussed in the text, assume that a total of 43 Ss participated, that $SST = 2131$, and that $SSW = 1834$. Is F significant at $\alpha = .05$?
5. Recall from Chapter Seven that Jim owns an Italian sports car (Brand A) and Joe owns a German Sports car (Brand B). When they purchased cars from F.F.I.L. (Foolish Foreign Investments, Ltd.), Jim's brother-in-law, Hogan, purchased a British "Trouble" (abbreviated Tr.). After one year of recording maintenance costs, all three agree that the dealership is appropriately named, but there is some argument as to whether their expenses are due to automotive design or piracy in F.F.I.L.'s service department. If it is the latter, the average maintenance cost per month should be identical for all three, but if it is design, average maintenance costs should differ from car to car. They therefore decide to compare maintenance bills for six one-month periods selected at random from the preceding year.

A	B	Tr
$73	$47	$58
55	30	47
90	70	73
70	45	55
85	55	65
77	53	62

Test the hypothesis that average monthly maintenance costs are identical for the three makes of automobile. What is the significance level of your result?

6. Every night for a week record to the nearest second the time devoted to national and international news (not including sports and weather) on the late news (e.g., 11:00 P.M.) broadcast of your local ABC television station. During the same week have two other persons make similar recordings for your local CBS and NBC affiliates. Then test the hypothesis that the three major networks devote the same amount of coverage to national and international news.

F. Summary Table

Distribution of X_j	Sample Size(s)	H_0	H_1	Test Statistic T	Distribution of T
A. One-way Fixed Effects Analysis of Variance					
Normal	$N_1 = \ldots = N_J{}^a$	$\mu_j = \mu$	$\mu_j \neq \mu$	$\dfrac{MSA^b}{MSW}$	$F(J - 1, N - J)$

Assumptions: All observations x_{ij} independent of one another

$\qquad$ X_j normally distributed

$\qquad$ $\sigma_1{}^2 = \ldots = \sigma_J{}^2 = \sigma_e{}^2$

[a] Required only if assumption of equal variance *may* be inappropriate.

[b] $MSA = \dfrac{B - C}{J - 1}$

$\quad MSW = \dfrac{A - B}{N - J}$

where

$$A = \sum_j \sum_i x_{ij}{}^2$$

$$B = \sum_j^J N_j \bar{x}_j{}^2$$

$$C = N \bar{x}^2$$

appendices

summation

Use of the symbol Σ to represent summation was first introduced in Chapter Five and subsequently elaborated in Chapters Fourteen and Fifteen. To facilitate references to these various conventions, we have assembled them in a single appendix.

A. Subscripts and Limits of Summation

As indicated in Chapter Four, a subscript may be used to indicate a particular element in the value set of a random variable

$$x_i \in \{x_1, x_2, \ldots x_n\}$$

or it may be used to indicate the value associated with a particular observation in a sample

$$x_i \in \{x_1, x_2, \ldots x_N\}$$

In this regard, we used the notation

$$\sum_{i=1}^{n} x_i$$

to represent the sum of all n values that the random variable may possibly assume. That is,

$$\sum_{i=1}^{n} x_i = x_1 + x_2 + \ldots + x_n$$

In this notation, 1 is the smallest value that the subscript i may assume and is therefore called the *lower limit of summation*. Likewise, i cannot exceed n, which is therefore called the *upper limit of summation*. Similarly, we use the notation.

$$\sum_{i=1}^{N} x_i$$

to represent the sum of all N values actually observed in one's sample:

$$\sum_{i=1}^{N} x_i = x_1 + x_2 + \ldots + x_N$$

In this notation the limits of summation are 1 and N.

If it may be assumed that the first value in the sum is x_1, then the lower limit of summation may be omitted. Thus

$$\sum_{i=1}^{n} x_i = \sum_{i}^{n} x_i$$

and

$$\sum_{i=1}^{N} x_i = \sum_{i}^{N} x_i$$

Because the upper limit distinguishes the sum of the value set from the sum of the observations, however, it is seldom omitted unless uniquivocally established by the context.

Very frequently N observations will be distributed among J mutually exclusive groups such that the first group includes N_1 observations, the second group includes N_2 observations, . . . and the last group includes N_J observations. In such instances, each observation will be identified by two subscripts x_{ij} much as as a person has two names. The second subscript, like a surname, indicates group (or family) membership and may therefore assume any value from 1 to J, the number of groups. The first subscript distinguishes observations within the same group much as a given name distinguishes family members from one another and therefore runs from 1 to N_j, the number of observations in Group j.

When double subscripts are employed, the sum of observations in any particular group, say, Group j, is represented as

$$\sum_{i=1}^{Nj} x_{ij}$$

or

$$\sum_{i}^{N_j} x_{ij}$$

To represent the sum of all $N_1 + N_2 + \ldots + N_J = N$ observations, however, we customarily use double summation signs:

$$\sum_{i=1}^{N} x_i = \sum_{j}^{J} \sum_{i}^{N_j} x_{ij}$$

This notation simply indicates that the summation is to be computed by adding together the sums of observations in each group. That is,

$$\sum_{j}^{J} \sum_{i}^{N_j} x_{ij} = \sum_{i}^{N_1} x_{i1} + \sum_{i}^{N_2} x_{i2} + \ldots + \sum_{i}^{N_J} x_{iJ}$$

In order to accommodate the limitations of typography or, more frequently, just to simplify notation, many authors will omit both the upper and lower limits from a double summation. Thus, when the reader encounters the expression

$$\sum_{j} \sum_{i} x_{ij}$$

he must remember that $\sum_{j}$ runs from 1 to J, the number of groups, and that $\sum_{i}$ runs from 1 to N_j, the number of observations in Group j.

B. The Algebra of Summations

In the following derivations we shall omit all limits of summation except where necessary for emphasis or to avoid confusion. Thus, the expression

$$\sum x_i$$

Should be assumed to indicate

$$\sum_{i=1}^{N} x_i$$

except when otherwise indicated.

Rule 1   If c is a constant,

$$\sum c = Nc$$

To prove this rule, suppose that we have taken N observations on a random variable X and that every observation assumes the same value c. That is,

$$x_1 = x_2 = \ldots = x_N = c$$

Proof By the preceding expression

$$\sum c = \sum x_i$$

$$= x_1 + x_2 + \ldots + x_N$$

$$= \underbrace{c + c + \ldots + c}_{N \text{ times}}$$

$$= Nc$$

Rule 2 If c is a constant and X is any set of N real numbers,

$$\sum cx_i = c \sum x_i$$

Proof

$$\sum cx_i = cx_1 + cx_2 + \ldots + cx_N$$

$$= c(x_1 + x_2 + \ldots + x_N)$$

$$= c \sum x_i$$

Rule 3 If X is a set of N real numbers and Y is a set of N real numbers,

$$\sum_i^N (x_i y_i) = x_1 y_1 + x_2 y_2 + \ldots + x_N y_N$$

This is true by definition and requires no proof.

Rule 4 If X is a set of N real numbers and Y is a set of N real numbers,

$$\sum_i^N (x_i + y_i) = \sum_i^N x_i + \sum_i^N y_i$$

Proof

$$\sum_i^N (x_i + y_i) = (x_1 + y_1) + (x_2 + y_2) + \ldots + (x_N + y_N)$$

$$= x_1 + y_1 + x_2 + y_2 + \ldots + x_N + y_N$$

Rearranging terms,

$$\sum_i^N (x_i + y_i) = (x_1 + x_2 + \ldots + x_N) + (y_1 + y_2 + \ldots + y_N)$$

$$\sum_i^N x_i + \sum_i^N y_i$$

Rule 5 If N observations x_{ij} are distributed among J mutually exclusive groups with $N_1, N_2, \ldots N_J$ observations, respectively, and c is a constant across all J groups,

$$\sum_j \sum_i cx_{ij} = c \sum_j \sum_i x_{ij}$$

Proof If $N_1 + N_2 + \ldots + N_J = N$, then

$$\sum_j \sum_i x_{ij} = \sum_i^N x_i$$

Thus

$$\sum_j \sum_i c x_{ij} = \sum_i^N c x_i$$

By Rule 2 of the algebra of summations, however,

$$\sum_i^N c x_i = c \sum_i^N x_i$$

Thus

$$\sum_j \sum_i c x_{ij} = c \sum_j \sum_i x_{ij}$$

Rule 6 If N observations x_{ij} are distributed among J mutually exclusive groups with $N_1, N_2, \ldots N_J$ observations, respectively, and c_j is a constant for all observations in Group j, then

$$\sum_j \sum_i c_j x_{ij} = \sum_j^J c_j \sum_i^{Nj} x_{ij}$$

Proof By definition

$$\sum_j \sum_i x_{ij} = \sum_i^{N_1} x_{i1} + \sum_i^{N_2} x_{i2} + \ldots + \sum_i^{N_J} x_{iJ}$$

Similarly

$$\sum_j \sum_i c_j x_{ij} = \sum_i^{N_1} c_1 x_{i1} + \sum_i^{N_2} c_2 x_{i2} + \ldots + \sum_i^{N_J} c_J x_{iJ}$$

If, however, c_j is a constant in Group j, then by Rule 2 of the algebra of summations,

$$\sum_i^{Nj} c_j x_{ij} = c_j \sum_i^{Nj} x_{ij}$$

Thus

$$\sum_j \sum_i c_j x_{ij} = c_1 \sum_i^{N_1} x_{i1} + c_2 \sum_i^{N_2} x_{i2} + \ldots + c_J \sum_i^{N_J} x_{iJ}$$

$$= \sum_j^J c_j \sum_i^{Nj} x_{ij}$$

the algebra of expectations

In order to streamline our notation we shall omit all limits of summation except in instances where they must be indicated to avoid confusion. Therefore, the expression

$$\sum x_i$$

should be assumed equivalent to

$$\sum_{i=1}^{n} x_i$$

except when otherwise indicated.

Rule 1 If X is a random variable with expectation $E(X)$ and Y is a random variable with expectation $E(Y)$,

$$E(X + Y) = E(X) + E(Y)$$

Recall from Chapter two that a random variable may be thought of as a function on a random experiment. We may therefore consider X and Y as two different functions on our set of results $(e_1, e_2, \ldots e_m)$. That is,

$$x_i = X(e_i)$$

$$y_i = Y(e_i)$$

Proof By the preceding definitions,

$$E(X + Y) = \sum_{i=1}^{m} (x_i + y_i)\, p(x_i + y_i)$$

$$= \sum_{i=1}^{m} [X(e_i) + Y(e_i)]\, p[X(e_i) + Y(e_i)]$$

The numerical value $(x_i + y_i) = [X(e_i) + Y(e_i)]$ occurs only when our experiment yields the result e_i. Thus

$$p[X(e_i) + Y(e_i)] = p(e_i)$$

Therefore

$$E(X + Y) = \sum_{i=1}^{m} [X(e_i) + Y(e_i)]\, p[X(e_i) + Y(e_i)]$$

$$= \sum_{i=1}^{m} [X(e_i) + Y(e_i)]\, p(e_i)$$

If we expand this summation, we obtain

$$[X(e_i) + Y(e_i)]\, p(e_i) + \ldots + [X(e_m) + Y(e_m)]\, p(e_m)$$
$$= X(e_1)\, p(e_1) + Y(e_1)\, p(e_1) + \ldots + X(e_m)\, p(e_m) + Y(e_m)\, p(e_m)$$

If we now collect all X terms into a single summation and all Y terms into single summation, our expression may be rearranged to read

$$[X(e_1)\, p(e_1) + \ldots + X(e_m)\, p(e_m)] + [Y(e_1)\, p(e_1) + \ldots + Y(e_m)\, p(e_m)]$$

Recall, however, that our m results may be collected into n subsets $E_1, E_2, \ldots E_n$, where

$$E_i = \{e \,|\, X(e) = x_i\}$$

and

$$p(E_i) = p(x_i)$$

where i now runs from 1 to n, the number of subsets, or events. Hence, the expression

$$[X(e_1)\, p(e_1) + \ldots + X(e_m)\, p(e_m)]$$

becomes

$$[x_1\, p(E_1) + \ldots + x_n\, p(E_n)]$$

$$= \sum_{i=1}^{n} x_i\, p(E_i)$$

$$= \sum_{i=1}^{n} x_i\, p(x_i)$$

Similarly

$$[Y(e_1)\,p(e_1) + \ldots + Y(e_m)\,p(e_m)] = \sum_{i=1}^{n} y_i\,p(y_i)$$

Thus

$$E(X + Y) = \sum_{i=1}^{n} x_i\,p(x_i) + \sum_{i=1}^{n} y_i\,p(y_i)$$

$$= E(X) + E(Y)$$

Rule 2 If c is a constant,

$$E(c) = c$$

In general we are concerned only with the expected values of variables, and the expectation of a constant is therefore something of a degenerate case. Nonetheless, Rule 2 is amenable to proof if we consider that our constant c is a set X in which every element $\{x_1, x_2 \ldots x_n\}$ is equal to the same value c.

Proof

$$E(c) = E(X) \qquad \text{as defined above}$$

$$= \sum x_i\,p(x_i)$$

However

$$x_1 = x_2 = \ldots = x_n = c$$

Therefore

$$E(c) = \sum c\,p(x_i)$$

By the rules of summation given in Appendix I, the constant c may be factored out of the summation. Therefore

$$E(c) = c \sum p(x_i)$$

$$= c\,(1.0)$$

$$= c$$

Rule 3 If c is a constant and X is a random variable,

$$E(X + c) = E(X) + c$$

Proof If we treat the constant c as if it were a random variable

$$E(X + c) = E(X) + E(c)$$

by Rule 1 given above. However, by Rule 2 we know that

$$E(c) = c$$

Therefore

$$E(X + c) = E(X) + c$$

370

Rule 4 If c is a constant and X is a random variable,

$$E(cX) = c\,E(X)$$

Proof

$$E(cX) = \sum cx_i\,p(cx_i)$$

By the rules of summation given in Appendix I, the constant c may by factored out of the summation. Thus

$$E(cX) = c\sum x_i\,p(cx_i)$$

Because c is a constant, however, the product cx_i can occur only when $x = x_i$. Thus

$$p(cx_i) = p(x_i)$$

and therefore

$$\sum x_i\,p(cx_i) = \sum x_i\,p(x_i)$$

and

$$c\sum x_i\,p(cx_i) = c\sum x_i\,p(x_i)$$

by definition

$$\sum x_i\,p(x_i) = E(X)$$

Therefore

$$E(cX) = c\sum x_i\,p(x_i)$$
$$= c\,E(X)$$

Rule 5 If X and Y are independent random variables with variances $V(X)$ and $V(Y)$, respectively,

$$V(X + Y) = V(X) + V(Y)$$

Proof

$$V(X + Y) = E(X + Y)^2 - [E(X + Y)]^2$$

If we expand the first term this becomes

$$V(X + Y) = E(X^2 + 2XY + Y^2) - [E(X + Y)]^2$$

By Rules 1 and 4 of the algebra of expectations, the first term is therefore equal to $E(X^2) + 2\,E(XY) + E(Y^2)$, and the entire expression becomes

$$V(X + Y) = [E(X^2) + 2E(XY) + E(Y^2)] - [E(X + Y)]^2$$

By Rule 1 of the algebra of expectations the second term is equal to $[E(X) + E(Y)]^2$, and our entire expression is thus given by

$$V(X + Y) = [E(X^2) + 2E(XY) + E(Y^2)] - [E(X) + E(Y)]^2$$

If we expand the second term, $V(X + Y)$ is equal to

$$[E(X^2) + 2E(XY) + E(Y^2)] - \{[E(X)]^2 + 2E(X)E(Y) + [E(Y)]^2\}$$

If we remove parentheses and collect terms, this becomes

$$\{E(X^2) - [E(X)]^2\} + \{E(Y^2) - [E(Y)]^2\} + 2[E(XY) - E(X)E(Y)]$$

The first term, $\{E(X^2) - [E(X)]^2\}$, is equal to $V(X)$, and the second term is similarly equal to $V(Y)$. Our expression thus becomes

$$V(X) + V(Y) + 2[E(XY) - E(X)E(Y)]$$

In the last term, $[E(XY) - (EX)E(Y)]$ is the population covariance of X and Y, $\mathrm{cov}(XY)$. It is seen in Chapter Nine that when two variables are independent, their covariance is equal to zero. The last term therefore drops out, leaving

$$V(X + Y) = V(X) + V(Y)$$

Rule 6 If c is a constant,

$$V(c) = 0$$

Proof

$$V(c) = E(c^2) - [E(c)]^2$$

By Rule 1 of the algebra of expectations

$$E(c) = c$$

Therefore

$$[E(c)]^2 = c^2$$

Hence

$$V(c) = E(c^2) - [E(c)]^2$$
$$= c^2 - c^2$$
$$= 0$$

Rule 7 If c is a constant and X is a random variable,

$$V(X + c) = V(X)$$

Proof 1

$$V(X + c) = E(X + c)^2 - [E(X + c)]^2$$

Expanding the first term this becomes

$$V(X + c) = E(X^2 + 2Xc + c^2) - [E(X + c)]^2$$

By Rules 1, 2, and 4 of the algebra of expectations, the first term is equal to $E(X^2) + 2cE(X) + c^2$, and the entire expression is thus given by

$$V(X + c) = [E(X^2) + 2cE(X) + c^2] - [E(X + c)]^2$$

From Rule 3 of the algebra of expectations the second term is equal to $[c + E(X)]^2$, and the entire expression becomes

$$V(X + c) = [E(X^2) + 2cE(X) + c^2] - [c + E(X)]^2$$

If we expand the second term, $V(X + c)$ is equal to

$$[E(X^2) + 2cE(X) + c^2] - \{c^2 + 2cE(X) + [E(X)]^2\}$$

When we remove parentheses the c^2 terms and the cross-product terms cancel out, leaving

$$V(X + c) = E(X^2) - [E(X)]^2$$
$$= V(X)$$

Proof 2   Recall from the proof of Rule 5 that in general

$$V(X + Y) = V(X) + V(Y) + 2 \operatorname{cov}(XY)$$

It can be demonstrated from the formula for population covariance given in Chapter Nine that $\operatorname{cov}(cX)$, the covariance of a constant c and a variable X, is always equal to zero. If the reader is either willing to accept this as an assumption or demonstrate it for himself, the proof for Rule 7 becomes trivial. Treating the constant c as a variable

$$V(X + c) = V(X) + V(c) + 2 \operatorname{cov}(cX)$$

As indicated above, $\operatorname{cov}(cX) = 0$. Therefore

$$V(X + c) = V(X) + V(c)$$

However, from Rule 6 of the algebra of expectations, we know that

$$V(c) = 0$$

Therefore

$$V(X + c) = V(X)$$

Rule 8   If c is a constant and X is a random variable,

$$V(cX) = c^2 V(X)$$

Proof
$$V(cX) = E(cX^2) - [E(cX)]^2$$
$$= E(c^2X^2) - [E(cX)]^2$$

Both c and c^2 are constants, and by Rule 4 of the algebra of expectations they may be factored out of their respective expectations. Thus

$$V(cX) = c^2E(X^2) - [cE(X)]^2$$
$$= c^2E(X^2) - c^2[E(X)]^2$$

If we factor out the common factor, c^2, this becomes

$$V(cX) = c^2 (E(X^2) - [E(X)]^2)$$

However

$$E(X^2) - [E(X)]^2 = V(X)$$

Therefore

$$V(cX) = c^2 V(X)$$

sums of squares

A. Computational Formulas

By definition, the total sums of squares is given by the expression

$$SST = \sum_j \sum_i (x_{ij} - \bar{x})^2$$

Expanding the square and distributing the sum we obtain

$$SST = \sum \sum (x_{ij}^2 - 2x_{ij}\bar{x} + \bar{x}^2)$$
$$= \sum \sum x_{ij}^2 - 2\bar{x} \sum \sum x_{ij} + N\bar{x}^2$$

By definition, however,

$$\frac{\sum_j \sum_i x_{ij}}{N} = \bar{x}$$

and it must therefore be true that

$$\sum_j \sum_i x_{ij} = N\bar{x}$$

Making this substitution in the second term of our expanded sum, the expression becomes

$$SST = \sum_j \sum_i x_{ij}^2 - 2\bar{x}N\bar{x} + N\bar{x}^2$$

$$= \sum_j \sum_i x_{ij}^2 - 2N\bar{x}^2 + N\bar{x}^2$$

$$= \sum_j \sum_i x_{ij}^2 - N\bar{x}^2$$

The sum of squares within groups may also be expressed in a more computationally convenient form.

$$SSW = \sum_j \sum_i (x_{ij} - \bar{x}_j)^2$$

$$= \sum_j \sum_i (x_{ij}^2 - 2x_{ij}\bar{x}_j + \bar{x}_j^2)$$

In distributing the sums, however, it must be remembered that $\bar{x}_j$ is a constant with respect to the within-group sum $\sum_i$. Thus

$$SSW = \sum_j \sum_i x_{ij}^2 - 2\sum_j^J \bar{x}_j \sum_i^{N_j} x_{ij} + \sum_j^J N_j \bar{x}_j^2$$

To simplify this expression, we note that by definition

$$\frac{\sum_i^{N_j} x_{ij}}{N_j} = \bar{x}_j$$

Thus

$$\sum_j^{N_j} x_{ij} = N_j \bar{x}_j$$

Making this substitution in the second term,

$$SSW = \sum_j \sum_i x_{ij}^2 - 2\sum_j^J \bar{x}_j N_j \bar{x}_j + \sum_j^J N_j \bar{x}_j^2$$

$$= \sum_j \sum_i x_{ij}^2 - 2\sum_j^J N_j \bar{x}_j^2 + \sum_j^J N_j \bar{x}_j^2$$

$$= \sum_j \sum_i x_{ij}^2 - \sum_j^J N_j \bar{x}_j^2$$

To obtain the computational formula for SSA we simply take

$$SST - SSW = \sum_j \sum_i x_{ij}^2 - N\bar{x}^2 - \sum_j \sum_i x_{ij}^2 - \sum_j^J N_j \bar{x}_j^2$$

$$= \sum_j^J N_j \bar{x}_j^2 - N\bar{x}^2$$

B. Expectations of Basic Quantities Comprising Sums of Squares

If we compare the three formulas we have just derived, we see that the various sums of squares may be expressed as

$$SSW = A - B$$

$$SSA = B - C$$

and

$$SST = SSW + SSA$$

$$= (A - B) + (B - C)$$

$$= A - C$$

where

$$A = \sum_j \sum_i x_{ij}{}^2$$

$$B = \sum_j N_j \bar{x}_j{}^2$$

$$C = N\bar{x}^2$$

The expected values of these three quantities are derived below.

1. $E(A)$

$$E(A) = E(\sum_j \sum_i x_{ij}{}^2)$$

The expectation of a sum is equal to the sum of the expectations, and this expression may therefore be rewritten as

$$E(A) = \sum_j^J E(\sum_i^{N_j} x_{ij}{}^2)$$

If we consider that our observations x_{ij} are yet to be taken, this becomes

$$E(A) = \sum_j^J E(X_{1j}{}^2 + X_{2j}{}^2 + \ldots + X_{Nj}{}^2)$$

$$= \sum_j^J [E(X_{1j}{}^2) + E(X_{2j}{}^2) + \ldots + E(X_{Nj}{}^2)]$$

To determine $E(X^2)$, recall the computational formula for variance derived in Chapter Six:

$$V(X) = E(X^2) - [E(X)]^2$$

Thus

$$E(X^2) = V(X) + [E(X)]^2$$
$$= \sigma^2 + \mu^2$$

Therefore

$$E(A) = \sum_j^J \underbrace{[(\sigma_j{}^2 + \mu_j{}^2) + (\sigma_j{}^2 + \mu_j{}^2) + \ldots + (\sigma_j{}^2 + \mu_j{}^2)]}_{N_j \text{ times}}$$

$$= \sum_j^J N_j (\sigma_j{}^2 + \mu_j{}^2)$$

$$= \sum_j^J N_j\sigma_j{}^2 + \sum_j^J N_j\mu_j{}^2$$

However, we are assuming that every population variance $\sigma_j{}^2$ is equal to $\sigma_e{}^2$. Thus

$$E(A) = \sum_j^J N_j\sigma_e{}^2 + \sum_j^J N_j\mu_j{}^2$$

$$= \sigma_e{}^2 \sum_j^J N_j + \sum_j^J N_j\mu_j{}^2$$

$$= N\sigma_e{}^2 + \sum_j^J N_j\mu_j{}^2$$

2. $E(B)$

$$E(B) = E(\sum_j^J N_j\bar{x}_j{}^2)$$

If we again assume that our observations are yet to be taken, the values $\bar{x}_j$ become random variables $\bar{X}_j$ and our expression becomes

$$E(B) = E(N_1\bar{X}_1{}^2 + N_2\bar{X}_2{}^2 + \ldots + N_J\bar{X}_J{}^2)$$
$$= N_1E(\bar{X}_1{}^2) + N_2E(\bar{X}_2{}^2) + \ldots + N_JE(\bar{X}_J{}^2)$$
$$= \sum_j^J N_J E(\bar{X}_J{}^2)$$

Again, by the computational formula for variance, we know that

$$E(\bar{X}^2) = V(\bar{X}) + [E(\bar{X})]^2$$
$$= \frac{\sigma^2}{N} + \mu^2$$

Thus

$$E(B) = \sum_j^J N_j \left(\frac{\sigma_j^2}{N_j} + \mu_j^2 \right)$$

$$= \sum_j^J (\sigma_j^2 + N_j \mu_j^2)$$

$$= \sum_j^J \sigma_j^2 + \sum_j^J N_j \mu_j^2$$

Because σ_j^2 is assumed equal to σ_e^2 in all of our J populations, however, this becomes

$$E(B) = J\sigma_e^2 + \sum_j^J N_j \mu_j^2$$

3. $E(C)$

By the definition given above,

$$C = N\bar{x}^2$$

is a constant, but before our N observations are taken, C is a variable equal to $N\bar{X}^2$. Thus

$$E(C) = E(N\bar{X}^2)$$

$$= NE(\bar{X}^2)$$

We have already demonstrated that

$$E(\bar{X}^2) = \frac{\sigma^2}{N} + \mu^2$$

and with variance in all cases assumed equal to σ_e^2 it must be true that

$$E(C) = N\left(\frac{\sigma_e^2}{N} + \mu^2 \right)$$

$$= \sigma_e^2 + N\mu^2$$

378

solutions to problems

Chapter 1

1a p(red ball) = .25 **1b** $f = 2$ **1c** $m = 8$ **1d** p(black ball) = 0

1e p(white ball) = 1.00 **2a** p(beer) = .167 **2b** Yes

2c $S = \{s_1\ s_2\ s_3\ s_4\ s_5\ b_1\}$ where s_i is a soft drink and b_i is a beer; $p(s_i) = .167$, $p(b_i) = .167$

3a p(white ball) = .50 p(red ball) = .25 p(green ball) = .25

3b No **3c** Yes **4a** p(soft drink) = .833 p(beer) = .167 **4b** No

4c Yes

5a p(S lives on 3rd floor) = .333

5b p(S possesses marijuana and lives on 3rd floor) = 7(1/30) = .233

5c p(S possess marijuana | S lives on 3rd floor) =

$$\frac{p(S \text{ possesses marijuana and lives on 3rd floor})}{p(S \text{ lives on 3rd floor})} = \frac{.233}{.333} = .70$$

6a $p(5) = .167$ **6b** Prime numbers are 1, 2, 3, and 5. Therefore, p(prime number) = 4(1/6) = .667

6c If we stipulate that a prime number has appeared or is certain to appear, then the sample space includes only four possible results, all of which are equally likely. Obtaining a 5 is one of these results. Thus, $p(5|\text{prime number}) = .25$

6d No. $p(5) \neq p(5|\text{prime number})$

6e p(prime number and 5) = p(prime number) $p(5|\text{prime number})$ = (.67)(.25) = .167

Chapter 2

1 There are 36 possible results   **2** The number of spots   **3** Counting

11	21	31	41	51	61
12	22	32	42	52	62
13	23	33	43	53	63
14	24	34	44	54	64
15	25	35	45	55	65
16	26	36	46	56	66

4 The set includes 11 values. $\{2, 3, 4, 5, 6, \underline{7, 8, 9}, 10, 11, 12\}$

5

x_i	2	3	4	5	6	7	8	9	10	11	12
$p(x_i)$	.028	.056	.083	.111	.139	.167	.139	.111	.083	.056	.028

6

x_i	2	3	4	5	6	7	8	9	10	11	12
$F(x_i)$	.028	.084	.167	.278	.417	.584	.723	.834	.917	.973	1.00

Chapter 3

1 There are 4 ways to get the first ace, 3 ways to get the second ace, 4 ways to get the first king, 3 ways to get the second king, and 11 other cards to fill the hand. By the basic rules of counting, the answer is $(4)(3)(4)(3)(11) = 1584$

2 $\quad _5P_5 = \dfrac{5!}{(5 - 5)!} = 5! = 120$

3 The number of permutations in which men occupy the end seats is equal to the number of permutations of the three men seated two at a time multiplied by the number of permutations of the occupants of the three remaining seats. $(_3P_2)(_3P_3) = (6)(6) = 36$

4 No. The number of possible solutions is equal to the number of permutations of the 24 people taken in groups of 20.

$$_{24}P_{20} = \frac{24!}{(24 - 20)!} = \frac{24!}{4!} = 2.59(10^{22})$$

5 The easiest way to solve this problem is to take $1 - p(7$ or more hits$) =$

$$1 - \left[\frac{10!}{7!3!}(.9)^7\,(.1)^3 + \frac{10!}{8!2!}(.9)^8\,(.1)^2 + \frac{10!}{9!1!}(.9)^9\,(.1) + (.9)^{10} \right] = .0128$$

Chapter 4

1a Number of heads is a discrete random variable

1b However you group your data (in unit-width intervals or otherwise) your apparent limits should be integers

1c Your real limits are identical with your apparent limits

2a Time is a continuous random variable

2b Because the problem specified that time was to be recorded to the nearest second, your apparent limits (however you group your data) are expressed in whole seconds

2c Your real limits should be expressed in half-seconds

3a The sample space is every possible *pair* of family members

3b $_NC_2$ where N is the number of members in your family

3c The members of your family   **3d**   Select a single slip of paper

4   No. Your sample space in **3a** is the same whether your two slips are drawn simultaneously or consecutively without replacement

Chapter 5

1c $E(X) = 0(.1) + 1(.1) + 2(.1) + \ldots + 9(.1) = 4.5$

1d As N increases, your sample mean $\bar{x}$ should become closer to μ. That is, $|\bar{x} - \mu|$ should decrease as N increases

2 $E(X) = \sum_i^n x_i\,p(x_i) = 7$ **3** $E(X) = Np = 10(.5) = 5$

4a

x_i	9999.50	$-.50$
$p(x_i)$	.000017	.999983

4b $E(X) = \sum_{i}^{n} x_i \, p(x_i) = .000017(9,999.50) + .999983(-.50) = -.33$

Chapter 6

2a $\sigma^2 = (0 - 4.5)^2 (.1) + (1 - 4.5)^2 (.1) + \ldots + (9 - 4.5)^2 (.1) = 8.25$

2b As N increases, your sample variance s^2 should become closer to σ^2, That is, $|s^2 - \sigma^2|$ should decrease as N increases

3 $\sigma^2 = \sum_{i}^{n} (x_i - \mu)^2 \, p(x_i) = 5.83$ **4** $\sigma^2 = Npq = 10(.5)(.5) = 2.5$

Chapter 7

1a $z_1 = \dfrac{630 - 480}{85} = 1.76$ $z_2 = \dfrac{580 - 450}{80} = 1.63$

1b $\mu_Y = 500$ **1c** $\sigma_Y = 100$ **1d** $y_1 = 676$ $y_2 = 663$

2a It establishes the zero-point of the scale at the freezing point of water

2b It changes the size of the unit. Each degree Farenheit is $\frac{5}{9}$ as large as a degree centigrade. Or, there are $\frac{9}{5}$ of a Farenheit degree per centigrade degree

3 $z = \dfrac{8 - 7}{\sqrt{5.83}} = .414$ **4** $z = \dfrac{8 - 5}{\sqrt{2.5}} = 1.897$

Chapter 8

1 $F(1.76) = .961$. Thus $p(z \geq 1.76) = 1 - .961 = .039$
$F(1.63) = .948$. Thus $p(z \geq 1.63) = 1 - .948 = .052$

382

2 $F(1.48) = .93$. Thus $Y = 500 + 1.48(100) = 648$

$F(1.34) = .91$. Thus $Y = 500 + 1.34(100) = 634$

3a $p[x > 7 \mid X{:}B(10, 0.5)] = .055$

3b $z = \dfrac{7.5 - 5}{\sqrt{2.5}} = 1.58$. $F(1.58) = .943$. Thus $p(z \geq 1.58) = 1 - .943 = .057$

3c Yes. The difference between the exact probability in **2a** and the approximated probability in **2b** is less than 4 percent of the exact probability.

4a $z = \dfrac{6.5 - 9}{\sqrt{.90}} = -2.64$. $F(-2.64) = .0041$

4b No. The approximate value in **3a** underestimates the exact probability (See answer to problem **5** in Chapter Three) by about 68 percent. Sample size is not large enough to compensate for the asymmetry produced by such an extreme value of p

Chapter 9

1a $C_{XY} = 89.31$ **1b** $C_{XY} = 7.44$

1c Covariance is influenced by the units of measurement

2a $C_{Z_X Z_Y} = .35$ **2b** $C_{Z_X Z_Y} = .35$

2c Covariance for standardized variables is not influenced by the units of measurement

3 $r_{XY} = .35$ **4** r is the covariance of standardized variables

Chapter 10

1a $\mu = 2.0000$ $\sigma^2 = .6667$

1b $p(x > 2) = p\left(z > \dfrac{2.5 - 2}{\sqrt{.6667}}\right) = p(z > .61) = 1 - F(.61) = .271$

1c $.333 - .271 = .062$

2a

$\bar{x}_i$	1.0	1.5	2.0	2.5	3.0
$p(\bar{x}_i)$	.111	.222	.333	.222	.111

2b $\mu_{\bar{X}} = \sum_{i}^{n} \bar{x}_i \, p(\bar{x}_i) = 2.0000 \qquad \sigma_{\bar{X}}^2 = \sum_{i}^{n} (\bar{x}_i - \mu_{\bar{X}})^2 \, p(\bar{x}_i) = .3333$

2c $\mu_{\bar{X}} = \mu \qquad \sigma_{\bar{X}}^2 = \dfrac{\sigma^2}{N}$

2d $p(\bar{x} > 2.5) = p\!\left(z > \dfrac{2.75 - 2}{\sqrt{.3333}}\right) = p(z > 1.30) = 1 - F(1.30) = .097$

2e $.111 - .097 = .014$

3a

$\bar{x}_i$	1.00	1.33	1.67	2.00	2.33	2.67	3.00
$p(\bar{x}_i)$	.037	.111	.222	.259	.222	.111	.037

3b $\mu_{\bar{X}} = \sum_{i}^{n} \bar{x}_i \, p(\bar{x}_i) = 2.0000 \qquad \sigma_{\bar{X}}^2 = \sum_{i}^{n} (\bar{x}_i - \mu_{\bar{X}})^2 \, p(\bar{x}_i) = .2222$

3c $\mu_{\bar{X}} = \mu \qquad \sigma_{\bar{X}}^2 = \dfrac{\sigma^2}{N}$

3d $p(x > 2.67) = p\!\left(z > \dfrac{2.835 - 2}{\sqrt{.2222}}\right) = p(z > 1.77) = 1 - F(1.77) = .038$

3e $.037 - .038 = -.001$

4 As N increases, the probability distribution of $\bar{X}$ approaches the normal

5 $\bar{x} = 2.67 \qquad \sigma_{\bar{X}} = \sqrt{.2222}$. Therefore, a 99 percent confidence interval for μ is $2.67 - 2.58 \sqrt{.2222}$ to $2.67 + 2.58 \sqrt{.2222}$, or 1.45 to 3.90.

6 $p = 1.0$ that the interval 1.45 to 3.90 spans the population mean μ

Chapter 11

1a The theory is that the coin is fair

1b If the coin is fair, then the number of heads in 10 tosses should be distributed binomially with $N = 10$ and $p = .50$. Your model is therefore

$$X : B(10, .50)$$

1c Under your model, you should expect $Np = (10)(.50) = 5$ heads in 10 tosses. Thus

$$H_0 : E(X) = 5$$

You have decided that even if the coin is biased (i.e., even if $p > .50$) the bias is of no real importance unless p is greater than .85. Given this decision, you may formulate a simple alternative based on $p = .85$:

$$H_1: E(X) = (10)(.85) = 8.5$$

2a The appropriate units of observation are the results of 10 tosses of the coin
2b An appropriate test statistic is the number of heads in 10 tosses
2c To select your rejection rule, you must first compute

$$\alpha = p(x \geq x_c | H_0 \text{ correct})$$

and

$$\beta = p(x < x_c | H_1 \text{ correct})$$

for each rule. You then multiply α by \$90 and β by \$25 to find the expected losses under each rule:

Rule 1	(.0010)(\$90.00) = \$ .09	(.8031)(\$25.00) = \$20.08
Rule 2	(.0108)(\$90.00) = \$ .97	(.4557)(\$25.00) = \$11.39
Rule 3	(.0547)(\$90.00) = \$ 4.92	(.1798)(\$25.00) = \$ 4.50
Rule 4	(.1719)(\$90.00) = \$15.47	(.0500)(\$25.00) = \$ 1.25

By the minimax criterion you should use Rule 3, under which your maximum expected loss is \$4.92

3a If the coin lands "heads" seven times, Rule 3 requires that H_1 be rejected
3b $p(x = 7 | H_0 \text{ correct}) = .1172$
3c $p(x = 7 | H_1 \text{ correct}) = .1298$
3d The data are even less likely under H_0 than under H_1 and it would therefore make no sense at all to accept the test hypothesis

Chapter 12

1a The theory is that the test is not biased
1b Because intelligence is largely polygenic, he may assume that it is normally distributed. Given the parameters of his standardization population, then, the model is

$$X: N(100, 225)$$

1c
$$H_0: \mu_B = 100$$
$$H_1: \mu_B < 100$$

2a I.Q. scores of black subjects

2b $Z_{\bar{X}} = \dfrac{\bar{X} - \mu_0}{\sigma_{\bar{X}}}$ **2c** $\{z | z \leq -1.64\}$

4a Yes. $z_{\bar{X}} = \dfrac{98.25 - 100}{\sqrt{225/200}} = -1.65$. Therefore $z_{\bar{X}} \in \{z \mid z \leq -1.64\}$

4b It suggests that the I.Q. test does indeed exhibit cultural bias

5a $\delta = \dfrac{|\mu_B - \mu_0|}{\sigma} = \dfrac{5}{15} = .333$ **5b** $H_1: \mu_B = \mu_0 - \delta\sigma = \mu_0 - .333(15)$
$$= 95$$

5c $z_\beta = 1.96$

5d $N = \left[\dfrac{\sigma(z_\beta - z_\alpha)}{\mu_0 - \mu_1}\right]^2 = \left[\dfrac{15(1.96 + 1.64)}{5}\right]^2 = 117$

5e $-1.64 = \dfrac{\bar{x}_c - 100}{\sqrt{225/117}}$. Thus, $\bar{x}_c = 100 - 1.64\sqrt{225/117} = 97.72$

5f He should reject H_1 **5g** Yes. The degree of bias in the test is not scientifically important

6a Given only 15 Ss and an estimate of σ, the psychologist must use a t test. For $v = N - 1 = 14$, the critical value of his test statistic is
$$t_\alpha = -1.76 = \frac{\bar{x}_c - 100}{18/\sqrt{15}}$$
Therefore $\bar{x}_c = 100 - 1.76(18/\sqrt{15}) = 91.82$

6b If he knows that $\sigma = 18$, he could use the normal distribution to test his hypothesis, and the critical value of his test statistic would be
$$z_\alpha = -1.64 = \frac{\bar{x}_c - 100}{18/\sqrt{15}}$$
Therefore, $\bar{x}_c = 100 - 1.64(18/\sqrt{15}) = 92.37$

6c The Z-test will yield significance for smaller departures from μ_0 and is therefore more powerful. It should be apparent from **6a** and **6b**, however, that $t(14)$ is already very nearly normal.

7a $H_0: \mu_W - \mu_B = \Delta_0 = 0$ **7b** $H_1: \mu_W - \mu_B = \Delta_1 = \Delta_0 + \delta\sigma_{x_1 - x_2}$
$$= 0 + .5\sqrt{169 + 256}$$
$$= 10.31$$

7c $N_W = \hat{\sigma}_W(\hat{\sigma}_W + \hat{\sigma}_B)\left[\dfrac{z_\beta - z_\alpha}{\Delta_0 - \Delta_1}\right]^2 = 13(13 + 16)\left[\dfrac{-1.64 - 1.64}{10.31}\right]^2 = 38$

$N_B = \hat{\sigma}_B(\hat{\sigma}_W + \hat{\sigma}_B)\left[\dfrac{z_\beta - z_\alpha}{\Delta_0 - \Delta_1}\right]^2 = 16(13 + 16)\left[\dfrac{-1.64 - 1.64}{10.31}\right]^2 = 47$

7d
$$\frac{\bar{x}_W - \bar{x}_B}{\sqrt{(\hat{\sigma}_W^2/N_W) + (\hat{\sigma}_B^2/N_B)}} = \frac{102.75 - 97.50}{\sqrt{9.894}} = \frac{5.25}{3.145} = 1.67$$

The psychologist should therefore reject H_0

8a With $N_W = N_B$ the psychologist can put his tongue in his cheek and assume that $\sigma_W^2 = \sigma_B^2$. The appropriate test statistic is therefore

$$\frac{\bar{X}_W - \bar{X}_B}{\sqrt{\hat{\sigma}^2_{pooled}\left[\dfrac{N_W + N_B}{N_W N_B}\right]}}$$

and is distributed as $t(N_W + N_B - 2)$. The critical value $t_\alpha = 1.72$

8b
$$\hat{\sigma}^2_{pooled} = \frac{(N_W - 1)\hat{\sigma}_W^2 + (N_B - 1)\hat{\sigma}_B^2}{N_W + N_B - 2} = 212.5$$

Therefore

$$t = \frac{102.75 - 97.50}{\sqrt{212.5\left(\dfrac{24}{144}\right)}} = \frac{5.25}{5.95} = .882$$

This difference is less than 1.72 and is therefore nonsignificant at $\alpha = .05$

Chapter 13

1 $\chi^2 = 21.06$

2 $E[\chi^2 | \chi^2 : \chi^2(14)] = 14$
$V[\chi^2 | \chi^2 : \chi^2(14)] = 28$

3 $\chi_b^2 = 12.59$
$\chi_a^2 = 1.64$

4 $E[\chi^2 | \chi^2 : \chi^2(6)] = 6$

5 No. $\chi_b^2 - E(\chi^2) = 6.59$ and $E(\chi^2) - \chi_a^2 = 4.36$. $\chi^2(6)$ is skewed

6a No   **6b** $\dfrac{\chi_a^2(225)}{14} = \dfrac{23.68(225)}{14} = 380.57$ **6c** Yes

6d $\sqrt{380.57} = 19.51$. This value is not surprisingly larger than $\sigma_0 = 15$

7 $F(1.78) = .75$ **8** $F(2.72) = .95$. Thus, $F_\alpha = 2.72$

9a $H_1 : \sigma_B^2 > \sigma_W^2$ or $H_1 : \sigma_W^2 < \sigma_B^2$

9b (Under $H_1 : \sigma_B^2 > \sigma_W^2$) $\hat{\sigma}_B^2/\hat{\sigma}_W^2 = 1.51$ and $F_\alpha = 2.82$
(Under $H_1 : \sigma_W^2 < \sigma_B^2$) $\hat{\sigma}_W^2/\hat{\sigma}_B^2 = .66$ and $F_\alpha = .355$
Whichever alternative hypothesis is used, the result is not significant at $\alpha = .05$

9c The result would be significant for approximately $\alpha = .74$

9d Even if the assumption is incorrect, little error is likely to result

Chapter 14

1a/1b

Δz_j	$Np_j = \mathbf{F}_j$
-3.00 to -2.40	$50(.008) = .40$
-2.40 to -1.80	$50(.028) = 1.40$
-1.80 to -1.20	$50(.08) = 4.00$
-1.20 to $-.60$	$50(.16) = 8.00$
$-.60$ to 0.00	$50(.23) = 11.50$
0.00 to $.60$	$50(.23) = 11.50$
$.60$ to 1.20	$50(.16) = 8.00$
1.20 to 1.80	$50(.08) = 4.00$
1.80 to 2.40	$50(.028) = 1.40$
2.40 to 3.00	$50(.008) = .40$

1c No

1b The smallest probability p_j is .008. Thus, if $N(.008)$ is to equal 5, then $N = 5/.008 = 625$

1e/1f

Δz_j	$Np_j = \mathbf{F}_j$
-3.00 to -1.28	$50(.10) = 5$
-1.28 to $-.84$	$50(.10) = 5$
$-.84$ to $-.52$	$50(.10) = 5$
$-.52$ to $-.25$	$50(.10) = 5$
$-.25$ to 0.00	$50(.10) = 5$
0.00 to $.25$	$50(.10) = 5$
$.25$ to $.52$	$50(.10) = 5$
$.52$ to $.84$	$50(.10) = 5$
$.84$ to 1.28	$50(.10) = 5$
1.28 to 3.00	$50(.10) = 5$

1i $v = C - 1 = 9$

1j $\tilde{\chi}^2$ will probably be significant because values in Table **B** are uniformly distributed

2a If you construct an expected frequency distribution based on five cells, your expected frequencies for D and F will be less than 5. You should therefore collapse these two cells, yielding the following values $\mathbf{F}_{jk}$:

Grade	A	B	C	D or F
Lower division	5	10	12	5
Upper division	5	10	12	5

388

2b No

3a The violation appears in Table 14.5. Several of the expected frequencies are less than 5

3b The data must be grouped into two classes (scores above the median and scores below the median) instead of five classes, yielding the following values of F_{jk}:

Δx_i	0–29	30–49
0–1 year	7.5	7.5
2—years	17.5	17.5

3c With Yates' correction for continuity $\tilde{\chi}^2 = 6.10$

3d Yes

4 Any $\tilde{\chi}^2$ test is necessarily one-tailed, but a test of association in which $R = C = 2$ can yield significance from either of *two* patterns of differences $(f_{jk} - F_{jk})$:

<table>
<tr><td align="center">A</td><td></td><td align="center">B</td></tr>
<tr><td>$f_{11} > F_{11}$ $f_{21} < F_{21}$</td><td></td><td>$f_{11} < F_{11}$ $f_{21} > F_{21}$</td></tr>
<tr><td colspan="3" align="center">or</td></tr>
<tr><td>$f_{12} < F_{12}$ $f_{22} > F_{22}$</td><td></td><td>$f_{12} > F_{12}$ $f_{22} < F_{22}$</td></tr>
</table>

If the experimenter makes no prediction about the pattern to be expected if H_0 is incorrect, therefore, $p(A \mid H_0 \text{ correct}) = \alpha/2$ and $p(B \mid H_0 \text{ correct}) = \alpha/2$. In problem 3, however, the teacher expects that B is the *only* possible alternative to H_0, and his total probability of a Type I error must therefore be $\alpha/2$

Chapter 15

1 *MSW* taps only error variance. From the definitional formula

$$MSW = \frac{\sum_j \sum_i (x_{ij} - \bar{x}_j)^2}{N - J}$$

it is apparent that *MSW* reflects only differences among individuals within the same treatment group. Because all such Ss receive the same experimental treatment, any variability must be entirely attributable to error. This is confirmed by deriving $E(MSW)$, which equals σ_e^2

2   *MSA* taps both error variance and treatment variance. From the definitional formula

$$MSA = \frac{\sum_{j}^{J} N_j(\bar{x}_j - \bar{x})^2}{J - 1}$$

we see that *MSA* reflects variability in our group means $\bar{x}_j$. We know that $\overline{X}$ is a random variable, and even if H_0 were correct, therefore, we would not expect $\bar{x}_1 = \bar{x}_2 = \ldots \bar{x}_J$. Thus, *MSA* will reflect error variance. If, however, H_0 is not correct and $\mu_j \neq \mu$, the differences $(\mu_j - \mu)$, which can be attributed only to experimental treatments, will also be reflected in *MSA*. This is confirmed by deriving $E(MSA)$, which equals

$$\sigma_e^2 + \frac{\sum_{j}^{J} N_j\mu_j^2 - N\mu^2}{J - 1}$$

Clearly, the bias factor reflects differences between the population means μ_j and their hypothetical common value μ

3   From problem **1** we know that $E(MSW) = \sigma_e^2$, and from problem **2** we know that $E(MSA) = \sigma_e^2$ plus a bias term that equals zero if H_0 is correct. When H_0 is correct, therefore, *MSW* and *MSA* are both unbiased estimators of σ_e^2, and from Chapter thirteen we know that the ratio of any two such estimators is distributed as Fisher's *F*

4   Yes. In $F(2, 40)$ $F_{.95} = 3.23$; in problem **4** $F = 3.24$

5 $MSA = 950.00$ $MSW = 134.8$ $F = 7.05$

In $F(2, 15)$ $F_{.99} = 6.36$. Therefore *F* is significant at $\alpha = .01$.

appendix V

tables

Table **A**. Squares and Square Roots of Numbers from 1 to 1000

Number	Square	Square Root	Number	Square	Square Root
1	1	1.0000	41	16 81	6.4031
2	4	1.4142	42	17 64	6.4807
3	9	1.7321	43	18 49	6.5574
4	16	2.0000	44	19 36	6.6332
5	25	2.2361	45	20 25	6.7082
6	36	2.4495	46	21 16	6.7823
7	49	2.6458	47	22 09	6.8557
8	64	2.8284	48	23 04	6.9282
9	81	3.0000	49	24 01	7.0000
10	1 00	3.1623	50	25 00	7.0711
11	1 21	3.3166	51	26 01	7.1414
12	1 44	3.4641	52	27 04	7.2111
13	1 69	3.6056	53	28 09	7.2801
14	1 96	3.7417	54	29 16	7.3485
15	2 25	3.8730	55	30 25	7.4162
16	2 56	4.0000	56	31 36	7.4833
17	2 89	4.1231	57	32 49	7.5498
18	3 24	4.2426	58	33 64	7.6158
19	3 61	4.3589	59	34 81	7.6811
20	4 00	4.4721	60	36 00	7.7460
21	4 41	4.5826	61	37 21	7.8102
22	4 84	4.6904	62	38 44	7.8740
23	5 29	4.7958	63	39 69	7.9373
24	5 76	4.8990	64	40 96	8.0000
25	6 25	5.0000	65	42 25	8.0623
26	6 76	5.0990	66	43 56	8.1240
27	7 29	5.1962	67	44 89	8.1854
28	7 84	5.2915	68	46 24	8.2462
29	8 41	5.3852	69	47 61	8.3066
30	9 00	5.4772	70	49 00	8.3666
31	9 61	5.5678	71	50 41	8.4261
32	10 24	5.6569	72	51 84	8.4853
33	10 89	5.7446	73	53 29	8.5440
34	11 56	5.8310	74	54 76	8.6023
35	12 25	5.9161	75	56 25	8.6603
36	12 96	6.0000	76	57 76	8.7178
37	13 69	6.0828	77	59 29	8.7750
38	14 44	6.1644	78	60 84	8.8318
39	15 21	6.2450	79	62 41	8.8882
40	16 00	6.3246	80	64 00	8.9443

Source. Table **A** is taken from Sorenson, H., *Statistics for Students of Psychology and Education*, McGraw-Hill Book Company, 1936, used with permission.

Number	Square	Square Root	Number	Square	Square Root
81	65 61	9.0000	121	1 46 41	11.0000
82	67 24	9.0554	122	1 48 84	11.0454
83	68 89	9.1104	123	1 51 29	11.0905
84	70 56	9.1652	124	1 53 76	11.1355
85	72 25	9.2195	125	1 56 25	11.1803
86	73 96	9.2736	126	1 58 76	11.2250
87	75 69	9.3274	127	1 61 29	11.2694
88	77 44	9.3808	128	1 63 84	11.3137
89	79 21	9.4340	129	1 66 41	11.3578
90	81 00	9.4868	130	1 69 00	11.4018
91	82 81	9.5394	131	1 71 61	11.4455
92	84 64	9.5917	132	1 74 24	11.4891
93	86 49	9.6437	133	1 76 89	11.5326
94	88 36	9.6954	134	1 79 56	11.5758
95	90 25	9.7468	135	1 82 25	11.6190
96	92 16	9.7980	136	1 84 96	11.6619
97	94 09	9.8489	137	1 87 69	11.7047
98	96 04	9.8995	138	1 90 44	11.7473
99	98 01	9.9499	139	1 93 21	11.7898
100	1 00 00	10.0000	140	1 96 00	11.8322
101	1 02 01	10.0499	141	1 98 81	11.8743
102	1 04 04	10.0995	142	2 01 64	11.9164
103	1 06 09	10.1489	143	2 04 49	11.9583
104	1 08 16	10.1980	144	2 07 36	12.0000
105	1 10 25	10.2470	145	2 10 25	12.0416
106	1 12 36	10.2956	146	2 13 16	12.0830
107	1 14 49	10.3441	147	2 16 09	12.1244
108	1 16 64	10.3923	148	2 19 04	12.1655
109	1 18 81	10.4403	149	2 22 01	12.2066
110	1 21 00	10.4881	150	2 25 00	12.2474
111	1 23 21	10.5357	151	2 28 01	12.2882
112	1 25 44	10.5830	152	2 31 04	12.3288
113	1 27 69	10.6301	153	2 34 09	12.3693
114	1 29 96	10.6771	154	2 37 16	12.4097
115	1 32 25	10.7238	155	2 40 25	12.4499
116	1 34 56	10.7703	156	2 43 36	12.4900
117	1 36 89	10.8167	157	2 46 49	12.5300
118	1 39 24	10.8628	158	2 49 64	12.5698
119	1 41 61	10.9087	159	2 52 81	12.6095
120	1 44 00	10.9545	160	2 56 00	12.6491

Table A (*continued*)

Number	Square	Square Root	Number	Square	Square Root
161	2 59 21	12.6886	201	4 04 01	14.1774
162	2 62 44	12.7279	202	4 08 04	14.2127
163	2 65 69	12.7671	203	4 12 09	14.2478
164	2 68 96	12.8062	204	4 16 16	14.2829
165	2 72 25	12.8452	205	4 20 25	14.3178
166	2 75 56	12.8841	206	4 24 36	14.3527
167	2 78 89	12.9228	207	4 28 49	14.3875
168	2 82 24	12.9615	208	4 32 64	14.4222
169	2 85 61	13.0000	209	4 36 81	14.4568
170	2 89 00	13.0384	210	4 41 00	14.4914
171	2 92 41	13.0767	211	4 45 21	14.5258
172	2 95 84	13.1149	212	4 49 44	14.5602
173	2 99 29	13.1529	213	4 53 69	14.5945
174	3 02 76	13.1909	214	4 57 96	14.6287
175	3 06 25	13.2288	215	4 62 25	14.6629
176	3 09 76	13.2665	216	4 66 56	14.6969
177	3 13 29	13.3041	217	4 70 89	14.7309
178	3 16 84	13.3417	218	4 75 24	14.7648
179	3 20 41	13.3791	219	4 79 61	14.7986
180	3 24 00	13.4164	220	4 84 00	14.8324
181	3 27 61	13.4536	221	4 88 41	14.8661
182	3 31 24	13.4907	222	4 92 84	14.8997
183	3 34 89	13.5277	223	4 97 29	14.9332
184	3 38 56	13.5647	224	5 01 76	14.9666
185	3 42 25	13.6015	225	5 06 25	15.0000
186	3 45 96	13.6382	226	5 10 76	15.0333
187	3 49 69	13.6748	227	5 15 29	15.0665
188	3 53 44	13.7113	228	5 19 84	15.0997
189	3 57 21	13.7477	229	5 24 41	15.1327
190	3 61 00	13.7840	230	5 29 00	15.1658
191	3 64 81	13.8203	231	5 33 61	15.1987
192	3 68 64	13.8564	232	5 38 24	15.2315
193	3 72 49	13.8924	233	5 42 89	15.2643
194	3 76 36	13.9284	234	5 47 56	15.2971
195	3 80 25	13.9642	235	5 52 25	15.3297
196	3 84 16	14.0000	236	5 56 96	15.3623
197	3 88 09	14.0357	237	5 61 69	15.3948
198	3 92 04	14.0712	238	5 66 44	15.4272
199	3 96 01	14.1067	239	5 71 21	15.4596
200	4 00 00	14.1421	240	5 76 00	15.4919

Table A (*continued*)

Number	Square	Square Root	Number	Square	Square Root
241	5 80 81	15.5242	281	7 89 61	16.7631
242	5 85 64	15.5563	282	7 95 24	16.7929
243	5 90 49	15.5885	283	8 00 89	16.8226
244	5 95 36	15.6205	284	8 06 56	16.8523
245	6 00 25	15.6525	285	8 12 25	16.8819
246	6 05 16	15.6844	286	8 17 96	16.9115
247	6 10 09	15.7162	287	8 23 69	16.9411
248	6 15 04	15.7480	288	8 29 44	16.9706
249	6 20 01	15.7797	289	8 35 21	17.0000
250	6 25 00	15.8114	290	8 41 00	17.0294
251	6 30 01	15.8430	291	8 46 81	17.0587
252	6 35 04	15.8745	292	8 52 64	17.0880
253	6 40 09	15.9060	293	8 58 49	17.1172
254	6 45 16	15.9374	294	8 64 36	17.1464
255	6 50 25	15.9687	295	8 70 25	17.1756
256	6 55 36	16.0000	296	8 76 16	17.2047
257	6 60 49	16.0312	297	8 82 09	17.2337
258	6 65 64	16.0624	298	8 88 04	17.2627
259	6 70 81	16.0935	299	8 94 01	17.2916
260	6 76 00	16.1245	300	9 00 00	17.3205
261	6 81 21	16.1555	301	9 06 01	17.3494
262	6 86 44	16.1864	302	9 12 04	17.3781
263	6 91 69	16.2173	303	9 18 09	17.4069
264	6 96 96	16.2481	304	9 24 16	17.4356
265	7 02 25	16.2788	305	9 30 25	17.4642
266	7 07 56	16.3095	306	9 36 36	17.4929
267	7 12 89	16.3401	307	9 42 49	17.5214
268	7 18 24	16.3707	308	9 48 64	17.5499
269	7 23 61	16.4012	309	9 54 81	17.5784
270	7 29 00	16.4317	310	9 61 00	17.6068
271	7 34 41	16.4621	311	9 67 21	17.6352
272	7 39 84	16.4924	312	9 73 44	17.6635
273	7 45 29	16.5227	313	9 79 69	17.6918
274	7 50 76	16.5529	314	9 85 96	17.7200
275	7 56 25	16.5831	315	9 92 25	17.7482
276	7 61 76	16.6132	316	9 98 56	17.7764
277	7 67 29	16.6433	317	10 04 89	17.8045
278	7 72 84	16.6733	318	10 11 24	17.8326
279	7 78 41	16.7033	319	10 17 61	17.8606
280	7 84 00	16.7332	320	10 24 00	17.8885

Table A (*continued*)

Number	Square	Square Root	Number	Square	Square Root
321	10 30 41	17.9165	361	13 03 21	19.0000
322	10 36 84	17.9444	362	13 10 44	19.0263
323	10 43 29	17.9722	363	13 17 69	19.0526
324	10 49 76	18.0000	364	13 24 96	19.0788
325	10 56 25	18.0278	365	13 32 25	19.1050
326	10 62 76	18.0555	366	13 39 56	19.1311
327	10 69 29	18.0831	367	13 46 89	19.1572
328	10 75 84	18.1108	368	13 54 24	19.1833
329	10 82 41	18.1384	369	13 61 61	19.2094
330	10 89 00	18.1659	370	13 69 00	19.2354
331	10 95 61	18.1934	371	13 76 41	19.2614
332	11 02 24	18.2209	372	13 83 84	19.2873
333	11 08 89	18.2483	373	13 91 29	19.3132
334	11 15 56	18.2757	374	13 98 76	19.3391
335	11 22 25	18.3030	375	14 06 25	19.3649
336	11 28 96	18.3303	376	14 13 76	19.3907
337	11 35 69	18.3576	377	14 21 29	19.4165
338	11 42 44	18.3848	378	14 28 84	19.4422
339	11 49 21	18.4120	379	14 36 41	19.4679
340	11 56 00	18.4391	380	14 44 00	19.4936
341	11 62 81	18.4662	381	14 51 61	19.5192
342	11 69 64	18.4932	382	14 59 24	19.5448
343	11 76 49	18.5203	383	14 66 89	19.5704
344	11 83 36	18.5472	384	14 74 56	19.5959
345	11 90 25	18.5742	385	14 82 25	19.6214
346	11 97 16	18.6011	386	14 89 96	19.6469
347	12 04 09	18.6279	387	14 97 69	19.6723
348	12 11 04	18.6548	388	15 05 44	19.6977
349	12 18 01	18.6815	389	15 13 21	19.7231
350	12 25 00	18.7083	390	15 21 00	19.7484
351	12 32 01	18.7350	391	15 28 81	19.7737
352	12 39 04	18.7617	392	15 36 64	19.7990
353	12 46 09	18.7883	393	15 44 49	19.8242
354	12 53 16	18.8149	394	15 52 36	19.8494
355	12 60 25	18.8414	395	15 60 25	19.8746
356	12 67 36	18.8680	396	15 68 16	19.8997
357	12 74 49	18.8944	397	15 76 09	19.9249
358	12 81 64	18.9209	398	15 84 04	19.9499
359	12 88 81	18.9473	399	15 92 01	19.9750
360	12 96 00	18.9737	400	16 00 00	20.0000

Table A (*continued*)

Number	Square	Square Root	Number	Square	Square Root
401	16 08 01	20.0250	441	19 44 81	21.0000
402	16 16 04	20.0499	442	19 53 64	21.0238
403	16 24 09	20.0749	443	19 62 49	21.0476
404	16 32 16	20.0998	444	19 71 36	21.0713
405	16 40 25	20.1246	445	19 80 25	21.0950
406	16 48 36	20.1494	446	19 89 16	21.1187
407	16 56 49	20.1742	447	19 98 09	21.1424
408	16 64 64	20.1990	448	20 07 04	21.1660
409	16 72 81	20.2237	449	20 16 01	21.1896
410	16 81 00	20.2485	450	20 25 00	21.2132
411	16 89 21	20.2731	451	20 34 01	21.2368
412	16 97 44	20.2978	452	20 43 04	21.2603
413	17 05 69	20.3224	453	20 52 09	21.2838
414	17 13 96	20.3470	454	20 61 16	21.3073
415	17 22 25	20.3715	455	20 70 25	21.3307
416	17 30 56	20.3961	456	20 79 36	21.3542
417	17 38 89	20.4206	457	20 88 49	21.3776
418	17 47 24	20.4450	458	20 97 64	21.4009
419	17 55 61	20.4695	459	21 06 81	21.4243
420	17 64 00	20.4939	460	21 16 00	21.4476
421	17 72 41	20.5183	461	21 25 21	21.4709
422	17 80 84	20.5426	462	21 34 44	21.4942
423	17 89 29	20.5670	463	21 43 69	21.5174
424	17 97 76	20.5913	464	21 52 96	21.5407
425	18 06 25	20.6155	465	21 62 25	21.5639
426	18 14 76	20.6398	466	21 71 56	21.5870
427	18 23 29	20.6640	467	21 80 89	21.6102
428	18 31 84	20.6882	468	21 90 24	21.6333
429	18 40 41	20.7123	469	21 99 61	21.6564
430	18 49 00	20.7364	470	22 09 00	21.6795
431	18 57 61	20.7605	471	22 18 41	21.7025
432	18 66 24	20.7846	472	22 27 84	21.7256
433	18 74 89	20.8087	473	22 37 29	21.7486
434	18 83 56	20.8327	474	22 46 76	21.7715
435	18 92 25	20.8567	475	22 56 25	21.7945
436	19 00 96	20.8806	476	22 65 76	21.8174
437	19 09 69	20.9045	477	22 75 29	21.8403
438	19 18 44	20.9284	478	22 84 84	21.8632
439	19 27 21	20.9523	479	22 94 41	21.8861
440	19 36 00	20.9762	480	23 04 00	21.9089

Table A (*continued*)

Number	Square	Square Root	Number	Square	Square Root
481	23 13 61	21.9317	521	27 14 41	22.8254
482	23 23 24	21.9545	522	27 24 84	22.8473
483	23 32 89	21.9773	523	27 35 29	22.8692
484	23 42 56	22.0000	524	27 45 76	22.8910
485	23 52 25	22.0227	525	27 56 25	22.9129
486	23 61 96	22.0454	526	27 66 76	22.9347
487	23 71 69	22.0681	527	27 77 29	22.9565
488	23 81 44	22.0907	528	27 87 84	22.9783
489	23 91 21	22.1133	529	27 98 41	23.0000
490	24 01 00	22.1359	530	28 09 00	23.0217
491	24 10 81	22.1585	531	28 19 61	23.0434
492	24 20 64	22.1811	532	28 30 24	23.0651
493	24 30 49	22.2036	533	28 40 89	23.0868
494	24 40 36	22.2261	534	28 51 56	23.1084
495	24 50 25	22.2486	535	28 62 25	23.1301
496	24 60 16	22.2711	536	28 72 96	23.1517
497	24 70 09	22.2935	537	28 83 69	23.1733
498	24 80 04	22.3159	538	28 94 44	23.1948
499	24 90 01	22.3383	539	29 05 21	23.2164
500	25 00 00	22.3607	540	29 16 00	23.2379
501	25 10 01	22.3830	541	29 26 81	23.2594
502	25 20 04	22.4054	542	29 37 64	23.2809
503	25 30 09	22.4277	543	29 48 49	23.3024
504	25 40 16	22.4499	544	29 59 36	23.3238
505	25 50 25	22.4722	545	29 70 25	23.3452
506	25 60 36	22.4944	546	29 81 16	23.3666
507	25 70 49	22.5167	547	29 92 09	23.3880
508	25 80 64	22.5389	548	30 03 04	23.4094
509	25 90 81	22.5610	549	30 14 01	23.4307
510	26 01 00	22.5832	550	30 25 00	23.4521
511	26 11 21	22.6053	551	30 36 01	23.4734
512	26 21 44	22.6274	552	30 47 04	23.4947
513	26 31 69	22.6495	553	30 58 09	23.5160
514	26 41 96	22.6716	554	30 69 16	23.5372
515	26 52 25	22.6936	555	30 80 25	23.5584
516	26 62 56	22.7156	556	30 91 36	23.5797
517	26 72 89	22.7376	557	31 02 49	23.6008
518	26 83 24	22.7596	558	31 13 64	23.6220
519	26 93 61	22.7816	559	31 24 81	23.6432
520	27 04 00	22.8035	560	31 36 00	23.6643

Number	Square	Square Root	Number	Square	Square Root
561	31 47 21	23.6854	601	36 12 01	24.5153
562	31 58 44	23.7065	602	36 24 04	24.5357
563	31 69 69	23.7276	603	36 36 09	24.5561
564	31 80 96	23.7487	604	36 48 16	24.5764
565	31 92 25	23.7697	605	36 60 25	24.5967
566	32 03 56	23.7908	606	36 72 36	24.6171
567	32 14 89	23.8118	607	36 84 49	24.6374
568	32 26 24	23.8328	608	36 96 64	24.6577
569	32 37 61	23.8537	609	37 08 81	24.6779
570	32 49 00	23.8747	610	37 21 00	24.6982
571	32 60 41	23.8956	611	37 33 21	24.7184
572	32 71 84	23.9165	612	37 45 44	24.7385
573	32 83 29	23.9374	613	37 57 69	24.7588
574	32 94 76	23.9583	614	37 69 96	24.7790
575	33 06 25	23.9792	615	37 82 25	24.7992
576	33 17 76	24.0000	616	37 94 56	24.8193
577	33 29 29	24.0208	617	38 06 89	24.8395
578	33 40 84	24.0416	618	38 19 24	24.8596
579	33 52 41	24.0624	619	38 31 61	24.8797
580	33 64 00	24.0832	620	38 44 00	24.8998
581	33 75 61	24.1039	621	38 56 41	24.9199
582	33 87 24	24.1247	622	38 68 84	24.9399
583	33 98 89	24.1454	623	38 81 29	24.9600
584	34 10 56	24.1661	624	38 93 76	24.9800
585	34 22 25	24.1868	625	39 06 25	25.0000
586	34 33 96	24.2074	626	39 18 76	25.0200
587	34 45 69	24.2281	627	39 31 29	25.0400
588	34 57 44	24.2487	628	39 43 84	25.0599
589	34 69 21	24.2693	629	39 56 41	25.0799
590	34 81 00	24.2899	630	39 69 00	25.0998
591	34 92 81	24.3105	631	39 81 61	25.1197
592	35 04 64	24.3311	632	39 94 24	25.1396
593	35 16 49	24.3516	633	40 06 89	25.1595
594	35 28 36	24.3721	634	40 19 56	25.1794
595	35 40 25	24.3926	635	40 32 25	25.1992
596	35 52 16	24.4131	636	40 44 96	25.2190
597	35 64 09	24.4336	637	40 57 69	25.2389
598	35 76 04	24.4540	638	40 70 44	25.2587
599	35 88 01	24.4745	639	40 83 21	25.2784
600	36 00 00	24.4949	640	40 96 00	25.2982

Number	Square	Square Root	Number	Square	Square Root
641	41 08 81	25.3180	681	46 37 61	26.0960
642	41 21 64	25.3377	682	46 51 24	26.1151
643	41 34 49	25.3574	683	46 64 89	26.1343
644	41 47 36	25.3772	684	46 78 56	26.1534
645	41 60 25	25.3969	685	46 92 25	26.1725
646	41 73 16	25.4165	686	47 05 96	26.1916
647	41 86 09	25.4362	687	47 19 69	26.2107
648	41 99 04	25.4558	688	47 33 44	26.2298
649	42 12 01	25.4755	689	47 47 21	26.2488
650	42 25 00	25.4951	690	47 61 00	26.2679
651	42 38 01	25.5147	691	47 74 81	26.2869
652	42 51 04	25.5343	692	47 88 64	26.3059
653	42 64 09	25.5539	693	48 02 49	26.3249
654	42 77 16	25.5734	694	48 16 36	26.3439
655	42 90 25	25.5930	695	48 30 25	26.3629
656	43 03 36	25.6125	696	48 44 16	26.3818
657	43 16 49	25.6320	697	48 58 09	26.4008
658	43 29 64	25.6515	698	48 72 04	26.4197
659	43 42 81	25.6710	699	48 86 01	26.4386
660	43 56 00	25.6905	700	49 00 00	26.4575
661	43 69 21	25.7099	701	49 14 01	26.4764
662	43 82 44	25.7294	702	49 28 04	26.4953
663	43 95 69	25.7488	703	49 42 09	26.5141
664	44 08 96	25.7682	704	49 56 16	26.5330
665	44 22 25	25.7876	705	49 70 25	26.5518
666	44 35 56	25.8070	706	49 84 36	26.5707
667	44 48 89	25.8263	707	49 98 49	26.5895
668	44 62 24	25.8457	708	50 12 64	26.6083
669	44 75 61	25.8650	709	50 26 81	26.6271
670	44 89 00	25.8844	710	50 41 00	26.6458
671	45 02 41	25.9037	711	50 55 21	26.6646
672	45 15 84	25.9230	712	50 69 44	26.6833
673	45 29 29	25.9422	713	50 83 69	26.7021
674	45 42 76	25.9615	714	50 97 96	26.7208
675	45 56 25	25.9808	715	51 12 25	26.7395
676	45 69 76	26.0000	716	51 26 56	26.7582
677	45 83 29	26.0192	717	51 40 89	26.7769
678	45 96 84	26.0384	718	51 55 24	26.7955
679	46 10 41	26.0576	719	51 69 61	26.8142
680	46 24 00	26.0768	720	51 84 00	26.8328

Number	Square	Square Root	Number	Square	Square Root
721	51 98 41	26.8514	761	57 91 21	27.5862
722	52 12 84	26.8701	762	58 06 44	27.6043
723	52 27 29	26.8887	763	58 21 69	27.6225
724	52 41 76	26.9072	764	58 36 96	27.6405
725	52 56 25	26.9258	765	58 52 25	27.6586
726	52 70 76	26.9444	766	58 67 56	27.6767
727	52 85 29	29.9629	767	58 82 89	27.6948
728	52 99 84	26.9815	768	58 98 24	27.7128
729	53 14 41	27.0000	769	59 13 61	27.7308
730	53 29 00	27.0185	770	59 29 00	27.7489
731	53 43 61	27.0370	771	59 44 41	27.7669
732	53 58 24	27.0555	772	59 59 84	27.7849
733	53 72 89	27.0740	773	59 75 29	27.8029
734	53 87 56	27.0924	774	59 90 76	27.8209
735	54 02 25	27.1109	775	60 06 25	27.8388
736	54 16 96	27.1293	776	60 21 76	27.8568
737	54 31 69	27.1477	777	60 37 29	27.8747
738	54 46 44	27.1662	778	60 52 84	27.8927
739	54 61 27	27.1846	779	60 68 41	27.9106
740	54 76 00	27.2029	780	60 84 00	27.9285
741	54 90 81	27.2213	781	60 99 61	27.9464
742	55 05 64	27.2397	782	61 15 24	27.9643
743	55 20 49	27.2580	783	61 30 89	27.9821
744	55 35 36	27.2764	784	61 46 56	28.0000
745	55 50 25	27.2947	785	61 62 25	28.0179
746	55 65 16	27.3130	786	61 77 96	28.0357
747	55 80 09	27.3313	787	61 93 69	28.0535
748	55 95 04	27.3496	788	62 09 44	28.0713
749	56 10 01	27.3679	789	62 25 21	28.0891
750	56 25 00	27.3861	790	62 41 00	28.1069
751	56 40 01	27.4044	791	62 56 81	28.1247
752	56 55 04	27.4226	792	62 72 64	28.1425
753	56 70 09	27.4408	793	62 88 49	28.1603
754	56 85 16	27.4591	794	63 04 36	28.1780
755	57 00 25	27.4773	795	63 20 25	28.1957
756	57 15 36	27.4955	796	63 36 16	28.2135
757	57 30 49	27.5136	797	63 52 09	28.2312
758	57 45 64	27.5318	798	63 68 04	28.2489
759	57 60 81	27.5500	799	63 84 01	28.2666
760	57 76 00	27.5681	800	64 00 00	28.2843

Table A (*continued*)

Number	Square	Square Root	Number	Square	Square Root
801	64 16 01	28.3019	841	70 72 81	29.0000
802	64 32 04	28.3196	842	70 89 64	29.0172
803	64 48 09	28.3373	843	71 06 49	29.0345
804	64 64 16	28.3049	844	71 23 36	29.0517
805	64 80 25	28.3725	845	71 40 25	29.0689
806	64 96 36	28.3901	846	71 57 16	29.0861
807	65 12 49	28.4077	847	71 74 09	29.1033
808	65 28 64	28.4253	848	71 91 04	29.1204
809	65 44 81	28.4429	849	72 08 01	29.1376
810	65 61 00	28.4605	850	72 25 00	29.1548
811	65 77 21	28.4781	851	72 42 01	29.1719
812	65 93 44	28.4956	852	72 59 04	29.1890
813	66 09 69	28.5132	853	72 76 09	29.2062
814	66 25 96	28.5307	854	72 93 16	29.2233
815	66 42 25	28.5482	855	73 10 25	29.2404
816	66 58 56	28.5657	856	73 27 36	29.2575
817	66 74 89	28.5832	857	73 44 49	29.2746
818	66 91 24	28.6007	858	73 61 64	29.2916
819	67 07 61	28.6082	859	73 78 81	29.3087
820	67 24 00	28.6356	860	73 96 00	29.3258
821	67 40 41	28.6531	861	74 13 21	29.3428
822	67 56 84	28.6705	862	74 30 44	29.3598
823	67 73 29	28.6880	863	74 47 69	29.3769
824	67 89 76	28.7054	864	74 64 96	29.3939
825	68 06 25	28.7228	865	74 82 25	29.4109
826	68 22 76	28.7402	866	74 99 56	29.4279
827	68 39 29	28.7576	867	75 16 89	29.4449
828	68 55 84	28.7750	868	75 34 24	29.4618
829	68 72 41	28.7924	869	75 51 61	29.4788
830	68 89 00	28.8097	870	75 69 00	29.4958
831	69 05 61	28.8271	871	75 86 41	29.5127
832	69 22 24	28.8444	872	76 03 84	29.5296
833	69 38 89	28.8617	873	76 21 29	29.5466
834	69 55 56	28.8791	874	76 38 76	29.5635
835	69 72 25	28.8964	875	76 56 25	29.5804
836	69 88 96	28.9137	876	76 73 76	29.5973
837	70 05 69	28.9310	877	76 91 29	29.6142
838	70 22 44	28.9482	878	77 08 84	29.6311
839	70 39 21	28.9655	879	77 26 41	29.6479
840	70 56 00	28.9828	880	77 44 00	29.6648

402

Table A (*continued*)

Number	Square	Square Root	Number	Square	Square Root
881	77 61 61	29.6816	921	84 82 41	30.3480
882	77 79 24	29.6985	922	85 00 84	30.3645
883	77 96 89	29.7153	923	85 19 29	30.3809
884	78 14 56	29.7321	924	85 37 76	30.3974
885	78 32 25	29.7489	925	85 56 25	30.4138
886	78 49 96	29.7658	926	85 74 76	30.4302
887	78 67 69	29.7825	927	85 93 29	30.4467
888	78 85 44	29.7993	928	86 11 84	30.4631
889	79 03 21	29.8161	929	86 30 41	30.4795
890	79 21 00	29.8329	930	86 49 00	30.4959
891	79 38 81	29.8496	931	86 67 61	30.5123
892	79 56 64	29.8664	932	86 86 24	30.5287
893	79 74 49	29.8831	933	87 04 89	30.5450
894	79 92 36	29.8998	934	87 23 56	30.5614
895	80 10 25	29.9166	935	87 42 25	30.5778
896	80 28 16	29.9333	936	87 60 96	30.5941
897	80 46 09	29.9500	937	87 79 69	30.6105
898	80 64 04	29.9666	938	87 98 44	30.6268
899	80 82 01	29.9833	939	88 17 21	30.6431
900	81 00 00	30.0000	940	88 36 00	30.6594
901	81 18 01	30.0167	941	88 54 81	30.6757
902	81 36 04	30.0333	942	88 73 64	30.6920
903	81 54 09	30.0500	943	88 92 49	30.7083
904	81 72 16	30.0666	944	89 11 36	30.7246
905	81 90 25	30.0832	945	89 30 25	30.7409
906	82 08 36	30.0998	946	89 49 16	30.7571
907	82 26 49	30.1164	947	89 68 09	30.7734
908	82 44 64	30.1330	948	89 87 04	30.7896
909	82 62 81	30.1496	949	90 06 01	30.8058
910	82 81 00	30.1662	950	90 25 00	30.8221
911	82 99 21	30.1828	951	90 44 01	30.8383
912	83 17 44	30.1993	952	90 63 04	30.8545
913	83 35 69	30.2159	953	90 82 09	30.8707
914	83 53 96	30.2324	954	91 01 16	30.8869
915	83 72 25	30.2490	955	91 20 25	30.9031
916	83 90 56	30.2655	956	91 39 26	30.9192
917	84 08 89	30.2820	957	91 58 49	30.9354
918	84 27 24	30.2985	958	91 77 64	30.9516
919	84 45 61	30.3150	959	91 96 81	30.9677
920	84 64 00	30.3315	960	92 16 00	30.9839

Number	Square	Square Root	Number	Square	Square Root
961	92 35 21	31.0000	981	96 23 61	31.3209
962	92 54 44	31.0161	982	96 43 24	31.3369
963	92 73 69	31.0322	983	96 62 89	31.3528
964	92 92 96	31.0483	984	96 82 56	31.3688
965	93 12 25	31.0644	985	97 02 25	31.3847
966	93 31 56	31.0805	986	97 21 96	31.4006
967	93 50 89	31.0966	987	97 41 69	31.4166
968	93 70 24	31.1127	988	97 61 44	31.4325
969	93 89 61	31.1288	989	97 81 21	31.4484
970	94 09 00	31.1448	990	98 01 00	31.4643
971	94 28 41	31.1609	991	98 20 81	31.4802
972	94 47 84	31.1769	992	98 40 64	31.4960
973	94 67 29	31.1929	993	98 60 49	31.5119
974	94 86 76	31.2090	994	98 80 36	31.5278
975	95 06 25	31.2250	995	99 00 25	31.5436
976	95 25 76	31.2410	996	99 20 16	31.5595
977	95 45 29	31.2570	997	99 40 09	31.5753
978	95 64 84	31.2730	998	99 60 04	31.5911
979	95 84 41	31.2890	999	99 80 01	31.6070
980	96 04 00	31.3050	1000	100 00 00	31.6228

Line/Col.	(1)	(2)	(3)	(4)	(5)	(6)	(7)	(8)	(9)	(10)	(11)	(12)	(13)	(14)
1	10480	15011	01536	02011	81647	91646	69179	14194	62590	36207	20969	99570	91291	90700
2	22368	46573	25595	85393	30995	89198	27982	53402	93965	34095	52666	19174	39615	99505
3	24130	48360	22527	97265	76393	64809	15179	23830	49340	32081	30680	19655	63348	58629
4	42167	93093	06243	61680	07856	16376	39440	53537	71341	57004	00849	74917	97758	16379
5	37570	39975	81837	16656	06121	91782	60468	81305	49684	60672	14110	06927	01263	54613
6	77921	06907	11008	42751	27756	53498	18602	70659	90655	15053	21916	81825	44394	42880
7	99562	72905	56420	69994	98872	31016	71194	18738	44013	48840	63213	21069	10634	12952
8	96301	91977	05463	07972	18876	20922	94595	56869	69014	60045	18425	84903	42508	32307
9	89579	14342	63661	10281	17453	18103	57740	84378	25331	12566	58678	44947	05585	56941
10	85475	36857	43342	53988	53060	59533	38867	62300	08158	17983	16439	11458	18593	64952
11	28918	69578	88231	33276	70997	79936	56865	05859	90106	31595	01547	85590	91610	78188
12	63553	40961	48235	03427	49626	69445	18663	72695	52180	29847	12234	90511	33703	90322
13	09429	93969	52636	92737	88974	33488	36320	17617	30015	08272	84115	27156	30613	74952
14	10365	61129	87529	85689	48237	52267	67689	93394	01511	26358	85104	20285	29975	89868
15	07119	97336	71048	08178	77233	13916	47564	81056	97735	85977	29372	74461	28551	90707
16	51085	12765	51821	51259	77452	16308	60756	92144	49442	53900	70960	63990	75601	40719
17	02368	21382	52404	60268	89368	19885	55322	44819	01188	65255	68435	44919	05944	55157
18	01011	54092	33362	94904	31273	04146	18594	29852	71585	85030	51132	01915	92747	64951
19	52162	53916	46369	58586	23216	14513	83149	98736	23495	64350	94738	17752	35156	35749
20	07056	97628	33787	09998	42698	06691	76988	13602	51851	46104	88916	19509	25625	58104
21	48663	91245	85828	14346	09272	30168	90229	04734	59193	22178	30421	61666	99904	32812
22	54164	58492	22421	74103	47070	25306	76468	26384	58151	06646	21524	15227	96909	44592
23	23639	32363	05597	24200	13363	38005	94342	28728	35806	06912	17012	64161	18296	22851
24	29334	27001	87637	87308	58731	00256	45834	15398	46557	41135	10367	07684	36188	18510
25	02488	33062	28834	07351	19731	92420	60952	61280	50001	67658	32586	86679	50720	94953
26	81525	72295	04839	96423	24788	82651	66566	14778	76797	14780	13300	87074	79666	95725
27	29676	20591	68086	26432	46901	20849	89768	81536	86645	12659	92259	57102	80428	25280
28	00742	57392	39064	66432	84673	40027	32832	61362	98947	96067	64760	64584	96096	98253
29	05366	04213	25669	26422	44407	44048	37937	63904	45766	66134	75470	66520	34693	90449
30	91921	26418	64117	94305	26766	25940	39972	22209	71500	64568	91402	42416	07844	69618
31	00582	04711	87917	77341	42206	35136	74087	99547	81817	42607	43808	76655	62028	76630
32	00725	69884	62797	56170	86324	88072	76222	36086	84637	93161	76038	65855	77919	88006
33	69011	65797	95876	55293	18988	27354	26575	08625	40801	59920	29841	80150	12777	48501
34	25976	57948	29888	88604	67917	48708	18912	82271	65424	69774	33611	54262	85963	03547
35	09763	83473	73577	12908	30883	18317	28290	35797	05998	41688	34952	37888	38917	88050
36	91567	42595	27958	30134	04024	86385	29880	99730	55536	84855	29080	09250	79656	73211
37	17955	56349	90999	49127	20044	59931	06115	29542	18059	02008	73708	83517	31603	42791
38	46503	18584	18845	49618	02304	51038	20655	58727	28168	15475	56942	53389	20562	87338
39	92157	89634	94824	78171	84610	82834	09922	25417	44137	48413	25555	21246	35509	20468
40	14577	62765	35605	81263	39667	47358	56873	56307	61607	49518	89656	20103	77490	18062

Source. Table **B** is reprinted, with permission, from the *Handbook of tables for probability and statistics*, 2nd edition, 1968, The Chemical Rubber Co., Cleveland, Ohio.

Line/Col.	(1)	(2)	(3)	(4)	(5)	(6)	(7)	(8)	(9)	(10)	(11)	(12)	(13)	(14)
41	98427	07523	33362	64270	01638	92477	66969	98420	04480	45585	46565	04102	46880	45709
42	34914	63976	88720	82765	34476	17032	87589	40836	32427	70002	70663	88863	77775	69348
43	70060	28277	39475	46473	23219	53416	94970	25832	69975	94884	19661	72828	00102	66794
44	53976	54914	06990	67245	68350	82948	11398	42878	80287	88267	47363	46634	06541	97809
45	76072	28515	40980	07391	58745	25774	22987	80059	39911	96189	41151	14222	60697	59583
46	90725	52210	83974	29992	65831	38857	50490	83765	55657	14361	31720	57375	56228	41546
47	64364	67412	33339	31926	14883	24413	59744	92351	97473	89286	35931	04110	23726	51900
48	08962	00358	31662	25388	61642	34072	81249	35648	56891	69352	48373	45578	78547	81788
49	95012	68379	93526	70765	10593	04542	76463	54328	02349	17247	28865	14777	62730	92277
50	15664	10493	20492	38391	91132	21999	59516	81652	27195	48223	46751	22923	32261	85653
51	16408	81899	04153	53381	79401	21438	83035	92350	36693	31238	59649	91754	72772	02338
52	18629	81953	05520	91962	04739	13092	97662	24822	94730	06496	35090	04822	86772	98299
53	73115	35101	47498	87637	99016	71060	88824	71013	18735	20286	23153	72924	35165	43040
54	57491	16703	23167	49323	45021	33132	12544	41035	80780	45393	44812	12515	98931	91202
55	30405	83946	23792	14422	15059	45799	22716	19792	09983	74353	68668	30429	70735	25499
56	16631	35006	85900	98275	32388	52390	16815	69298	82732	38480	73817	32523	41961	44437
57	96773	20206	42559	78985	05300	22164	24369	54224	35983	19687	11052	91491	60383	19746
58	38935	64202	14349	82674	66523	44133	00697	35552	35970	19124	63318	29686	03387	59846
59	31624	76384	17403	53363	44167	64486	64758	75366	76554	31601	12614	33072	60332	92325
60	78919	19474	23632	27889	47914	02584	37680	20801	72152	39339	34806	08930	85001	87820
61	03931	33309	57047	74211	63445	17361	62825	39908	05607	91284	68833	25570	38818	46920
62	74426	33278	43972	10119	89917	15665	52872	73823	73144	88662	88970	74492	51805	99378
63	09066	00903	20795	95452	92648	45454	09552	88815	16553	51125	79375	97596	16296	66092
64	42238	12426	87025	14267	20979	04508	64535	31355	86064	29472	47689	05974	52468	16834
65	16153	08002	26504	41744	81959	65642	74240	56302	00033	67107	77510	70625	28725	34191
66	21457	40742	29820	96783	29400	21840	15035	34537	33310	06116	95240	15957	16572	06004
67	21581	57802	02050	89728	17937	37621	47075	42080	96403	46826	68995	43805	33386	21597
68	55612	78095	83197	33732	05810	24813	86902	60397	16489	03264	88525	42786	05269	92532
69	44657	66999	99324	51281	84463	60563	79312	93454	68876	24571	93911	25650	12682	73572
70	91340	84979	46949	81973	37949	61023	43997	15263	80644	43942	89203	71795	99533	50501
71	91227	21199	31935	27022	84067	05462	35216	14486	29891	68607	41867	14951	91696	85065
72	50001	38140	66321	19924	72163	09538	12151	06878	91903	18749	34405	56087	82790	70925
73	65390	05224	72958	28609	81406	39147	25549	48542	42627	45233	57202	94617	23772	07896
74	27504	96131	83944	41575	10573	08619	64482	73923	36152	05184	94142	25299	84387	34925
75	37169	94851	39117	89632	00959	16487	65536	49071	39782	17095	02330	74301	00275	48280
76	11508	70225	51111	38351	19444	66499	71945	05422	13442	78675	84081	66938	93654	59894
77	37449	30362	06694	54690	04052	53115	62757	95348	78662	11163	81651	50245	34971	52924
78	46515	70331	85922	38329	57015	15765	97161	17869	45349	61796	66345	81073	49106	79860
79	30986	81223	42416	58353	21532	30502	32305	86482	05174	07901	54339	58861	74818	46942
80	63798	64995	46583	09765	44160	78218	83991	42865	92520	83531	80377	35909	81250	54238

Line/Col.	(1)	(2)	(3)	(4)	(5)	(6)	(7)	(8)	(9)	(10)	(11)	(12)	(13)	(14)
81	82486	84846	99254	67632	43218	50076	21361	64816	51202	88124	41870	52689	51275	83556
82	21885	32906	92431	09060	64297	51674	64126	62570	26123	05155	59194	52799	28225	85762
83	60336	98782	07408	53458	13564	59089	26445	29789	85205	41001	12535	12133	14645	23541
84	43937	46891	24010	25560	86355	33941	35786	54990	71899	15475	95434	98227	21824	19585
85	97656	63175	89303	16275	07100	92063	21942	18611	47348	20203	18534	03862	78095	50136
86	03299	01221	05418	38982	55758	92237	26759	86367	21216	98442	08303	56613	91511	75928
87	79626	06486	03574	17668	07785	76020	79924	25651	83325	88428	85076	72811	22717	50585
88	85636	68335	47539	03129	65651	11977	02510	26113	99447	68645	34327	15152	55230	93448
89	18039	14367	61337	06177	12143	46609	32989	74014	64708	00533	35398	58408	13261	47908
90	08362	15656	60627	36478	65648	16764	53412	09013	07832	41574	17639	82163	60859	75567
91	79556	29068	04142	16268	15387	12856	66227	38358	22478	73373	88732	09443	82558	05250
92	92608	82674	27072	32534	17075	27698	98204	63863	11951	34648	88022	56148	34925	57031
93	23982	25835	40055	67006	12293	02753	14827	22235	35071	99704	37543	11601	35503	85171
94	09915	96306	05908	97901	28395	14186	00821	80703	70426	74647	76310	88717	37890	40129
95	50937	33300	26695	62247	69927	76123	50842	43834	86654	70959	79725	93872	28117	19233
96	42488	78077	69882	61657	34136	79180	97526	43992	04098	73571	80799	76536	71255	64239
97	46764	86273	63003	93017	31204	36692	40202	35275	57306	55543	53203	18098	47635	88684
98	03237	45430	55417	63282	90816	17349	88298	90183	36600	78406	06216	95787	42579	90730
99	86591	81482	52667	61583	14972	90053	89534	76036	49199	43716	97548	04379	46370	28672
100	38534	01715	94964	87288	65680	43772	39560	12918	86537	62738	19636	51132	25739	56947

Table C. Cumulative Probabilities for the Standard Normal Random Variable. Entry: $F(z_j) = p(z \leq z_j)$

z_j	.00	.01	.02	.03	.04	.05	.06	.07	.08	.09
−3.0	$.0^2 1350$	$.0^2 1306$	$.0^2 1264$	$.0^2 1223$	$.0^2 1183$	$.0^2 1114$	$.0^2 1107$	$.0^2 1070$	$.0^2 1035$	$.0^2 1001$
−2.9	$.0^2 1866$	$.0^2 1807$	$.0^2 1750$	$.0^2 1695$	$.0^2 1641$	$.0^2 1589$	$.0^2 1538$	$.0^2 1489$	$.0^2 1441$	$.0^2 1395$
−2.8	$.0^2 2555$	$.0^2 2477$	$.0^2 2401$	$.0^2 2327$	$.0^2 2256$	$.0^2 2186$	$.0^2 2118$	$.0^2 2052$	$.0^2 1988$	$.0^2 1926$
−2.7	$.0^2 3467$	$.0^2 3364$	$.0^2 3264$	$.0^2 3167$	$.0^2 3072$	$.0^2 2980$	$.0^2 2890$	$.0^2 2803$	$.0^2 2718$	$.0^2 2635$
−2.6	$.0^2 4661$	$.0^2 4527$	$.0^2 4396$	$.0^2 4269$	$.0^2 4145$	$.0^2 4025$	$.0^2 3907$	$.0^2 3793$	$.0^2 3681$	$.0^2 3573$
−2.5	$.0^2 6210$	$.0^2 6037$	$.0^2 5868$	$.0^2 5703$	$.0^2 5543$	$.0^2 5386$	$.0^2 5234$	$.0^2 5085$	$.0^2 4940$	$.0^2 4799$
−2.4	$.0^2 8198$	$.0^2 7976$	$.0^2 7760$	$.0^2 7549$	$.0^2 7344$	$.0^2 7143$	$.0^2 6947$	$.0^2 6756$	$.0^2 6569$	$.0^2 6387$
−2.3	$.01072$	$.01044$	$.01017$	$.0^2 9903$	$.0^2 9642$	$.0^2 9387$	$.0^2 9137$	$.0^2 8894$	$.0^2 8656$	$.0^2 8424$
−2.2	$.01390$	$.01355$	$.01321$	$.01287$	$.01255$	$.01222$	$.01191$	$.01160$	$.01130$	$.01101$
−2.1	$.01786$	$.01743$	$.01700$	$.01659$	$.01616$	$.01578$	$.01539$	$.01500$	$.01463$	$.01426$
−2.0	$.02275$	$.02222$	$.02169$	$.02118$	$.02068$	$.02018$	$.01970$	$.01923$	$.01876$	$.01831$
−1.9	$.02872$	$.02807$	$.02743$	$.02680$	$.02619$	$.02559$	$.02500$	$.02442$	$.02385$	$.02330$
−1.8	$.03593$	$.03515$	$.03438$	$.03362$	$.03288$	$.03216$	$.03144$	$.03074$	$.03005$	$.02938$
−1.7	$.04457$	$.04363$	$.04272$	$.04182$	$.04093$	$.04006$	$.03920$	$.03836$	$.03754$	$.03673$
−1.6	$.05480$	$.05370$	$.05262$	$.05155$	$.05050$	$.04947$	$.04846$	$.04746$	$.04648$	$.04551$
−1.5	$.06681$	$.06552$	$.06426$	$.06301$	$.06178$	$.06057$	$.05938$	$.05821$	$.05705$	$.05592$
−1.4	$.08076$	$.07927$	$.07780$	$.07636$	$.07493$	$.07353$	$.07215$	$.07078$	$.06944$	$.06811$
−1.3	$.09680$	$.09510$	$.09342$	$.09176$	$.09012$	$.08851$	$.08691$	$.08534$	$.08379$	$.08226$
−1.2	$.1151$	$.1131$	$.1112$	$.1093$	$.1075$	$.1056$	$.1038$	$.1020$	$.1003$	$.09853$
−1.1	$.1375$	$.1335$	$.1314$	$.1292$	$.1271$	$.1251$	$.1230$	$.1210$	$.1190$	$.1170$
−1.0	$.1587$	$.1562$	$.1539$	$.1515$	$.1492$	$.1469$	$.1446$	$.1423$	$.1401$	$.1379$
−0.9	$.1841$	$.1814$	$.1788$	$.1762$	$.1736$	$.1711$	$.1685$	$.1660$	$.1635$	$.1611$
−0.8	$.2119$	$.2090$	$.2061$	$.2033$	$.2005$	$.1977$	$.1949$	$.1922$	$.1894$	$.1867$
−0.7	$.2420$	$.2389$	$.2358$	$.2327$	$.2297$	$.2266$	$.2236$	$.2206$	$.2177$	$.2148$
−0.6	$.2743$	$.2709$	$.2676$	$.2643$	$.2611$	$.2578$	$.2546$	$.2514$	$.2483$	$.2451$
−0.5	$.3085$	$.3050$	$.3015$	$.2981$	$.2946$	$.2912$	$.2877$	$.2843$	$.2810$	$.2776$
−0.4	$.3446$	$.3409$	$.3372$	$.3336$	$.3300$	$.3264$	$.3228$	$.3192$	$.3156$	$.3121$
−0.3	$.3821$	$.3783$	$.3745$	$.3707$	$.3669$	$.3632$	$.3594$	$.3557$	$.3520$	$.3483$
−0.2	$.4207$	$.4168$	$.4129$	$.4090$	$.4052$	$.4013$	$.3974$	$.3936$	$.3897$	$.3859$
−0.1	$.4602$	$.4562$	$.4522$	$.4483$	$.4443$	$.4404$	$.4364$	$.4325$	$.4286$	$.4247$
−0.0	$.5000$	$.4960$	$.4920$	$.4880$	$.4840$	$.4801$	$.4761$	$.4721$	$.4681$	$.4641$

	0	1	2	3	4	5	6	7	8	9
.0	.5000	.5040	.5080	.5120	.5160	.5199	.5239	.5279	.5319	.5359
.1	.5398	.5438	.5478	.5517	.5557	.5596	.5636	.5675	.5714	.5753
.2	.5793	.5832	.5871	.5910	.5948	.5987	.6026	.6064	.6103	.6141
.3	.6179	.6217	.6255	.6293	.6331	.6368	.6406	.6443	.6480	.6517
.4	.6554	.6591	.6628	.6664	.6700	.6736	.6772	.6808	.6844	.6879
.5	.6915	.6950	.6985	.7019	.7054	.7088	.7123	.7157	.7190	.7224
.6	.7257	.7291	.7324	.7357	.7389	.7422	.7454	.7486	.7517	.7549
.7	.7580	.7611	.7642	.7673	.7703	.7734	.7764	.7794	.7823	.7852
.8	.7881	.7910	.7939	.7967	.7995	.8023	.8051	.8078	.8106	.8133
.9	.8159	.8186	.8212	.8238	.8264	.8289	.8315	.8340	.8365	.8389
1.0	.8413	.8438	.8461	.8485	.8508	.8531	.8554	.8577	.8599	.8621
1.1	.8643	.8665	.8686	.8708	.8729	.8749	.8770	.8790	.8810	.8830
1.2	.8849	.8869	.8888	.8907	.8925	.8944	.8962	.8980	.8997	.90147
1.3	.90320	.90490	.90658	.90824	.90988	.91149	.91309	.91466	.91621	.91774
1.4	.91924	.92073	.92220	.92364	.92507	.92647	.92785	.92922	.93056	.93189
1.5	.93319	.93448	.93574	.93699	.93822	.93943	.94062	.94179	.94295	.94408
1.6	.94520	.94630	.94738	.94845	.94950	.95053	.95154	.95254	.95352	.95449
1.7	.95543	.95637	.95728	.95818	.95907	.95994	.96080	.96164	.96246	.96327
1.8	.96407	.96485	.96562	.96638	.96712	.96784	.96856	.96926	.96995	.97062
1.9	.97128	.97193	.97257	.97320	.97381	.97441	.97500	.97558	.97615	.97670
2.0	.97725	.97778	.97831	.97882	.97932	.97982	.98030	.98077	.98124	.98169
2.1	.98214	.98257	.98300	.98341	.98382	.98422	.98461	.98500	.98537	.98574
2.2	.98610	.98645	.98679	.98713	.98745	.98778	.98809	.98840	.98870	.98899
2.3	.98928	.98956	.98983	$.9^{2}0097$	$.9^{2}0358$	$.9^{2}0613$	$.9^{2}0863$	$.9^{2}1106$	$.9^{2}1344$	$.9^{2}1576$
2.4	$.9^{2}1802$	$.9^{2}2024$	$.9^{2}2240$	$.9^{2}2451$	$.9^{2}2656$	$.9^{2}2857$	$.9^{2}3053$	$.9^{2}3244$	$.9^{2}3431$	$.9^{2}3613$
2.5	$.9^{2}3790$	$.9^{2}3963$	$.9^{2}4132$	$.9^{2}4297$	$.9^{2}4457$	$.9^{2}4614$	$.9^{2}4766$	$.9^{2}4915$	$.9^{2}5060$	$.9^{2}5201$
2.6	$.9^{2}5339$	$.9^{2}5473$	$.9^{2}5604$	$.9^{2}5731$	$.9^{2}5855$	$.9^{2}5975$	$.9^{2}6093$	$.9^{2}6207$	$.9^{2}6319$	$.9^{2}6427$
2.7	$.9^{2}6533$	$.9^{2}6636$	$.9^{2}6736$	$.9^{2}6833$	$.9^{2}6928$	$.9^{2}7020$	$.9^{2}7110$	$.9^{2}7197$	$.9^{2}7282$	$.9^{2}7365$
2.8	$.9^{2}7445$	$.9^{2}7523$	$.9^{2}7599$	$.9^{2}7673$	$.9^{2}7744$	$.9^{2}7814$	$.9^{2}7882$	$.9^{2}7948$	$.9^{2}8012$	$.9^{2}8074$
2.9	$.9^{2}8134$	$.9^{2}8193$	$.9^{2}8250$	$.9^{2}8305$	$.9^{2}8359$	$.9^{2}8411$	$.9^{2}8462$	$.9^{2}8511$	$.9^{2}8559$	$.9^{2}8605$
3.0	$.9^{2}8650$	$.9^{2}8694$	$.9^{2}8736$	$.9^{2}8777$	$.9^{2}8817$	$.9^{2}8856$	$.9^{2}8893$	$.9^{2}8930$	$.9^{2}8965$	$.9^{2}8999$

Source. Table **C** is taken with permission from Hald, A. *Statistical tables and formulas*, John Wiley & Sons, Inc., New York, 1952.
Note: $.0^{1}1350 = .001350$. $.9^{2}8650 = .998650$.

Table **D.** Student's *t*-Values for Selected Cumulative Probabilities

ν \ F	.60	.75	.90	.95	.975	.99	.995	.9995
1	.325	1.000	3.078	6.314	12.706	31.821	63.657	636.619
2	.289	.816	1.886	2.920	4.303	6.965	9.925	31.598
3	.277	.765	1.638	2.353	3.182	4.541	5.841	12.941
4	.271	.741	1.533	2.132	2.776	3.747	4.604	8.610
5	.267	.727	1.476	2.015	2.571	3.365	4.032	6.859
6	.265	.718	1.440	1.943	2.447	3.143	3.707	5.959
7	.263	.711	1.415	1.895	2.365	2.998	3.499	5.405
8	.262	.706	1.397	1.860	2.306	2.896	3.355	5.041
9	.261	.703	1.383	1.833	2.262	2.821	3.250	4.781
10	.260	.700	1.372	1.812	2.228	2.764	3.169	4.587
11	.260	.697	1.363	1.796	2.201	2.718	3.106	4.437
12	.259	.695	1.356	1.782	2.179	2.681	3.055	4.318
13	.259	.694	1.350	1.771	2.160	2.650	3.012	4.221
14	.258	.692	1.345	1.761	2.145	2.624	2.977	4.140
15	.258	.691	1.341	1.753	2.131	2.602	2.947	4.073
16	.258	.690	1.337	1.746	2.120	2.583	2.921	4.015
17	.257	.689	1.333	1.740	2.110	2.567	2.898	3.965
18	.257	.688	1.330	1.734	2.101	2.552	2.878	3.922
19	.257	.688	1.328	1.729	2.093	2.539	2.861	3.883
20	.257	.687	1.325	1.725	2.086	2.528	2.845	3.850
21	.257	.686	1.323	1.721	2.080	2.518	2.831	3.819
22	.256	.686	1.321	1.717	2.074	2.508	2.819	3.792
23	.256	.685	1.319	1.714	2.069	2.500	2.807	3.767
24	.256	.685	1.318	1.711	2.064	2.492	2.797	3.745
25	.256	.684	1.316	1.708	2.060	2.485	2.787	3.725
26	.256	.684	1.315	1.706	2.056	2.479	2.779	3.707
27	.256	.684	1.314	1.703	2.052	2.473	2.771	3.690
28	.256	.683	1.313	1.701	2.048	2.467	2.763	3.674
29	.256	.683	1.311	1.699	2.045	2.462	2.756	3.659
30	.256	.683	1.310	1.697	2.042	2.457	2.750	3.646
40	.255	.681	1.303	1.684	2.021	2.423	2.704	3.551
60	.254	.679	1.296	1.671	2.000	2.390	2.660	3.460
120	.254	.677	1.289	1.658	1.980	2.358	2.617	3.373
∞	.253	.674	1.282	1.645	1.960	2.326	2.576	3.291

Source. Table **D** is taken from Fisher and Yates: *Statistical tables for biological, agricultural and medical research*, published by Oliver and Boyd, Edinburgh, and by permission of the authors and publishers.

Table E. Chi Square Values for Selected Cumulative Probabilities

v / F	.005	.010	.025	.050	.100	.250	.500	.750	.900	.950	.975	.990	.995	.999
1	.0000393	.000157	.000982	.00393	.0158	.102	.455	1.32	2.71	3.84	5.02	6.63	7.88	10.8
2	.0100	.0201	.0506	.103	.211	.575	1.39	2.77	4.61	5.99	7.38	9.21	10.6	13.8
3	.0717	.115	.216	.352	.584	1.21	2.37	4.11	6.25	7.81	9.35	11.3	12.8	16.3
4	.207	.297	.484	.711	1.06	1.92	3.36	5.39	7.78	9.49	11.1	13.3	14.9	18.5
5	.412	.554	.831	1.15	1.61	2.67	4.35	6.63	9.24	11.1	12.8	15.1	16.7	20.5
6	.676	.872	1.24	1.64	2.20	3.45	5.35	7.84	10.6	12.6	14.4	16.8	18.5	22.5
7	.989	1.24	1.69	2.17	2.83	4.25	6.35	9.04	12.0	14.1	16.0	18.5	20.3	24.3
8	1.34	1.65	2.18	2.73	3.49	5.07	7.34	10.2	13.4	15.5	17.5	20.1	22.0	26.1
9	1.73	2.09	2.70	3.33	4.17	5.90	8.34	11.4	14.7	16.9	19.0	21.7	23.6	27.9
10	2.16	2.56	3.25	3.94	4.87	6.74	9.34	12.5	16.0	18.3	20.5	23.2	25.2	29.6
11	2.60	3.05	3.82	4.57	5.58	7.58	10.3	13.7	17.3	19.7	21.9	24.7	26.8	31.3
12	3.07	3.57	4.40	5.23	6.30	8.44	11.3	14.8	18.5	21.0	23.3	26.2	28.3	32.9
13	3.57	4.11	5.01	5.89	7.04	9.30	12.3	16.0	19.8	22.4	24.7	27.7	29.8	34.5
14	4.07	4.66	5.63	6.57	7.79	10.2	13.3	17.1	21.1	23.7	26.1	29.1	31.3	36.1
15	4.60	5.23	6.26	7.26	8.55	11.0	14.3	18.2	22.3	25.0	27.5	30.6	32.8	37.7
16	5.14	5.81	6.91	7.96	9.31	11.9	15.3	19.4	23.5	26.3	28.8	32.0	34.3	39.3
17	5.70	6.41	7.56	8.67	10.1	12.8	16.3	20.5	24.8	27.6	30.2	33.4	35.7	40.8
18	6.26	7.01	8.23	9.39	10.9	13.7	17.3	21.6	26.0	28.9	31.5	34.8	37.2	42.3
19	6.84	7.63	8.91	10.1	11.7	14.6	18.3	22.7	27.2	30.1	32.9	36.2	38.6	43.8
20	7.43	8.26	9.59	10.9	12.4	15.5	19.3	23.8	28.4	31.4	34.2	37.6	40.0	45.3
21	8.03	8.90	10.3	11.6	13.2	16.3	20.3	24.9	29.6	32.7	35.5	38.9	41.4	46.8
22	8.64	9.54	11.0	12.3	14.0	17.2	21.3	26.0	30.8	33.9	36.8	40.3	42.8	48.3
23	9.26	10.2	11.7	13.1	14.8	18.1	22.3	27.1	32.0	35.2	38.1	41.6	44.2	49.7
24	9.89	10.9	12.4	13.8	15.7	19.0	23.3	28.2	33.2	36.4	39.4	43.0	45.6	51.2
25	10.5	11.5	13.1	14.6	16.5	19.9	24.3	29.3	34.4	37.7	40.6	44.3	46.9	52.6
26	11.2	12.2	13.8	15.4	17.3	20.8	25.3	30.4	35.6	38.9	41.9	45.6	48.3	54.0
27	11.8	12.9	14.6	16.2	18.1	21.7	26.3	31.5	36.7	40.1	43.2	47.0	49.6	55.5
28	12.5	13.6	15.3	16.9	18.9	22.7	27.3	32.6	37.9	41.3	44.5	48.3	51.0	56.9
29	13.1	14.3	16.0	17.7	19.8	23.6	28.3	33.7	39.1	42.6	45.7	49.6	52.3	58.3
30	13.8	15.0	16.8	18.5	20.6	24.5	29.3	34.8	40.3	43.8	47.0	50.9	53.7	59.7

Source. Table **E** is reprinted, with permission, from Pearson, E. S., and Hartley, H. O. (1962), *Biometrika tables for statisticians*, Vol. 1, Biometrika Trustees, London.
Column .999 is reprinted abridged from R. A. Fisher and F. Yates: *Statistical tables for biological, agricultural and medical research*, published by Oliver Boyd, Edinburgh, and by permission of the authors and publishers.

Table F. Fisher's F-Values for Selected Cumulative Probabilities

ν_1, Degrees of Freedom for Numerator

ν_2	F	1	2	3	4	5	6	7	8	9	10	11	12	F
1	.0005	$.0^662$	$.0^350$	$.0^238$	$.0^294$	.016	.022	.027	.032	.036	.039	.042	.045	.0005
	.001	$.0^525$	$.0^210$	$.0^260$	.013	.021	.028	.034	.039	.044	.048	.051	.054	.001
	.005	$.0^462$	$.0^251$	.018	.032	.044	.054	.062	.068	.073	.078	.082	.085	.005
	.01	$.0^325$	.010	.029	.047	.062	.073	.082	.089	.095	.100	.104	.107	.01
	.025	$.0^215$	.026	.057	.082	.100	.113	.124	.132	.139	.144	.149	.153	.025
	.05	$.0^262$	.054	.099	.130	.151	.167	.179	.188	.195	.201	.207	.211	.05
	.10	.025	.117	.181	.220	.246	.265	.279	.289	.298	.304	.310	.315	.10
	.25	.172	.389	.494	.553	.591	.617	.637	.650	.661	.670	.680	.684	.25
	.50	1.00	1.50	1.71	1.82	1.89	1.94	1.98	2.00	2.03	2.04	2.05	2.07	.50
	.75	5.83	7.50	8.20	8.58	8.82	8.98	9.10	9.19	9.26	9.32	9.36	9.41	.75
	.90	39.9	49.5	53.6	55.8	57.2	58.2	58.9	59.4	59.9	60.2	60.5	60.7	.90
	.95	161	200	216	225	230	234	237	239	241	242	243	244	.95
	.975	648	800	864	900	922	937	948	957	963	969	973	977	.975
	.99	405^1	500^1	540^1	562^1	576^1	586^1	593^1	598^1	602^1	606^1	608^1	611^1	.99
	.995	162^2	200^2	216^2	225^2	231^2	234^2	237^2	239^2	241^2	242^2	243^2	244^2	.995
	.999	406^3	500^3	540^3	562^3	576^3	586^3	593^3	598^3	602^3	606^3	609^3	611^3	.999
	.9995	162^4	200^4	216^4	225^4	231^4	234^4	237^4	239^4	241^4	242^4	243^4	244^4	.9995
2	.0005	$.0^650$	$.0^350$	$.0^242$	.011	.020	.029	.037	.044	.050	.056	.061	.065	.0005
	.001	$.0^520$	$.0^210$	$.0^268$	.016	.027	.037	.046	.052	.061	.067	.072	.077	.001
	.005	$.0^450$	$.0^250$	.020	.038	.055	.069	.081	.091	.099	.106	.112	.118	.005
	.01	$.0^320$	.010	.032	.056	.075	.092	.105	.116	.125	.132	.139	.144	.01
	.025	$.0^213$	.026	.062	.094	.119	.138	.153	.165	.175	.183	.190	.196	.025
	.05	$.0^250$	.053	.105	.144	.173	.194	.211	.224	.235	.244	.251	.257	.05
	.10	.020	.111	.183	.231	.265	.289	.307	.321	.333	.342	.350	.356	.10
	.25	.133	.333	.439	.500	.540	.568	.588	.604	.616	.626	.633	.641	.25
	.50	.667	1.00	1.13	1.21	1.25	1.28	1.30	1.32	1.33	1.34	1.35	1.36	.50
	.75	2.57	3.00	3.15	3.23	3.28	3.31	3.34	3.35	3.37	3.38	3.39	3.39	.75
	.90	8.53	9.00	9.16	9.24	9.29	9.33	9.35	9.37	9.38	9.39	9.40	9.41	.90
	.95	18.5	19.0	19.2	19.2	19.3	19.3	19.4	19.4	19.4	19.4	19.4	19.4	.95
	.975	38.5	39.0	39.2	39.2	39.3	39.3	39.4	39.4	39.4	39.4	39.4	39.4	.975
	.99	98.5	99.0	99.2	99.2	99.3	99.3	99.4	99.4	99.4	99.4	99.4	99.4	.99
	.995	198	199	199	199	199	199	199	199	199	199	199	199	.995
	.999	998	999	999	999	999	999	999	999	999	999	999	999	.999
	.9995	200^1	200^1	200^1	200^1	200^1	200^1	200^1	200^1	200^1	200^1	200^1	200^1	.9995
3	.0005	$.0^646$	$.0^350$	$.0^244$	.012	.023	.033	.043	.052	.060	.067	.074	.079	.0005
	.001	$.0^519$	$.0^210$	$.0^271$	.018	.030	.042	.053	.063	.072	.079	.086	.093	.001
	.005	$.0^446$	$.0^250$	.021	.041	.060	.077	.092	.104	.115	.124	.132	.138	.005
	.01	$.0^319$	.010	.034	.060	.083	.102	.118	.132	.143	.153	.161	.168	.01
	.025	$.0^212$	.026	.065	.100	.129	.152	.170	.185	.197	.207	.216	.224	.025
	.05	$.0^246$	.052	.108	.152	.185	.210	.230	.246	.259	.270	.279	.287	.05
	.10	.019	.109	.185	.239	.276	.304	.325	.342	.356	.367	.376	.384	.10
	.25	.122	.317	.424	.489	.531	.561	.582	.600	.613	.624	.633	.641	.25
	.50	.585	.881	1.00	1.06	1.10	1.13	1.15	1.16	1.17	1.18	1.19	1.20	.50
	.75	2.02	2.28	2.36	2.39	2.41	2.42	2.43	2.44	2.44	2.44	2.45	2.45	.75
	.90	5.54	5.46	5.39	5.34	5.31	5.28	5.27	5.25	5.24	5.23	5.22	5.22	.90
	.95	10.1	9.55	9.28	9.12	9.01	8.94	8.89	8.85	8.81	8.79	8.76	8.74	.95
	.975	17.4	16.0	15.4	15.1	14.9	14.7	14.6	14.5	14.5	14.4	14.4	14.3	.975
	.99	34.1	30.8	29.5	28.7	28.2	27.9	27.7	27.5	27.3	27.2	27.1	27.1	.99
	.995	55.6	49.8	47.5	46.2	45.4	44.8	44.4	44.1	43.9	43.7	43.5	43.4	.995
	.999	167	149	141	137	135	133	132	131	130	129	129	128	.999
	.9995	266	237	225	218	214	211	209	208	207	206	204	204	.9995

ν_2, Degrees of Freedom for Denominator

Read $.0^356$ as .00056, 200^1 as 2,000, 162^4 as 1,620,000, and so on.

Source. Table **F** was reprinted with permission from W. J. Dixon and F. J. Massey, Jr., *Introduction to statistical analysis* (3rd ed.), McGraw-Hill Book Co., New York, 1969. Part of this table was extracted from (1) A. Hald, *Statistical tables and formulas*, John Wiley & Sons, Inc., New York, 1952; (2) M. Merrington and C. M. Thompson, *Biometrika*, *33*, 1943; (3) C. Colcord and L. S. Deming, *Sankhyā*, *2*, 1936. Permission to reprint the needed entries was granted in each case.

ν_1, Degrees of Freedom for Numerator

F	15	20	24	30	40	50	60	100	120	200	500	∞	F	
.0005	.051	.058	.062	.066	.069	.072	.074	.077	.078	.080	.081	.083	.0005	1
.001	.060	.067	.071	.075	.079	.082	.084	.087	.088	.089	.091	.092	.001	
.005	.093	.101	.105	.109	.113	.116	.118	.121	.122	.124	.126	.127	.005	
.01	.115	.124	.128	.132	.137	.139	.141	.145	.146	.148	.150	.151	.01	
.025	.161	.170	.175	.180	.184	.187	.189	.193	.194	.196	.198	.199	.025	
.05	.220	.230	.235	.240	.245	.248	.250	.254	.255	.257	.259	.261	.05	
.10	.325	.336	.342	.347	.353	.356	.358	.362	.364	.366	.368	.370	.10	
.25	.698	.712	.719	.727	.734	.738	.741	.747	.749	.752	.754	.756	.25	
.50	2.09	2.12	2.13	2.15	2.16	2.17	2.17	2.18	2.18	2.19	2.19	2.20	.50	
.75	9.49	9.58	9.63	9.67	9.71	9.74	9.76	9.78	9.80	9.82	9.84	9.85	.75	
.90	61.2	61.7	62.0	62.3	62.5	62.7	62.8	63.0	63.1	63.2	63.3	63.3	.90	
.95	246	248	249	250	251	252	252	253	253	254	254	254	.95	
.975	985	993	997	100^1	101^1	101^1	101^1	101^1	101^1	102^1	102^1	102^1	.975	
.99	616^2	621^1	623^1	626^1	629^1	630^1	631^1	633^1	634^1	635^1	636^1	637^1	.99	
.995	246^2	248^2	249^2	250^2	251^2	252^2	253^2	253^2	254^2	254^2	254^2	255^2	.995	
.999	616^3	621^3	623^3	626^3	629^3	630^3	631^3	633^3	634^3	635^3	636^3	637^3	.999	
.9995	246^4	248^4	249^4	250^4	251^4	252^4	252^4	253^4	253^4	253^4	254^4	254^4	.9995	
.0005	.076	.088	.094	.101	.108	.113	.116	.122	.124	.127	.130	.132	.0005	2
.001	.088	.100	.107	.114	.121	.126	.129	.135	.137	.140	.143	.145	.001	
.005	.130	.143	.150	.157	.165	.169	.173	.179	.181	.184	.187	.189	.005	
.01	.157	.171	.178	.186	.193	.198	.201	.207	.209	.212	.215	.217	.01	
.025	.210	.224	.232	.239	.247	.251	.255	.261	.263	.266	.269	.271	.025	
.05	.272	.286	.294	.302	.309	.314	.317	.324	.326	.329	.332	.334	.05	
.10	.371	.386	.394	.402	.410	.415	.418	.424	.426	.429	.433	.434	.10	
.25	.657	.672	.680	.689	.697	.702	.705	.711	.713	.716	.719	.721	.25	
.50	1.38	1.39	1.40	1.41	1.42	1.42	1.43	1.43	1.43	1.44	1.44	1.44	.50	
.75	3.41	3.43	3.43	3.44	3.45	3.45	3.46	3.47	3.47	3.48	3.48	3.48	.75	
.90	9.42	9.44	9.45	9.46	9.47	9.47	9.47	9.48	9.48	9.49	9.49	9.49	.90	
.95	19.4	19.4	19.5	19.5	19.5	19.5	19.5	19.5	19.5	19.5	19.5	19.5	.95	
.975	39.4	39.4	39.5	39.5	39.5	39.5	39.5	39.5	39.5	39.5	39.5	39.5	.975	
.99	99.4	99.4	99.5	99.5	99.5	99.5	99.5	99.5	99.5	99.5	99.5	99.5	.99	
.995	199	199	199	199	199	199	199	199	199	199	199	200	.995	
.999	999	999	999	999	999	999	999	999	999	999	999	999	.999	
.9995	200^1	200^1	200^1	200^1	200^1	200^1	200^1	200^1	200^1	200^1	200^1	200^1	.9995	
.0005	.093	.109	.117	.127	.136	.143	.147	.156	.158	.162	.166	.169	.0005	3
.001	.107	.123	.132	.142	.152	.158	.162	.171	.173	.177	.181	.184	.001	
.005	.154	.172	.181	.191	.201	.207	.211	.220	.222	.227	.231	.234	.005	
.01	.185	.203	.212	.222	.232	.238	.242	.251	.253	.258	.262	.264	.01	
.025	.241	.259	.269	.279	.289	.295	.299	.308	.310	.314	.318	.321	.025	
.05	.304	.323	.332	.342	.352	.358	.363	.370	.373	.377	.382	.384	.05	
.10	.402	.420	.430	.439	.449	.455	.459	.467	.469	.474	.476	.480	.10	
.25	.658	.675	.684	.693	.702	.708	.711	.719	.721	.724	.728	.730	.25	
.50	1.21	1.23	1.23	1.24	1.25	1.25	1.25	1.26	1.26	1.26	1.27	1.27	.50	
.75	2.46	2.46	2.46	2.47	2.47	2.47	2.47	2.47	2.47	2.47	2.47	2.47	.75	
.90	5.20	5.18	5.18	5.17	5.16	5.15	5.15	5.14	5.14	5.14	5.14	5.13	.90	
.95	8.70	8.66	8.63	8.62	8.59	8.58	8.57	8.55	8.55	8.54	8.53	8.53	.95	
.975	14.3	14.2	14.1	14.1	14.0	14.0	14.0	14.0	13.9	13.9	13.9	13.9	.975	
.99	26.9	26.7	26.6	26.5	26.4	26.4	26.3	26.2	26.2	26.2	26.1	26.1	.99	
.995	43.1	42.8	42.6	42.5	42.3	42.2	42.1	42.0	42.0	41.9	41.9	41.8	.995	
.999	127	126	126	125	125	125	124	124	124	124	124	123	.999	
.9995	203	201	200	199	199	199	198	198	197	197	196	196	.9995	

ν_2, Degrees of Freedom for Denominator

v_1, Degrees of Freedom for Numerator

v_2, Degrees of Freedom for Denominator

	F	1	2	3	4	5	6	7	8	9	10	11	12	F
4	.0005	$.0^644$	$.0^550$	$.0^246$	.013	.024	.036	.047	.057	.066	.075	.082	.089	.0005
	.001	$.0^518$	$.0^210$	$.0^273$	.019	.032	.046	.058	.069	.079	.089	.097	.104	.001
	.005	$.0^444$	$.0^250$	.022	.043	.064	.083	.100	.114	.126	.137	.145	.153	.005
	.01	$.0^318$	.010	.035	.063	.088	.109	.127	.143	.156	.167	.176	.185	.01
	.025	$.0^211$	.026	.066	.104	.135	.161	.181	.198	.212	.224	.234	.243	.025
	.05	$.0^244$	.052	.110	.157	.193	.221	.243	.261	.275	.288	.298	.307	.05
	.10	.018	.108	.187	.243	.284	.314	.338	.356	.371	.384	.394	.403	.10
	.25	.117	.309	.418	.484	.528	.560	.583	.601	.615	.627	.637	.645	.25
	.50	.549	.828	.941	1.00	1.04	1.06	1.08	1.09	1.10	1.11	1.12	1.13	.50
	.75	1.81	2.00	2.05	2.06	2.07	2.08	2.08	2.08	2.08	2.08	2.08	2.08	.85
	.90	4.54	4.32	4.19	4.11	4.05	4.01	3.98	3.95	3.94	3.92	3.91	3.90	.90
	.95	7.71	6.94	6.59	6.39	6.26	6.16	6.09	6.04	6.00	5.96	5.94	5.91	.95
	.975	12.2	10.6	9.98	9.60	9.36	9.20	9.07	8.98	8.90	8.84	8.79	8.75	.975
	.99	21.2	18.0	16.7	16.0	15.5	15.2	15.0	14.8	14.7	14.5	14.4	14.4	.99
	.995	31.3	26.3	24.3	23.2	22.5	22.0	21.6	21.4	21.1	21.0	20.8	20.7	.995
	.999	74.1	61.2	56.2	53.4	51.7	50.5	49.7	49.0	48.5	48.0	47.7	47.4	.999
	.9995	106	87.4	80.1	76.1	73.6	71.9	70.6	69.7	68.9	68.3	67.8	67.4	.9995
5	.0005	$.0^643$	$.0^550$	$.0^247$	.014	.025	.038	.050	.061	.070	.081	.089	.096	.0005
	.001	$.0^517$	$.0^210$	$.0^275$	.019	.034	.048	.062	.074	.085	.095	.104	.112	.001
	.005	$.0^443$	$.0^250$	.022	.045	.067	.087	.105	.120	.134	.146	.156	.165	.005
	.01	$.0^317$	.010	.035	.064	.091	.114	.134	.151	.165	.177	.188	.197	.01
	.025	$.0^211$	.025	.067	.107	.140	.167	.189	.208	.223	.236	.248	.257	.025
	.05	$.0^243$	.052	.111	.160	.198	.228	.252	.271	.287	.301	.313	.322	.05
	.10	.017	.108	.188	.247	.290	.322	.347	.367	.383	.397	.408	.418	.10
	.25	.113	.305	.415	.483	.528	.560	.584	.604	.618	.631	.641	.650	.25
	.50	.528	.799	.907	.965	1.00	1.02	1.04	1.05	1.06	1.07	1.08	1.09	.50
	.75	1.69	1.85	1.88	1.89	1.89	1.89	1.89	1.89	1.89	1.89	1.89	1.89	.75
	.90	4.06	3.78	3.62	3.52	3.45	3.40	3.37	3.34	3.32	3.30	3.28	3.27	.90
	.95	6.61	5.79	5.41	5.19	5.05	4.95	4.88	4.82	4.77	4.74	4.71	4.68	.95
	.975	10.0	8.43	7.76	7.39	7.15	6.98	6.85	6.76	6.68	6.62	6.57	6.52	.975
	.99	16.3	13.3	12.1	11.4	11.0	10.7	10.5	10.3	10.2	10.1	9.96	9.89	.99
	.995	22.8	18.3	16.5	15.6	14.9	14.5	14.2	14.0	13.8	13.6	13.5	13.4	.995
	.999	47.2	37.1	33.2	31.1	29.7	28.8	28.2	27.6	27.2	26.9	26.6	26.4	.999
	.9995	63.6	49.8	44.4	41.5	39.7	38.5	37.6	36.9	36.4	35.9	35.6	35.2	.9995
6	.0005	$.0^643$	$.0^550$	$.0^247$	.014	.026	.039	.052	.064	.075	.085	.094	.103	.0005
	.001	$.0^517$	$.0^210$	$.0^275$	.020	.035	.050	.064	.078	.090	.101	.111	.119	.001
	.005	$.0^443$	$.0^250$	.022	.045	.069	.090	.109	.126	.140	.153	.164	.174	.005
	.01	$.0^317$	.010	.036	.066	.094	.118	.139	.157	.172	.186	.197	.207	.01
	.025	$.0^211$	.025	.068	.109	.143	.172	.195	.215	.231	.246	.258	.268	.025
	.05	$.0^243$	.052	.112	.162	.202	.233	.259	.279	.296	.311	.324	.334	.05
	.10	.017	.107	.189	.249	.294	.327	.354	.375	.392	.406	.418	.429	.10
	.25	.111	.302	.413	.481	.524	.561	.586	.606	.622	.635	.645	.654	.25
	.50	.515	.780	.886	.942	.977	1.00	1.02	1.03	1.04	1.05	1.05	1.06	.50
	.75	1.62	1.76	1.78	1.79	1.79	1.78	1.78	1.78	1.77	1.77	1.77	1.77	.75
	.90	3.78	3.46	3.29	3.18	3.11	3.05	3.01	2.98	2.96	2.94	2.92	2.90	.90
	.95	5.99	5.14	4.76	4.53	4.39	4.28	4.21	4.15	4.10	4.06	4.03	4.00	.95
	.975	8.81	7.26	6.60	6.23	5.99	5.82	5.70	5.60	5.52	5.46	5.41	5.37	.975
	.99	13.7	10.9	9.78	9.15	8.75	8.47	8.26	8.10	7.98	7.87	7.79	7.72	.99
	.995	18.6	14.5	12.9	12.0	11.5	11.1	10.8	10.6	1.04	10.2	10.1	10.0	.995
	.999	35.5	27.0	23.7	21.9	20.8	20.0	19.5	19.0	18.7	18.4	18.2	18.0	.999
	.9995	46.1	34.8	30.4	28.1	26.6	25.6	24.9	24.3	23.9	23.5	23.2	23.0	.9995

Table **F** (*continued*)

ν_1, Degrees of Freedom for Numerator

F	15	20	24	30	40	50	60	100	120	200	500	∞	F	
.0005	.105	.125	.135	.147	.159	.166	.172	.183	.186	.191	.196	.200	.0005	4
.001	.121	.141	.152	.163	.176	.183	.188	.200	.202	.208	.213	.217	.001	
.005	.172	.193	.204	.216	.229	.237	.242	.253	.255	.260	.266	.269	.005	
.01	.204	.226	.237	.249	.261	.269	.274	.285	.287	.293	.298	.301	.01	
.025	.263	.284	.296	.308	.320	.327	.332	.342	.346	.351	.356	.359	.025	
.05	.327	.349	.360	.372	.384	.391	.396	.407	.409	.413	.418	.422	.05	
.10	.424	.445	.456	.467	.478	.485	.490	.500	.502	.508	.510	.514	.10	
.25	.664	.683	.692	.702	.712	.718	.722	.731	.733	.737	.740	.743	.25	
.50	1.14	1.15	1.16	1.16	1.17	1.18	1.18	1.18	1.18	1.19	1.19	1.19	.50	
.75	2.08	2.08	2.08	2.08	2.08	2.08	2.08	2.08	2.08	2.08	2.08	2.08	.75	
.90	3.87	3.84	3.83	3.82	3.80	3.80	3.79	3.78	3.78	3.77	3.76	3.76	.90	
.95	5.86	5.80	5.77	5.75	5.72	5.70	5.69	5.66	5.66	5.65	5.64	5.63	.95	
.975	8.66	8.56	8.51	8.46	8.41	8.38	8.36	8.32	8.31	8.29	8.27	8.26	.975	
.99	14.2	14.0	13.9	13.8	13.7	13.7	13.7	13.6	13.6	13.5	13.5	13.5	.99	
.995	20.4	20.2	20.0	19.9	19.8	19.7	19.6	19.5	19.5	19.4	19.4	19.3	.995	
.999	46.8	46.1	45.8	45.4	45.1	44.9	44.7	44.5	44.4	44.3	44.1	44.0	.999	
.9995	66.5	65.5	65.1	64.6	64.1	63.8	63.6	63.2	63.1	62.9	62.7	62.6	.9995	
.0005	.115	.137	.150	.163	.177	.186	.192	.205	.209	.216	.222	.226	.0005	5
.001	.132	.155	.167	.181	.195	.204	.210	.223	.227	.233	.239	.244	.001	
.005	.186	.210	.223	.237	.251	.260	.266	.279	.282	.288	.294	.299	.005	
.01	.219	.244	.257	.270	.285	.293	.299	.312	.315	.322	.328	.331	.01	
.025	.280	.304	.317	.330	.344	.353	.359	.370	.374	.380	.386	.390	.025	
.05	.345	.369	.382	.395	.408	.417	.422	.432	.437	.442	.448	.452	.05	
.10	.440	.463	.476	.488	.501	.508	.514	.524	.527	.532	.538	.541	.10	
.25	.669	.690	.700	.711	.722	.728	.732	.741	.743	.748	.752	.755	.25	
.50	1.10	1.11	1.12	1.12	1.13	1.13	1.14	1.14	1.14	1.15	1.15	1.15	.50	
.75	1.89	1.88	1.88	1.88	1.88	1.88	1.87	1.87	1.87	1.87	1.87	1.87	.75	
.90	3.24	3.21	3.19	3.17	3.16	3.15	3.14	3.13	3.12	3.12	3.11	3.10	.90	
.95	4.62	4.56	4.53	4.50	4.46	4.44	4.43	4.41	4.40	4.39	4.37	4.36	.95	
.975	6.43	6.33	6.28	6.23	6.18	6.14	6.12	6.08	6.07	6.05	6.03	6.02	.975	
.99	9.72	9.55	9.47	9.38	9.29	9.24	9.20	9.13	9.11	9.08	9.04	9.02	.99	
.995	13.1	12.9	12.8	12.7	12.5	12.5	12.4	12.3	12.3	12.2	12.2	12.1	.995	
.999	25.9	25.4	25.1	24.9	24.6	24.4	24.3	24.1	24.1	23.9	23.8	23.8	.999	
.9995	34.6	33.9	33.5	33.1	32.7	32.5	32.3	32.1	32.0	31.8	31.7	31.6	.9995	
.0005	.123	.148	.162	.177	.193	.203	.210	.225	.229	.236	.244	.249	.0005	6
.001	.141	.166	.180	.195	.211	.222	.229	.243	.247	.255	.262	.267	.001	
.005	.197	.224	.238	.253	.269	.279	.286	.301	.304	.312	.318	.324	.005	
.01	.232	.258	.273	.288	.304	.313	.321	.334	.338	.346	.352	.357	.01	
.025	.293	.320	.334	.349	.364	.375	.381	.394	.398	.405	.412	.415	.025	
.05	.358	.385	.399	.413	.428	.437	.444	.457	.460	.467	.472	.476	.05	
.10	.453	.478	.491	.505	.519	.526	.533	.546	.548	.556	.559	.564	.10	
.25	.675	.696	.707	.718	.729	.736	.741	.751	.753	.758	.762	.765	.25	
.50	1.07	1.08	1.09	1.10	1.10	1.11	1.11	1.11	1.12	1.12	1.12	1.12	.50	
.75	1.76	1.76	1.75	1.75	1.75	1.75	1.74	1.74	1.74	1.74	1.74	1.74	.75	
.90	2.87	2.84	2.82	2.80	2.78	2.77	2.76	2.75	2.74	2.73	2.73	2.72	.90	
.95	3.94	3.87	3.84	3.81	3.77	3.75	3.74	3.71	3.70	3.69	3.68	3.67	.95	
.975	5.27	5.17	5.12	5.07	5.01	4.98	4.96	4.92	4.90	4.88	4.86	4.85	.975	
.99	7.56	7.40	7.31	7.23	7.14	7.09	7.06	6.99	6.97	6.93	6.90	6.88	.99	
.995	9.81	9.59	9.47	9.36	9.24	9.17	9.12	9.03	9.00	8.95	8.91	8.88	.995	
.999	17.6	17.1	16.9	16.7	16.4	16.3	16.2	16.0	16.0	15.9	15.8	15.7	.999	
.9995	22.4	21.9	21.7	21.4	21.1	20.9	20.7	20.5	20.4	20.3	20.2	20.1	.9995	

ν_2, Degrees of Freedom for Denominator

ν_1, Degrees of Freedom for Numerator

ν_2, Degrees of Freedom for Denominator

F	1	2	3	4	5	6	7	8	9	10	11	12	F
7 .0005	$.0^642$	$.0^350$	$.0^248$	.014	.027	.040	.053	.066	.078	.088	.099	.108	.0005
.001	$.0^517$	$.0^210$	$.0^276$	.020	.035	.051	.067	.081	.093	.105	.115	.125	.001
.005	$.0^442$	$.0^250$	.023	.046	.070	.093	.113	.130	.145	.159	.171	.181	.005
.01	$.0^317$	.010	.036	.067	.096	.121	.143	.162	.178	.192	.205	.216	.01
.025	$.0^210$	.025	.068	.110	.146	.176	.200	.221	.238	.253	.266	.277	.025
.05	$.0^242$	.052	.113	.164	.205	.238	.264	.286	.304	.319	.332	.343	.05
.10	.017	.107	.190	.251	.297	.332	.359	.381	.399	.414	.427	.438	.10
.25	.110	.300	.412	.481	.528	.562	.588	.608	.624	.637	.649	.658	.25
.50	.506	.767	.871	.926	.960	.983	1.00	1.01	1.02	1.03	1.04	1.04	.50
.75	1.57	1.70	1.72	1.72	1.71	1.71	1.70	1.70	1.69	1.69	1.69	1.68	.75
.90	3.59	3.26	3.07	2.96	2.88	2.83	2.78	2.75	2.72	2.70	2.68	2.67	.90
.95	5.59	4.74	4.35	4.12	3.97	3.87	3.79	3.73	3.68	3.64	3.60	3.57	.95
.975	8.07	6.54	5.89	5.52	5.29	5.12	4.99	4.90	4.82	4.76	4.71	4.67	.975
.99	12.2	9.55	8.45	7.85	7.46	7.19	6.99	6.84	6.72	6.62	6.54	6.47	.99
.995	16.2	12.4	10.9	10.0	9.52	9.16	8.89	8.68	8.51	8.38	8.27	8.18	.995
.999	29.2	21.7	18.8	17.2	16.2	15.5	15.0	14.6	14.3	14.1	13.9	13.7	.999
.9995	37.0	27.2	23.5	21.4	20.2	19.3	18.7	18.2	17.8	17.5	17.2	17.0	.9995
8 .0005	$.0^642$	$.0^350$	$.0^248$	.014	.027	.041	.055	.068	.081	.092	.102	.112	.0005
.001	$.0^517$	$.0^210$	$.0^276$	.020	.036	.053	.068	.083	.096	.109	.120	.130	.001
.005	$.0^442$	$.0^250$	.027	.047	.072	.095	.115	.133	.149	.164	.176	.187	.005
.01	$.0^317$	.010	.036	.068	.097	.123	.146	.166	.183	.198	.211	.222	.01
.025	$.0^210$	.025	.069	.111	.148	.179	.204	.226	.244	.259	.273	.285	.025
.05	$.0^242$	.052	.113	.166	.208	.241	.268	.291	.310	.326	.339	.351	.05
.10	.017	.107	.190	.253	.299	.335	.363	.386	.405	.421	.435	.445	.10
.25	.109	.298	.411	.481	.529	.563	.589	.610	.627	.640	.654	.661	.25
.50	.499	.757	.860	.915	.948	.971	.988	1.00	1.01	1.02	1.02	1.03	.50
.75	1.54	1.66	1.67	1.66	1.66	1.65	1.64	1.64	1.64	1.63	1.63	1.62	.75
.90	3.46	3.11	2.92	2.81	2.73	2.67	2.62	2.59	2.56	2.54	2.52	2.50	.90
.95	5.32	4.46	4.07	3.84	3.69	3.58	3.50	3.44	3.39	3.35	3.31	3.28	.95
.975	7.57	6.06	5.42	5.05	4.82	4.65	4.53	4.43	4.36	4.30	4.24	4.20	.975
.99	11.3	8.65	7.59	7.01	6.63	6.37	6.18	6.03	5.91	5.81	5.73	5.67	.99
.995	14.7	11.0	9.60	8.81	8.30	7.95	7.69	7.50	7.34	7.21	7.10	7.01	.995
.999	25.4	18.5	15.8	14.4	13.5	12.9	12.4	12.0	11.8	11.5	11.4	11.2	.999
.9995	31.6	22.8	19.4	17.6	16.4	15.7	15.1	14.6	14.3	14.0	13.8	13.6	.9995
9 .0005	$.0^641$	$.0^350$	$.0^248$	.015	.027	.042	.056	.070	.083	.094	.105	.115	.0005
.001	$.0^517$	$.0^210$	$.0^277$	.021	.037	.054	.070	.085	.099	.112	.123	.134	.001
.005	$.0^442$	$.0^250$	.023	.047	.073	.096	.117	.136	.153	.168	.181	.192	.005
.01	$.0^317$	.010	.037	.068	.098	.125	.149	.169	.187	.202	.216	.228	.01
.025	$.0^210$	.025	.069	.112	.150	.181	.207	.230	.248	.265	.279	.291	.025
.05	$.0^240$	.052	.113	.167	.210	.244	.272	.296	.315	.331	.345	.358	.05
.10	.017	.107	.191	.254	.302	.338	.367	.390	.410	.426	.441	.452	.10
.25	.108	.297	.410	.480	.529	.564	.591	.612	.629	.643	.654	.664	.25
.50	.494	.749	.852	.906	.939	.962	.978	.990	1.00	1.01	1.01	1.02	.50
.75	1.51	1.62	1.63	1.63	1.62	1.61	1.60	1.60	1.59	1.59	1.58	1.58	.75
.90	3.36	3.01	2.81	2.69	2.61	2.55	2.51	2.47	2.44	2.42	2.40	2.38	.90
.95	5.12	4.26	3.86	3.63	3.48	3.37	3.29	3.23	3.18	3.14	3.10	3.07	.95
.975	7.21	5.71	5.08	4.72	4.48	4.32	4.20	4.10	4.03	3.98	3.91	3.87	.975
.99	10.6	8.02	6.99	6.42	6.06	5.80	5.61	5.47	5.35	5.36	5.18	5.11	.99
.995	13.6	10.1	8.72	7.96	7.47	7.13	6.88	6.69	6.54	6.42	6.31	6.23	.995
.999	22.9	16.4	13.9	12.6	11.7	11.1	10.7	10.4	10.1	9.89	9.71	9.57	.999
.9995	28.0	19.9	16.8	15.1	14.1	13.3	12.8	12.4	12.1	11.8	11.6	11.4	.9995

ν₁, Degrees of Freedom for Numerator

F	15	20	24	30	40	50	60	100	120	200	500	∞	F	
.0005	.130	.157	.172	.188	.206	.217	.225	.242	.246	.255	.263	.268	.0005	7
.001	.148	.176	.191	.208	.225	.237	.245	.261	.266	.274	.282	.288	.001	
.005	.206	.235	.251	.267	.285	.296	.304	.319	.324	.332	.340	.345	.005	
.01	.241	.270	.286	.303	.320	.331	.339	.355	.358	.366	.373	.379	.01	
.025	.304	.333	.348	.364	.381	.392	.399	.413	.418	.426	.433	.437	.025	
.05	.369	.398	.413	.428	.445	.455	.461	.476	.479	.485	.493	.498	.05	
.10	.463	.491	.504	.519	.534	.543	.550	.562	.566	.571	.578	.582	.10	
.25	.679	.702	.713	.725	.737	.745	.749	.760	.762	.767	.772	.775	.25	
.50	1.05	1.07	1.07	1.08	1.08	1.09	1.09	1.10	1.10	1.10	1.10	1.10	.50	
.75	1.68	1.67	1.67	1.66	1.66	1.66	1.65	1.65	1.65	1.65	1.65	1.65	.75	
.90	2.63	2.59	2.58	2.56	2.54	2.52	2.51	2.50	2.49	2.48	2.48	2.47	.90	
.95	3.51	3.44	3.41	3.38	3.34	3.32	3.30	3.27	3.27	3.25	3.24	3.23	.95	
.975	4.57	4.47	4.42	4.36	4.31	4.28	4.25	4.21	4.20	4.18	4.16	4.14	.975	
.99	6.31	6.16	6.07	5.99	5.91	5.86	5.82	5.75	5.74	5.70	5.67	5.65	.99	
.995	7.97	7.75	7.65	7.53	7.42	7.35	7.31	7.22	7.19	7.15	7.10	7.08	.995	
.999	13.3	12.9	12.7	12.5	12.3	12.2	12.1	11.9	11.9	11.8	11.7	11.7	.999	
.9995	16.5	16.0	15.7	15.5	15.2	15.1	15.0	14.7	14.7	14.6	14.5	14.4	.9995	
.0005	.136	.164	.181	.198	.218	.230	.239	.257	.262	.271	.281	.287	.0005	8
.001	.155	.184	.200	.218	.238	.250	.259	.277	.282	.292	.300	.306	.001	
.005	.214	.244	.261	.279	.299	.311	.319	.337	.341	.351	.358	.364	.005	
.01	.250	.281	.297	.315	.334	.346	.354	.372	.376	.385	.392	.398	.01	
.025	.313	.343	.360	.377	.395	.407	.415	.431	.435	.442	.450	.456	.025	
.05	.379	.409	.425	.441	.459	.469	.477	.493	.496	.505	.510	.516	.05	
.10	.472	.500	.515	.531	.547	.556	.563	.578	.581	.588	.595	.599	.10	
.25	.684	.707	.718	.730	.743	.751	.756	.767	.769	.775	.780	.783	.25	
.50	1.04	1.05	1.06	1.07	1.07	1.07	1.08	1.08	1.08	1.09	1.09	1.09	.50	
.75	1.62	1.61	1.60	1.60	1.59	1.59	1.59	1.58	1.58	1.58	1.58	1.58	.75	
.90	2.46	2.42	2.40	2.38	2.36	2.35	2.34	2.32	2.32	2.31	2.30	2.29	.90	
.95	3.22	3.15	3.12	3.08	3.04	3.02	3.01	2.97	2.97	2.95	2.94	2.93	.95	
.975	4.10	4.00	3.95	3.89	3.84	3.81	3.78	3.74	3.73	3.70	3.68	3.67	.975	
.99	5.52	5.36	5.28	5.20	5.12	5.07	5.03	4.96	4.95	4.91	4.88	4.86	.99	
.995	6.81	6.61	6.50	6.40	6.29	6.22	6.18	6.09	6.06	6.02	5.98	5.95	.995	
.999	10.8	10.5	10.3	10.1	9.92	9.80	9.73	9.57	9.54	9.46	9.39	9.34	.999	
.9995	13.1	12.7	12.5	12.2	12.0	11.8	11.8	11.6	11.5	11.4	11.4	11.3	.9995	
.0005	.141	.171	.188	.207	.228	.242	.251	.270	.276	.287	.297	.303	.0005	9
.001	.160	.191	.208	.228	.249	.262	.271	.291	.296	.307	.316	.323	.001	
.005	.220	.253	.271	.290	.310	.324	.332	.351	.356	.366	.376	.382	.005	
.01	.257	.289	.307	.326	.346	.358	.368	.386	.391	.400	.410	.415	.01	
.025	.320	.352	.370	.388	.408	.420	.428	.446	.450	.459	.467	.473	.025	
.05	.386	.418	.435	.452	.471	.483	.490	.508	.510	.518	.526	.532	.05	
.10	.479	.509	.525	.541	.558	.568	.575	.588	.594	.602	.610	.613	.10	
.25	.687	.711	.723	.736	.749	.757	.762	.773	.776	.782	.787	.791	.25	
.50	1.03	1.04	1.05	1.05	1.06	1.06	1.07	1.07	1.07	1.08	1.08	1.08	.50	
.75	1.57	1.56	1.56	1.55	1.55	1.54	1.54	1.53	1.53	1.53	1.53	1.53	.75	
.90	2.34	2.30	2.28	2.25	2.23	2.22	2.21	2.19	2.18	2.17	2.17	2.16	.90	
.95	3.01	2.94	2.90	2.86	2.83	2.80	2.79	2.76	2.75	2.73	2.72	2.71	.95	
.975	3.77	3.67	3.61	3.56	3.51	3.47	3.45	3.40	3.39	3.37	3.35	3.33	.975	
.99	4.96	4.81	4.73	4.65	4.57	4.52	4.48	4.42	4.40	4.36	4.33	4.31	.99	
.995	6.03	5.83	5.73	5.62	5.52	5.45	5.41	5.32	5.30	5.26	5.21	5.19	.995	
.999	9.24	8.90	8.72	8.55	8.37	8.26	8.19	8.04	8.00	7.93	7.86	7.81	.999	
.9995	11.0	10.6	10.4	10.2	9.94	9.80	9.71	9.53	9.49	9.40	9.32	9.26	.9995	

ν₂, Degrees of Freedom for Denominator

ν_1, Degrees of Freedom for Numerator

	F	1	2	3	4	5	6	7	8	9	10	11	12	F
10	.0005	$.0^641$	$.0^350$	$.0^249$	.015	.028	.043	.057	.071	.085	.097	.108	.119	.0005
	.001	$.0^517$	$.0^210$	$.0^277$	.021	.037	.054	.071	.087	.101	.114	.126	.137	.001
	.005	$.0^441$	$.0^250$	.023	.048	.073	.098	.119	.139	.156	.171	.185	.197	.005
	.01	$.0^317$	.010	.037	.069	.100	.127	.151	.172	.190	.206	.220	.233	.01
	.025	$.0^210$	.025	.069	.113	.151	.183	.210	.233	.252	.269	.283	.296	.025
	.05	$.0^241$	.052	.114	.168	.211	.246	.275	.299	.319	.336	.351	.363	.05
	.10	.017	.106	.191	.255	.303	.340	.370	.394	.414	.430	.444	.457	.10
	.25	.107	.296	.409	.480	.529	.565	.592	.613	.631	.645	.657	.667	.25
	.50	.490	.743	.845	.899	.932	.954	.971	.983	.992	1.00	1.01	1.01	.50
	.75	1.49	1.60	1.60	1.59	1.59	1.58	1.57	1.56	1.56	1.55	1.55	1.54	.75
	.90	3.28	2.92	2.73	2.61	2.52	2.46	2.41	2.38	2.35	2.32	2.30	2.28	.90
	.95	4.96	4.10	3.71	3.48	3.33	3.22	3.14	3.07	3.02	2.98	2.94	2.91	.95
	.975	6.94	5.46	4.83	4.47	4.24	4.07	3.95	3.85	3.78	3.72	3.66	3.62	.975
	.99	10.0	7.56	6.55	5.99	5.64	5.39	5.20	5.06	4.94	4.85	4.77	4.71	.99
	.995	12.8	9.43	8.08	7.34	6.87	6.54	6.30	6.12	5.97	5.85	5.75	5.66	.995
	.999	21.0	14.9	12.6	11.3	10.5	9.92	9.52	9.20	8.96	8.75	8.58	8.44	.999
	.9995	25.5	17.9	15.0	13.4	12.4	11.8	11.3	10.9	10.6	10.3	10.1	9.93	.9995
11	.0005	$.0^641$	$.0^350$	$.0^249$	.015	.028	.043	.058	.072	.086	.099	.111	.121	.0005
	.001	$.0^516$	$.0^210$	$.0^278$	.021	.038	.055	.072	.088	.103	.116	.129	.140	.001
	.005	$.0^440$	$.0^250$	.023	.048	.074	.099	.121	.141	.158	.174	.188	.200	.005
	.01	$.0^316$	.010	.037	.069	.100	.128	.153	.175	.193	.210	.224	.237	.01
	.025	$.0^210$	.025	.069	.114	.152	.185	.212	.236	.256	.273	.288	.301	.025
	.05	$.0^241$	.052	.114	.168	.212	.248	.278	.302	.323	.340	.355	.368	.05
	.10	.017	.106	.192	.256	.305	.342	.373	.397	.417	.435	.448	.461	.10
	.25	.107	.295	.408	.481	.529	.565	.592	.614	.633	.645	.658	.667	.25
	.50	.486	.739	.840	.893	.926	.948	.964	.977	.986	.994	1.00	1.01	.50
	.75	1.47	1.58	1.58	1.57	1.56	1.55	1.54	1.53	1.53	1.52	1.52	1.51	.75
	.90	3.23	2.86	2.66	2.54	2.45	2.39	2.34	2.30	2.27	2.25	2.23	2.21	.90
	.95	4.84	3.98	3.59	3.36	3.20	3.09	3.01	2.95	2.90	2.85	2.82	2.79	.95
	.975	6.72	5.26	4.63	4.28	4.04	3.88	3.76	3.66	3.59	3.53	3.47	3.43	.975
	.99	9.65	7.21	6.22	5.67	5.32	5.07	4.89	4.74	4.63	4.54	4.46	4.40	.99
	.995	12.2	8.91	7.60	6.88	6.42	6.10	5.86	5.68	5.54	5.42	5.32	5.24	.995
	.999	19.7	13.8	11.6	10.3	9.58	9.05	8.66	8.35	8.12	7.92	7.76	7.62	.999
	.9995	23.6	16.4	13.6	12.2	11.2	10.6	10.1	9.76	9.48	9.24	9.04	8.88	.9995
12	.0005	$.0^641$	$.0^350$	$.0^249$	.015	.028	.044	.058	.073	.087	.101	.113	.124	.0005
	.001	$.0^516$	$.0^210$	$.0^278$	.021	.038	.056	.073	.089	.104	.118	.131	.143	.001
	.005	$.0^439$	$.0^250$	.023	.048	.075	.100	.122	.143	.161	.177	.191	.204	.005
	.01	$.0^316$	.010	.037	.070	.101	.130	.155	.176	.196	.212	.227	.241	.01
	.025	$.0^210$	.025	.070	.114	.153	.186	.214	.238	.259	.276	.292	.305	.025
	.05	$.0^241$	.052	.114	.169	.214	.250	.280	.305	.325	.343	.358	.372	.05
	.10	.016	.106	.192	.257	.306	.344	.375	.400	.420	.438	.452	.466	.10
	.25	.106	.295	.408	.480	.530	.566	.594	.616	.633	.649	.662	.671	.25
	.50	.484	.735	.835	.888	.921	.943	.959	.972	.981	.989	.995	1.00	.50
	.75	1.46	1.56	1.56	1.55	1.54	1.53	1.52	1.51	1.51	1.50	1.50	1.49	.75
	.90	3.18	2.81	2.61	2.48	2.39	2.33	2.28	2.24	2.21	2.19	2.17	2.15	.90
	.95	4.75	3.89	3.49	3.26	3.11	3.00	2.91	2.85	2.80	2.75	2.72	2.69	.95
	.975	6.55	5.10	4.47	4.12	3.89	3.73	3.61	3.51	3.44	3.37	3.32	3.28	.975
	.99	9.33	6.93	5.95	5.41	5.06	4.82	4.64	4.50	4.39	4.30	4.22	4.16	.99
	.995	11.8	8.51	7.23	6.52	6.07	5.76	5.52	5.35	5.20	5.09	4.99	4.91	.995
	.999	18.6	13.0	10.8	9.63	8.89	8.38	8.00	7.71	7.48	7.29	7.14	7.01	.999
	.9995	22.2	15.3	12.7	11.2	10.4	9.74	9.28	8.94	8.66	8.43	8.24	8.08	.9995

ν_2, Degrees of Freedom for Denominator

ν₁, Degrees of Freedom for Numerator

F	15	20	24	30	40	50	60	100	120	200	500	∞	F	
.0005	.145	.177	.195	.215	.238	.251	.262	.282	.288	.299	.311	.319	.0005	10
.001	.164	.197	.216	.236	.258	.272	.282	.303	.309	.321	.331	.338	.001	
.005	.226	.260	.279	.299	.321	.334	.344	.365	.370	.380	.391	.397	.005	
.01	.263	.297	.316	.336	.357	.370	.380	.400	.405	.415	.424	.431	.01	
.025	.327	.360	.379	.398	.419	.431	.441	.459	.464	.474	.483	.488	.025	
.05	.393	.426	.444	.462	.481	.493	.502	.518	.523	.532	.541	.546	.05	
.10	.486	.516	.532	.549	.567	.578	.586	.602	.605	.614	.621	.625	.10	
.25	.691	.714	.727	.740	.754	.762	.767	.779	.782	.788	.793	.797	.25	
.50	1.02	1.03	1.04	1.05	1.05	1.06	1.06	1.06	1.06	1.07	1.07	1.07	.50	
.75	1.53	1.52	1.52	1.51	1.51	1.50	1.50	1.49	1.49	1.49	1.48	1.48	.75	
.90	2.24	2.20	2.18	2.16	2.13	2.12	2.11	2.09	2.08	2.07	2.06	2.06	.90	
.95	2.85	2.77	2.74	2.70	2.66	2.64	2.62	2.59	2.58	2.56	2.55	2.54	.95	
.975	3.52	3.42	3.37	3.31	3.26	3.22	3.20	3.15	3.14	3.12	3.09	3.08	.975	
.99	4.56	4.41	4.33	4.25	4.17	4.12	4.08	4.01	4.00	3.96	3.93	3.91	.99	
.995	5.47	5.27	5.17	5.07	4.97	4.90	4.86	4.77	4.75	4.71	4.67	4.64	.995	
.999	8.13	7.80	7.64	7.47	7.30	7.19	7.12	6.98	6.94	6.87	6.81	6.76	.999	
.9995	9.56	9.16	8.96	8.75	8.54	8.42	8.33	8.16	8.12	8.04	7.96	7.90	.9995	
.0005	.148	.182	.201	.222	.246	.261	.271	.293	.299	.312	.324	.331	.0005	11
.001	.168	.202	.222	.243	.266	.282	.292	.313	.320	.322	.343	.353	.001	
.005	.231	.266	.286	.308	.330	.345	.355	.376	.382	.394	.403	.412	.005	
.01	.268	.304	.324	.344	.366	.380	.391	.412	.417	.427	.439	.444	.01	
.025	.332	.368	.386	.407	.429	.442	.450	.472	.476	.485	.495	.503	.025	
.05	.398	.433	.452	.469	.490	.503	.513	.529	.535	.543	.552	.559	.05	
.10	.490	.524	.541	.559	.578	.588	.595	.614	.617	.625	.633	.637	.10	
.25	.694	.719	.730	.744	.758	.767	.773	.780	.788	.794	.799	.803	.25	
.50	1.02	1.03	1.03	1.04	1.05	1.05	1.05	1.06	1.06	1.06	1.06	1.06	.50	
.75	1.50	1.49	1.49	1.48	1.47	1.47	1.47	1.46	1.46	1.46	1.45	1.45	.75	
.90	2.17	2.12	2.10	2.08	2.05	2.04	2.03	2.00	2.00	1.99	1.98	1.97	.90	
.95	2.72	2.65	2.61	2.57	2.53	2.51	2.49	2.46	2.45	2.43	2.42	2.40	.95	
.975	3.33	3.23	3.17	3.12	3.06	3.03	3.00	2.96	2.94	2.92	2.90	2.88	.975	
.99	4.25	4.10	4.02	3.94	3.86	3.81	3. 8	3.71	3.69	3.66	3.62	3.60	.99	
.995	5.05	4.86	4.76	4.65	4.55	4.49	4.45	4.36	4.34	4.29	4.25	4.23	.996	
.999	7.32	7.01	6.85	6.68	6.52	6.41	6.35	6.21	6.17	6.10	6.04	6.00	.999	
.9995	8.52	8.14	7.94	7.75	7.55	7.43	7.35	7.18	7.14	7.06	6.98	6.93	.9995	
.0005	.152	.186	.206	.228	.253	.269	.280	.305	.311	.323	.337	.345	.0005	12
.001	.172	.207	.228	.250	.275	.291	.302	.326	.332	.344	.357	.365	.001	
.005	.235	.272	.292	.315	.339	.355	.365	.388	.393	.405	.417	.424	.005	
.01	.273	.310	.330	.352	.375	.391	.401	.422	.428	.441	.450	.458	.01	
.025	.337	.374	.394	.416	.437	.450	.461	.481	.487	.498	.508	.514	.025	
.05	.404	.439	.458	.478	.499	.513	.522	.541	.545	.556	.565	.571	.05	
.10	.496	.528	.546	.564	.583	.595	.604	.621	.625	.633	.641	.647	.10	
.25	.695	.721	.734	.748	.762	.771	.777	.789	.792	.799	.804	.808	.25	
.50	1.01	1.02	1.03	1.03	1.04	1.04	1.05	1.05	1.05	1.05	1.06	1.06	.50	
.75	1.48	1.47	1.46	1.45	1.45	1.44	1.44	1.43	1.43	1.43	1.42	1.42	.75	
.90	2.11	2.06	2.04	2.01	1.99	1.97	1.96	1.94	1.93	1.92	1.91	1.90	.90	
.95	2.62	2.54	2.51	2.47	2.43	2.40	2.38	2.35	2.34	2.32	2.31	2.30	.95	
.975	3.18	3.07	3.02	2.96	2.91	2.87	2.85	2.80	2.79	2.76	2.74	2.72	.975	
.99	4.01	3.86	3.78	3.70	3.62	3.57	3.54	3.47	3.45	3.41	3.38	3.36	.99	
.995	4.72	4.53	4.43	4.33	4.23	4.17	4.12	4.04	4.01	3.97	3.93	3.90	.995	
.999	6.71	6.40	6.25	6.09	5.93	5.83	5.76	5.63	5.59	5.52	5.46	5.42	.999	
.9995	7.74	7.37	7.18	7.00	6.80	6.68	6.61	6.45	6.41	6.33	6.25	6.20	.9995	

ν₂, Degrees of Freedom for Denominator

ν_1, Degrees of Freedom for Numerator

ν_2, Degrees of Freedom for Denominator

	F	1	2	3	4	5	6	7	8	9	10	11	12	F
15	.0005	$.0^641$	$.0^350$	$.0^249$	.015	.029	.045	.061	.076	.091	.105	.117	.129	.0005
	.001	$.0^516$	$.0^210$	$.0^279$	.021	.039	.057	.075	.092	.108	.123	.137	.149	.001
	.005	$.0^439$	$.0^250$	.023	.049	.076	.102	.125	.147	.166	.183	.198	.212	.003
	.01	$.0^316$	.010	.037	.070	.103	.132	.158	.181	.202	.219	.235	.249	.01
	.025	$.0^210$	.025	.070	.116	.156	.190	.219	.244	.265	.284	.300	.315	.025
	.05	$.0^241$	.051	.115	.170	.216	.254	.285	.311	.333	.351	.368	.382	.05
	.10	.016	.106	.192	.258	.309	.348	.380	.406	.427	.446	.461	.475	.10
	.25	.105	.293	.407	.480	.531	.568	.596	.618	.637	.652	.667	.676	.25
	.50	.478	.726	.826	.878	.911	.933	.948	.960	.970	.977	.984	.989	.50
	.75	1.43	1.52	1.52	1.51	1.49	1.48	1.47	1.46	1.46	1.45	1.44	1.44	.75
	.90	3.07	2.70	2.49	2.36	2.27	2.21	2.16	2.12	2.09	2.06	2.04	2.02	.90
	.95	4.54	3.68	3.29	3.06	2.90	2.79	2.71	2.64	2.59	2.54	2.51	2.48	.95
	.975	6.20	4.76	4.15	3.80	3.58	3.41	3.29	3.20	3.12	3.06	3.01	2.96	.975
	.00	8.68	6.36	5.42	4.89	4.56	4.32	4.14	4.00	3.89	3.80	3.73	3.67	.99
	.995	10.8	7.70	6.48	5.80	5.37	5.07	4.85	4.67	4.54	4.42	4.33	4.25	.995
	.999	16.6	11.3	9.34	8.25	7.57	7.09	6.74	6.47	6.26	6.08	5.93	5.81	.999
	.9995	19.5	13.2	10.8	9.48	8.66	8.10	7.68	7.36	7.11	6.91	6.75	6.60	.9995
20	.0005	$.0^640$	$.0^350$	$.0^250$	.015	.029	.046	.063	.079	.094	.109	.123	.136	.0005
	.001	$.0^516$	$.0^210$	$.0^279$	.022	.039	.058	.077	.095	.112	.128	.143	.156	.001
	.005	$.0^439$	$.0^250$	.023	.050	.077	.104	.129	.151	.171	.190	.206	.221	.005
	.01	$.0^316$	.010	.037	.071	.105	.135	.162	.187	.208	.227	.244	.259	.01
	.025	$.0^210$	.025	.071	.117	.158	.193	.224	.250	.273	.292	.310	.325	.025
	.05	$.0^240$	.051	.115	.172	.219	.258	.290	.318	.340	.360	.377	.393	.05
	.10	.016	.106	.193	.260	.312	.353	.385	.412	.435	.454	.472	.485	.10
	.25	.104	.292	.407	.480	.531	.569	.598	.622	.641	.656	.671	.681	.25
	.50	.472	.718	.816	.868	.900	.922	.938	.950	.959	.966	.972	.977	.50
	.75	1.40	1.49	1.48	1.47	1.45	1.44	1.43	1.42	1.41	1.40	1.39	1.39	.75
	.90	2.97	2.59	2.38	2.25	2.16	2.09	2.04	2.00	1.96	1.94	1.91	1.89	.90
	.95	4.35	3.49	3.10	2.87	2.71	2.60	2.51	2.45	2.39	2.35	2.31	2.28	.95
	.975	5.87	4.46	3.86	3.51	3.29	3.13	3.01	2.91	2.84	2.77	2.72	2.68	.975
	.99	8.10	5.85	4.94	4.43	4.10	3.87	3.70	3.56	3.46	3.37	3.29	3.23	.99
	.995	9.94	6.99	5.82	5.17	4.76	4.47	4.26	4.09	3.96	3.85	3.76	3.68	.995
	.999	14.8	9.95	8.10	7.10	6.46	6.02	5.69	5.44	5.24	5.08	4.94	4.82	.999
	.9995	17.2	11.4	9.20	8.02	7.28	6.76	6.38	6.08	5.85	5.66	5.51	5.38	.9995
24	.0005	$.0^640$	$.0^350$	$.0^250$	.015	.030	.046	.064	.080	.096	.112	.126	.139	.0005
	.001	$.0^516$	$.0^210$	$.0^279$	.022	.040	.059	.079	.097	.115	.131	.146	.160	.001
	.005	$.0^440$	$.0^250$	.023	.050	.078	.106	.131	.154	.175	.193	.210	.226	.005
	.01	$.0^316$	.010	.038	.072	.106	.137	.165	.189	.211	.231	.249	.264	.01
	.025	$.0^210$	.025	.071	.117	.159	.195	.227	.253	.277	.297	.315	.331	.035
	.05	$.0^240$	.051	.116	.173	.221	.260	.293	.321	.345	.365	.383	.399	.05
	.10	.016	.106	.193	.261	.313	.355	.388	.416	.439	.459	.476	.491	.10
	.25	.104	.291	.406	.480	.532	.570	.600	.623	.643	.659	.671	.684	.25
	.50	.469	.714	.812	.863	.895	.917	.932	.944	.953	.961	.967	.972	.50
	.75	1.39	1.47	1.46	1.44	1.43	1.41	1.40	1.39	1.38	1.38	1.37	1.36	.75
	.90	2.93	2.54	2.33	2.19	2.10	2.04	1.98	1.94	1.91	1.88	1.85	1.83	.90
	.95	4.26	3.40	3.01	2.78	2.62	2.51	2.42	2.36	2.30	2.25	2.21	2.18	.95
	.975	5.72	4.32	3.72	3.38	3.15	2.99	2.87	2.78	2.70	2.64	2.59	2.54	.975
	.99	7.82	5.61	4.72	4.22	3.90	3.67	3.50	3.36	3.26	3.17	3.09	3.03	.99
	.995	9.55	6.66	5.52	4.89	4.49	4.20	3.99	3.83	3.69	3.59	3.50	3.42	.995
	.999	14.0	9.34	7.55	6.59	5.98	5.55	5.23	4.99	4.80	4.64	4.50	4.39	.999
	.9995	16.2	10.6	8.52	7.39	6.68	6.18	5.82	5.54	5.31	5.13	4.98	4.85	.9995

Table **F** (*continued*)

ν_1, Degrees of Freedom for Numerator

F	15	20	24	30	40	50	60	100	120	200	500	∞	F	
.0005	.159	.197	.220	.244	.272	.290	.303	.330	.339	.353	.368	.377	.0005	15
.001	.181	.219	.242	.266	.294	.313	.325	.352	.360	.375	.388	.398	.001	
.005	.246	.286	.308	.333	.360	.377	.389	.415	.422	.435	.448	.457	.005	
.01	.284	.324	.346	.370	.397	.413	.425	.450	.456	.469	.483	.490	.01	
.025	.349	.389	.410	.433	.458	.474	.485	.508	.514	.526	.538	.546	.025	
.05	.416	.454	.474	.496	.519	.535	.545	.565	.571	.581	.592	.600	.05	
.10	.507	.542	.561	.581	.602	.614	.624	.641	.647	.658	.667	.672	.10	
.25	.701	.728	.742	.757	.772	.782	.788	.802	.805	.812	.818	.822	.25	
.50	1.00	1.01	1.02	1.02	1.03	1.03	1.03	1.04	1.04	1.04	1.04	1.05	.50	
.75	1.43	1.41	1.41	1.40	1.39	1.39	1.38	1.38	1.37	1.37	1.36	1.36	.75	
.90	1.97	1.92	1.90	1.87	1.85	1.83	1.82	1.79	1.79	1.77	1.76	1.76	.90	
.95	2.40	2.33	2.39	2.25	2.20	2.18	2.16	2.12	2.11	2.10	2.08	2.07	.95	
.975	2.86	2.76	2.70	2.64	2.59	2.55	2.52	2.47	2.46	2.44	2.41	2.40	.975	
.99	3.52	3.37	3.29	3.21	3.13	3.08	3.05	2.98	2.96	2.92	2.89	2.87	.99	
.995	4.07	3.88	3.79	3.69	3.59	3.52	3.48	3.39	3.37	3.33	3.29	3.26	.995	
.999	5.54	5.25	5.10	4.95	4.80	4.70	4.64	4.51	4.47	4.41	4.35	4.31	.999	
.9995	6.27	5.93	5.75	5.58	5.40	5.29	5.21	5.06	5.02	4.94	4.87	4.83	.9995	
.0005	.169	.211	.235	.263	.295	.316	.331	.364	.375	.391	.408	.422	.0005	20
.001	.191	.233	.258	.286	.318	.339	.354	.386	.395	.413	.429	.441	.001	
.005	.258	.301	.327	.354	.385	.405	.419	.448	.457	.474	.490	.500	.005	
.01	.297	.340	.365	.392	.422	.441	.455	.483	.491	.508	.521	.532	.01	
.025	.363	.406	.430	.456	.484	.503	.514	.541	.548	.562	.575	.585	.025	
.05	.430	.471	.493	.518	.544	.562	.572	.595	.603	.617	.629	.637	.05	
.10	.520	.557	.578	.600	.623	.637	.648	.671	.675	.685	.694	.704	.10	
.25	.708	.736	.751	.767	.784	.794	.801	.816	.820	.827	.835	.840	.25	
.50	.989	1.00	1.01	1.01	1.02	1.02	1.02	1.03	1.03	1.03	1.03	1.03	.50	
.75	1.37	1.36	1.35	1.34	1.33	1.33	1.32	1.31	1.31	1.30	1.30	1.29	.75	
.90	1.84	1.79	1.77	1.74	1.71	1.69	1.68	1.65	1.64	1.63	1.62	1.61	.90	
.95	2.20	2.12	2.08	2.04	1.99	1.97	1.95	1.91	1.90	1.88	1.86	1.84	.95	
.975	2.57	2.46	2.41	2.35	2.29	2.25	2.22	2.17	2.16	2.13	2.10	2.09	.975	
.99	3.09	2.94	2.86	2.78	2.69	2.64	2.61	2.54	2.52	2.48	2.44	2.42	.99	
.995	3.50	3.32	3.22	3.12	3.02	2.96	2.92	2.83	2.81	2.76	2.72	2.69	.995	
.999	4.56	4.29	4.15	4.01	3.86	3.77	3.70	3.58	3.54	3.48	3.42	3.38	.999	
.9995	5.07	4.75	4.58	4.42	4.24	4.15	4.07	3.93	3.90	3.82	3.75	3.70	.9995	
.0005	.174	.218	.244	.274	.309	.331	.349	.384	.395	.416	.434	.449	.0005	24
.001	.196	.241	.268	.298	.332	.354	.371	.405	.417	.437	.455	.469	.001	
.005	.264	.310	.337	.367	.400	.422	.437	.469	.478	.498	.515	.527	.005	
.01	.304	.350	.376	.405	.437	.459	.473	.505	.513	.529	.546	.558	.01	
.025	.370	.415	.441	.468	.498	.518	.531	.562	.568	.585	.599	.610	.025	
.05	.437	.480	.504	.530	.558	.575	.588	.613	.622	.637	.649	.659	.05	
.10	.527	.566	.588	.611	.635	.651	.662	.685	.691	.704	.715	.723	.10	
.25	.712	.741	.757	.773	.791	.802	.809	.825	.829	.837	.844	.850	.25	
.50	.983	.994	1.00	1.01	1.01	1.02	1.02	1.02	1.02	1.02	1.03	1.03	.50	
.75	1.35	1.33	1.32	1.31	1.30	1.29	1.29	1.28	1.28	1.27	1.27	1.26	.75	
.90	1.78	1.73	1.70	1.67	1.64	1.62	1.61	1.58	1.57	1.56	1.54	1.53	.90	
.95	2.11	2.03	1.98	1.94	1.89	1.86	1.84	1.80	1.79	1.77	1.75	1.73	.95	
.975	2.44	2.33	2.27	2.21	2.15	2.11	2.08	2.02	2.01	1.98	1.95	1.94	.975	
.99	2.89	2.74	2.66	2.58	2.49	2.44	2.40	2.33	2.31	2.27	2.24	2.21	.99	
.995	3.25	3.06	2.97	2.87	2.77	2.70	2.66	2.57	2.55	2.50	2.46	2.43	.995	
.999	4.14	3.87	3.74	3.59	3.45	3.35	3.29	3.16	3.14	3.07	3.01	2.97	.999	
.9995	4.55	4.25	4.09	3.93	3.76	3.66	3.59	3.44	3.41	3.33	3.27	3.22	.9995	

ν_2, Degrees of Freedom for Denominator

Table F (continued)

ν₂, Degrees of Freedom for Denominator

ν_1, Degrees of Freedom for Numerator

	F	1	2	3	4	5	6	7	8	9	10	11	12	F
30	.0005	$.0^6 40$	$.0^3 50$	$.0^2 50$	.015	.030	.047	.065	.082	.098	.114	.129	.143	.0005
	.001	$.0^5 16$	$.0^2 10$	$.0^2 80$	.022	.040	.060	.080	.099	.117	.134	.150	.164	.001
	.005	$.0^4 40$	$.0^2 50$	.024	.050	.079	.107	.133	.156	.178	.197	.215	.231	.005
	.01	$.0^3 16$	.010	.038	.072	.107	.138	.167	.192	.215	.235	.254	.270	.01
	.025	$.0^2 10$	.025	.071	.118	.161	.197	.229	.257	.281	.302	.321	.337	.025
	.05	$.0^2 40$	.051	.116	.174	.222	.263	.296	.325	.349	.370	.389	.406	.05
	.10	.016	.106	.193	.262	.315	.357	.391	.420	.443	.464	.481	.497	.10
	.25	.103	.290	.406	.480	.532	.571	.601	.625	.645	.661	.676	.688	.25
	.50	.466	.709	.807	.858	.890	.912	.927	.939	.948	.955	.961	.966	.50
	.75	1.38	1.45	1.44	1.42	1.41	1.39	1.38	1.37	1.36	1.35	1.35	1.34	.75
	.90	2.88	2.49	2.28	2.14	2.05	1.98	1.93	1.88	1.85	1.82	1.79	1.77	.90
	.95	4.17	3.32	2.92	2.69	2.53	2.42	2.33	2.27	2.21	2.16	2.13	2.09	.95
	.975	5.57	4.18	3.59	3.25	3.03	2.87	2.75	2.65	2.57	2.51	2.46	2.41	.975
	.99	7.56	5.39	4.51	4.02	3.70	3.47	3.30	3.17	3.07	2.98	2.91	2.84	.99
	.995	9.18	6.35	5.24	4.62	4.23	3.95	3.74	3.58	3.45	3.34	3.25	3.18	.995
	.999	13.3	8.77	7.05	6.12	5.53	5.12	4.82	4.58	4.39	4.24	4.11	4.00	.999
	.9995	15.2	9.90	7.90	6.82	6.14	5.66	5.31	5.04	4.82	4.65	4.51	4.38	.9995
40	.0005	$.0^6 40$	$.0^3 50$	$.0^2 50$	.016	.030	.048	.066	.084	.100	.117	.132	.147	.0005
	.001	$.0^5 16$	$.0^2 10$	$.0^2 80$	.022	.042	.061	.081	.101	.119	.137	.153	.169	.001
	.005	$.0^4 40$	$.0^2 50$	.024	.051	.080	.108	.135	.159	.181	.201	.220	.237	.005
	.01	$.0^3 16$	.010	.038	.073	.108	.140	.169	.195	.219	.240	.259	.276	.01
	.025	$.0^3 99$	.025	.071	.119	.162	.199	.232	.260	.285	.307	.327	.344	.025
	.05	$.0^2 40$	.051	.116	.175	.224	.265	.299	.329	.354	.376	.395	.412	.05
	.10	.016	.106	.194	.263	.317	.360	.394	.424	.448	.469	.488	.504	.10
	.25	.103	.290	.405	.480	.533	.572	.603	.627	.647	.664	.680	.691	.25
	.50	.463	.705	.802	.854	.885	.907	.922	.934	.943	.950	.956	.961	.50
	.75	1.36	1.44	1.42	1.40	1.39	1.37	1.36	1.35	1.34	1.33	1.32	1.31	.75
	.90	2.84	2.44	2.23	2.09	2.00	1.93	1.87	1.83	1.79	1.76	1.73	1.71	.90
	.95	4.08	3.23	2.84	2.61	2.45	2.34	2.25	2.18	2.12	2.08	2.04	2.00	.95
	.975	5.42	4.05	3.46	3.13	2.90	2.74	2.62	2.53	2.45	2.39	2.33	2.29	.975
	.99	7.31	5.18	4.31	3.83	3.51	3.29	3.12	2.99	2.89	2.80	2.73	2.66	.99
	.995	8.83	6.07	4.98	4.37	3.99	3.71	3.51	3.35	3.22	3.12	3.03	2.95	.995
	.999	12.6	8.25	6.60	5.70	5.13	4.73	4.44	4.21	4.02	3.87	3.75	3.64	.999
	.9995	14.4	9.25	7.33	6.30	5.64	5.19	4.85	4.59	4.38	4.21	4.07	3.95	.9995
60	.0005	$.0^6 40$	$.0^3 50$	$.0^2 51$	.016	.031	.048	.067	.085	.103	.120	.136	.152	.0005
	.001	$.0^5 16$	$.0^2 10$	$.0^2 80$	.022	.041	.062	.083	.103	.122	.140	.157	.174	.001
	.005	$.0^4 40$	$.0^2 50$	.024	.051	.081	.110	.137	.162	.185	.206	.225	.243	.005
	.01	$.0^3 16$	.010	.038	.073	.109	.142	.172	.199	.223	.245	.265	.283	.01
	.025	$.0^3 99$	.025	.071	.120	.163	.202	.235	.264	.290	.313	.333	.351	.025
	.05	$.0^2 40$	.051	.116	.176	.226	.267	.303	.333	.359	3.82	.402	.419	.05
	.10	.016	.106	.194	.264	.318	.362	.398	.428	.453	.475	.493	.510	.10
	.25	.102	.289	.405	.480	.534	.573	.604	.629	.650	.667	.680	.695	.25
	.50	.461	.701	.798	.849	.880	.901	.917	.928	.937	.945	.951	.956	.50
	.75	1.35	1.42	1.41	1.38	1.37	1.35	1.33	1.32	1.31	1.30	1.29	1.29	.75
	.90	2.79	2.39	2.18	2.04	1.95	1.87	1.82	1.77	1.74	1.71	1.68	1.66	.90
	.95	4.00	3.15	2.76	2.53	2.37	2.25	2.17	2.10	2.04	1.99	1.95	1.92	.95
	.975	5.29	3.93	3.34	3.01	2.79	2.63	2.51	2.41	2.33	2.27	2.22	2.17	.975
	.99	7.08	4.98	4.13	3.65	3.34	3.12	2.95	2.82	2.72	2.63	2.56	2.50	.99
	.995	8.49	5.80	4.73	4.14	3.76	3.49	3.29	3.13	3.01	2.90	2.82	2.74	.995
	.999	12.0	7.76	6.17	5.31	4.76	4.37	4.09	3.87	3.69	3.54	3.43	3.31	.999
	.9995	13.6	8.65	6.81	5.82	5.20	4.76	4.44	4.18	3.98	3.82	3.69	3.57	.9995

ν_1, Degrees of Freedom for Numerator

F	15	20	24	30	40	50	60	100	120	200	500	∞	F	
.0005	.179	.226	.254	.287	.325	.350	.369	.410	.420	.444	.467	.483	.0005	30
.001	.202	.250	.278	.311	.348	.373	.391	.431	.442	.465	.488	.503	.001	
.005	.271	.320	.349	.381	.416	.441	.457	.495	.504	.524	.543	.559	.005	
.01	.311	.360	.388	.419	.454	.476	.493	.529	.538	.559	.575	.590	.01	
.025	.378	.426	.453	.482	.515	.535	.551	.585	.592	.610	.625	.639	.025	
.05	.445	.490	.516	.543	.573	.592	.606	.637	.644	.658	.676	.685	.05	
.10	.534	.575	.598	.623	.649	.667	.678	.704	.710	.725	.735	.746	.10	
.25	.716	.746	.763	.780	.798	.810	.818	.835	.839	.848	.856	.862	.25	
.50	.978	.989	.994	1.00	1.01	1.01	1.01	1.02	1.02	1.02	1.02	1.02	.50	
.75	1.32	1.30	1.29	1.28	1.27	1.26	1.26	1.25	1.24	1.24	1.23	1.23	.75	
.90	1.72	1.67	1.64	1.61	1.57	1.55	1.54	1.51	1.50	1.48	1.47	1.46	.90	
.95	2.01	1.93	1.89	1.84	1.79	1.76	1.74	1.70	1.68	1.66	1.64	1.62	.95	
.975	2.31	2.20	2.14	2.07	2.01	1.97	1.94	1.88	1.87	1.84	1.81	1.79	.975	
.99	2.70	2.55	2.47	2.39	2.30	2.25	2.21	2.13	2.11	2.07	2.03	2.01	.99	
.995	3.01	2.82	2.73	2.63	2.52	2.46	2.42	2.32	2.30	2.25	2.21	2.18	.995	
.999	3.75	3.49	3.36	3.22	3.07	2.98	2.92	2.79	2.76	2.69	2.63	2.59	.999	
.9995	4.10	3.80	3.65	3.48	3.32	3.22	3.15	3.00	2.97	2.89	2.82	2.78	.9995	
.0005	.185	.236	.266	.301	.343	.373	.393	.441	.453	.480	.504	.525	.0005	40
.001	.209	.259	.290	.326	.367	.396	.415	.461	.473	.500	.524	.545	.001	
.005	.279	.331	.362	.396	.436	.463	.481	.524	.534	.559	.581	.599	.005	
.01	.319	.371	.401	.435	.473	.498	.516	.556	.567	.592	.613	.628	.01	
.025	.387	.437	.466	.498	.533	.556	.573	.610	.620	.641	.662	.674	.025	
.05	.454	.502	.529	.558	.591	.613	.627	.658	.669	.685	.704	.717	.05	
.10	.542	.585	.609	.636	.664	.683	.696	.724	.731	.747	.762	.772	.10	
.25	.720	.752	.769	.787	.806	.819	.828	.846	.851	.861	.870	.877	.25	
.50	.972	.983	.989	.994	1.00	1.00	1.01	1.01	1.01	1.01	1.02	1.02	.50	
.75	1.30	1.28	1.26	1.25	1.24	1.23	1.22	1.21	1.21	1.20	1.19	1.19	.75	
.90	1.66	1.61	1.57	1.54	1.51	1.48	1.47	1.43	1.42	1.41	1.39	1.38	.90	
.95	1.92	1.84	1.79	1.74	1.69	1.66	1.64	1.59	1.58	1.55	1.53	1.51	.95	
.975	2.18	2.07	2.01	1.94	1.88	1.83	1.80	1.74	1.72	1.69	1.66	1.64	.975	
.99	2.52	2.37	2.29	2.20	2.11	2.06	2.02	1.94	1.92	1.87	1.83	1.80	.99	
.995	2.78	2.60	2.50	2.40	2.30	2.23	2.18	2.09	2.06	2.01	1.96	1.93	.995	
.999	3.40	3.15	3.01	2.87	2.73	2.64	2.57	2.44	2.41	2.34	2.28	2.23	.999	
.9995	3.68	3.39	3.24	3.08	2.92	2.82	2.74	2.60	2.57	2.49	2.41	2.37	.9995	
.0005	.192	.246	.278	.318	.365	.398	.421	.478	.493	.527	.561	.585	.0005	60
.001	.216	.270	.304	.343	.389	.421	.444	.497	.512	.545	.579	.602	.001	
.005	.287	.343	.376	.414	.458	.488	.510	.559	.572	.602	.633	.652	.005	
.01	.328	.383	.416	.453	.495	.524	.545	.592	.604	.633	.658	.679	.01	
.025	.396	.450	.481	.515	.555	.581	.600	.641	.654	.680	.704	.720	.025	
.05	.463	.514	.543	.575	.611	.633	.652	.690	.700	.719	.746	.759	.05	
.10	.550	.596	.622	.650	.682	.703	.717	.750	.758	.776	.793	.806	.10	
.25	.725	.758	.776	.796	.816	.830	.840	.860	.865	.877	.888	.896	.25	
.50	.967	.978	.983	.989	.994	.998	1.00	1.00	1.01	1.01	1.01	1.01	.50	
.75	1.27	1.25	1.24	1.22	1.21	1.20	1.19	1.17	1.17	1.16	1.15	1.15	.75	
.90	1.60	1.54	1.51	1.48	1.44	1.41	1.40	1.36	1.35	1.33	1.31	1.29	.90	
.95	1.84	1.75	1.70	1.65	1.59	1.56	1.53	1.48	1.47	1.44	1.41	1.39	.95	
.975	2.06	1.94	1.88	1.82	1.74	1.70	1.67	1.60	1.58	1.54	1.51	1.48	.975	
.99	2.35	2.20	2.12	2.03	1.94	1.88	1.84	1.75	1.73	1.68	1.63	1.60	.99	
.995	2.57	2.39	2.29	2.19	2.08	2.01	1.96	1.86	1.83	1.78	1.73	1.69	.995	
.999	3.08	2.83	2.69	2.56	2.41	2.31	2.25	2.11	2.09	2.01	1.93	1.89	.999	
.9995	3.30	3.02	2.87	2.71	2.55	2.45	2.38	2.23	2.19	2.11	2.03	1.98	.9995	

ν_2, Degrees of Freedom for Denominator

ν_1, Degrees of Freedom for Numerator

	F	1	2	3	4	5	6	7	8	9	10	11	12	F
120	.0005	$.0^640$	$.0^350$	$.0^251$	.016	.031	.049	.067	.087	.105	.123	.140	.156	.0005
	.001	$.0^516$	$.0^210$	$.0^281$	.023	.042	.063	.084	.105	.125	.144	.162	.179	.001
	.005	$.0^439$	$.0^250$	.024	.051	.081	.111	.139	.165	.189	.211	.230	.249	.005
	.01	$.0^316$	.010	.038	.074	.110	.143	.174	.202	.227	.250	.271	.290	.01
	.025	$.0^399$	.025	.072	.120	.165	.204	.238	.268	.295	.318	.340	.359	.025
	.05	$.0^239$	.051	.117	.177	.227	.270	.306	.337	.364	.388	.408	.427	.05
	.10	.016	.105	.194	.265	.320	.365	.401	.432	.458	.480	.500	.518	.10
	.25	.102	.288	.405	.481	.534	.574	.606	.631	.652	.670	.685	.699	.25
	.50	.458	.697	.793	.844	.875	.896	.912	.923	.932	.939	.945	.950	.50
	.75	1.34	1.40	1.39	1.37	1.35	1.33	1.31	1.30	1.29	1.28	1.27	1.26	.75
	.90	2.75	2.35	2.13	1.99	1.90	1.82	1.77	1.72	1.68	1.65	1.62	1.60	.90
	.95	3.92	3.07	2.68	2.45	2.29	2.18	2.09	2.02	1.96	1.91	1.87	1.83	.95
	.975	5.15	3.80	3.23	2.89	2.67	2.52	2.39	2.30	2.22	2.16	2.10	2.05	.975
	.99	6.85	4.79	3.95	3.48	3.17	2.96	2.79	2.66	2.56	2.47	2.40	2.34	.99
	.995	8.18	5.54	4.50	3.92	3.55	3.28	3.09	2.93	2.81	2.71	2.62	2.54	.995
	.999	11.4	7.32	5.79	4.95	4.42	4.04	3.77	3.55	3.38	3.24	3.12	3.02	.999
	.9995	12.8	8.10	6.34	5.39	4.79	4.37	4.07	3.82	3.63	3.47	3.34	3.22	.9995
∞	.0005	$.0^639$	$.0^350$	$.0^251$	.016	.032	.050	.069	.088	.108	.127	.144	.161	.0005
	.001	$.0^516$	$.0^210$	$.0^281$	.023	.042	.063	.085	.107	.128	.148	.167	.185	.001
	.005	$.0^439$	$.0^250$	.024	.052	.082	.113	.141	.168	.193	.216	.236	.256	.005
	.01	$.0^316$	.010	.038	.074	.111	.145	.177	.206	.232	.256	.278	.298	.01
	.025	$.0^398$	.025	.072	.121	.166	.206	.241	.272	.300	.325	.347	.367	.025
	.05	$.0^239$	.051	.117	.178	.229	.273	.310	.342	.369	.394	.417	.436	.05
	.10	.016	.105	.195	.266	.322	.367	.405	.436	.463	.487	.508	.525	.10
	.25	.102	.288	.404	.481	.535	.576	.608	.634	.655	.674	.690	.703	.25
	.50	.455	.693	.789	.839	.870	.891	.907	.918	.927	.934	.939	.945	.50
	.75	1.32	1.39	1.37	1.35	1.33	1.31	1.29	1.28	1.27	1.25	1.24	1.24	.75
	.90	2.71	2.30	2.08	1.94	1.85	1.77	1.72	1.67	1.63	1.60	1.57	1.55	.90
	.95	3.84	3.00	2.60	2.37	2.21	2.10	2.01	1.94	1.88	1.83	1.79	1.75	.95
	.995	5.02	3.69	3.12	2.79	2.57	2.41	2.29	2.19	2.11	2.05	1.99	1.94	.975
	.99	6.63	4.61	3.78	3.32	3.02	2.80	2.64	2.51	2.41	2.32	2.25	2.18	.99
	.995	7.88	5.30	4.28	3.72	3.35	3.09	2.90	2.74	2.62	2.52	2.43	2.36	.995
	.999	10.8	6.91	5.42	4.62	4.10	3.74	3.47	3.27	3.10	2.96	2.84	2.74	.999
	.9995	12.1	7.60	5.91	5.00	4.42	4.02	3.72	3.48	3.30	3.14	3.02	2.90	.9995

ν_2, Degrees of Freedom for Denominator

For sample sizes larger than, say, 30, Critical Values of Fisher's F can be approximated from

$$\log F_\alpha \approx \left(\frac{a}{\sqrt{h - b}} \right) - cg$$

where $h = 2\nu_1\nu_2/(\nu_1 + \nu_2)$, $g = (\nu_2 - \nu_1)/\nu_1\nu_2$, and a, b, c are functions of F given below:

Values of F

	$F = .50$	.75	.90	.95	.975	.99	.995	.999	.9995
a	0	.5859	1.1131	1.4287	1.7023	2.0206	2.2373	2.6841	2.8580
b	—	.58	.77	.95	1.14	1.40	1.61	2.09	2.30
c	.290	.355	.527	.681	.846	1.073	1.250	1.672	1.857

424

Table F (*continued*)

F	15	20	24	30	40	50	60	100	120	200	500	∞	F	
.0005	.199	.256	.293	.338	.390	.429	.458	.524	.543	.578	.614	.676	.0005	120
.001	.223	.282	.319	.363	.415	.453	.480	.542	.568	.595	.631	.691	.001	
.005	.297	.356	.393	.434	.484	.520	.545	.605	.623	.661	.702	.733	.005	
.01	.338	.397	.433	.474	.522	.556	.579	.636	.652	.688	.725	.755	.01	
.025	.406	.464	.498	.536	.580	.611	.633	.684	.698	.729	.762	.789	.025	
.05	.473	.527	.559	.594	.634	.661	.682	.727	.740	.767	.785	.819	.05	
.10	.560	.609	.636	.667	.702	.726	.742	.781	.791	.815	.838	.855	.10	
.25	.730	.765	.784	.805	.828	.843	.853	.877	.884	.897	.911	.923	.25	
.50	.961	.972	.978	.983	.989	.992	.994	1.00	1.00	1.00	1.01	1.01	.50	
.75	1.24	1.22	1.21	1.19	1.18	1.17	1.16	1.14	1.13	1.12	1.11	1.10	.75	
.90	1.55	1.48	1.45	1.41	1.37	1.34	1.32	1.27	1.26	1.24	1.21	1.19	.90	
.95	1.75	1.66	1.61	1.55	1.50	1.46	1.43	1.37	1.35	1.32	1.28	1.25	.95	
.975	1.95	1.82	1.76	1.69	1.61	1.56	1.53	1.45	1.43	1.39	1.34	1.31	.975	
.99	2.19	2.03	1.95	1.86	1.76	1.70	1.66	1.56	1.53	1.48	1.42	1.38	.99	
.995	2.37	2.19	2.09	1.98	1.87	1.80	1.75	1.64	1.61	1.54	1.48	1.43	.995	
.999	2.78	2.53	2.40	2.26	2.11	2.02	1.95	1.82	1.76	1.70	1.62	1.54	.999	
.9995	2.96	2.67	2.53	2.38	2.21	2.11	2.01	1.88	1.84	1.75	1.67	1.60	.9995	
.0005	.207	.270	.311	.360	.422	.469	.505	.599	.624	.704	.804	1.00	.0005	∞
.001	.232	.296	.338	.386	.448	.493	.527	.617	.649	.719	.819	1.00	.001	
.005	.307	.372	.412	.460	.518	.559	.592	.671	.699	.762	.843	1.00	.005	
.01	.349	.413	.452	.499	.554	.595	.625	.699	.724	.782	.858	1.00	.01	
.025	.418	.480	.517	.560	.611	.645	.675	.741	.763	.813	.878	1.00	.025	
.05	.484	.543	.577	.617	.663	.694	.720	.781	.797	.840	.896	1.00	.05	
.10	.570	.622	.652	.687	.726	.752	.774	.826	.838	.877	.919	1.00	.10	
.25	.736	.773	.793	.816	.842	.860	.872	.901	.910	.932	.957	1.00	.25	
.50	.956	.967	.972	.978	.983	.987	.989	.993	.994	.997	.999	1.00	.50	
.75	1.22	1.19	1.18	1.16	1.14	1.13	1.12	1.09	1.08	1.07	1.04	1.00	.75	
.90	1.49	1.42	1.38	1.34	1.30	1.26	1.24	1.18	1.17	1.13	1.08	1.00	.90	
.95	1.67	1.57	1.52	1.46	1.39	1.35	1.32	1.24	1.22	1.17	1.11	1.00	.95	
.975	1.83	1.71	1.64	1.57	1.48	1.43	1.39	1.30	1.27	1.21	1.13	1.00	.975	
.99	2.04	1.88	1.79	1.70	1.59	1.52	1.47	1.36	1.32	1.25	1.15	1.00	.99	
.995	2.19	2.00	1.90	1.79	1.67	1.59	1.53	1.40	1.36	1.28	1.17	1.00	.995	
.999	2.51	2.27	2.13	1.99	1.84	1.73	1.66	1.49	1.45	1.34	1.21	1.00	.999	
.9995	2.65	2.37	2.22	2.07	1.91	1.79	1.71	1.53	1.48	1.36	1.22	1.00	.9995	

ν_2, Degrees of Freedom for Denominator

425

Index